AIChE Spring Meeting 2006

8th Process Plant Safety Symposium 2006

April 23-27, 2006
Orlando, Florida, USA

Printed from e-media with permission by:

Curran Associates, Inc.
57 Morehouse Lane
Red Hook, NY 12571
www.proceedings.com

ISBN: 978-1-60423-517-3

AIChE Spring Meeting 2006

8th Process Plant Safety Symposium 2006

April 23-27, 2006
Orlando, Florida, USA

PROCESS PLANT SAFETY SYMPOSIUM

Welcome to the 8th Process Plant Safety Symposium! This is now an annual symposium and is one of three topical conferences making up the AIChE 2nd Annual Global Congress on Process Safety (PPSS, LPS and CCPS). The Process Plant Safety Symposium includes:

- Global Congress Keynote Session--Process Safety in the Global Chemical and Petroleum Industries (Joint Session with LPS and CCPS)
- Loss Prevention: Past, Present, and Future (Joint Session with LPS)
- Improving Safety Culture
- Risk Assessment
- Facility Siting Issues
- Safety Instrumented Systems and Safety Critical Devices
- Case Histories and Lessons Learned (Joint Session with LPS and CCPS)

Over 30 technical papers will be presented during the Symposium. Copies of these papers are generally contained in this publication, including those from the Joint PPSS/LPS/CCPS sessions. The range and depth of the papers will extend the knowledge of all attendees. PPSS attendees will gain valuable knowledge and insight in the following PPSS focus areas:

- **Reviewing the Past, Present and Future of Process Safety & Loss Prevention** provides us insight on how far we have come, as well as how far we have to go in these vital arenas.

- **A strong Safety Culture is of critical importance** to safety over the long-term in any organization. Experiences and ideas for measuring and improving Safety Culture and the effect of Safety Culture on decision making will be presented.

- **Risk Assessment is key to evaluating and improving** process safety and managing business risk associated with process systems. Papers demonstrating application of Risk Assessment techniques will be shared in the areas of engineered systems, risk tolerability, utility failures, and modeling of rare events and domino effects.

- **Facility Siting remains critically important to Process Safety**. The BP Texas City explosion and Hurricane Katrina and Rita impacts will be discussed. Novel approaches and applications to Facility Siting will be presented, in addition to a review of lessons from OSHA enforcement.

- **Proper application of Safety Instrumented Systems and Safety Critical Devices** is important for process plants. Application of IEC 61511 (ANSI/ISA 84.00.01-2004) to various types of failures, application of safety life cycle methods, determination of requirements for equipment manufacturers, and the use of Layer of Protection Analysis (LOPA) to address hazardous scenarios are presented.

- **Sharing and learning from past failures** can help us to avoid repeat or similar incidents in the future.

We recognize that extra effort and dedication is required not only of the organizers and the contributors to the symposium but also of the attendees in these competitive and demanding times. We are grateful to all who have contributed their time and effort to make this a successful program, including the organizing committee, authors, and session chairs and co-chairs.

We hope you enjoy the conference, and that it improves your Process Safety and Risk knowledge and expertise--providing value to your organization.

Jim Thompson
PPSS Chair

Philip Myers
PPSS Co-Chair

8TH PROCESS PLANT SAFETY SYMPOSIUM
ORLANDO, FLORIDA
APRIL 23-27, 2006

Symposium Chair

James R. Thompson
PSM Coordinator
INVISTA S.à r.l.
P.O. Box 2626
Victoria, Texas 77902
Ph (361) 572-2032
Fax (361) 572-1515
james.r.thompson@invista.com

Symposium Co-Chair

Philip M. Myers
President
Advantage Risk Solutions, Inc.
P.O. Box 510
Sunbury, Ohio 43074
Ph (740) 965-6304
Fax (740) 965-9159
pmyers@arisksolution.com

Program	Session Chair	Session Co-Chair
Loss Prevention: Past, Present, and Future	David G. Clark E.I. du Pont de Nemours & Co., Inc. Wilmington, DE Ph (302) 774-8044 david.g.clark@usa.dupont.com	Walter L. Frank ABS Consulting Wilmington, DE Ph (302) 239-0496 wfrank@absconsulting.com
Improving Safety Culture	Dr. M. Sam Mannan Mary Kay O'Connor Process Safety Center, TAMU College Station, TX (979) 862-3985 mannan@tamu.edu	Dr. Stephanie C. Payne Department of Psychology Texas A&M University College Station, TX (979) 845-2090 scp@psyc.tamu.edu
Risk Assessment	Albert Ness Rohm & Haas Croydon, PA (215) 785-7567 aness@rohmhaas.com	Katherine Pearson Rohm & Haas Texas, Inc. Deer Park, TX (281) 228-8236 katherinepearson@rohmhaas.com
Facility Siting Issues	Jack Chosnek KnowledgeOne Houston, TX (281) 538-0220 jc@knowledge1.net	Michael Livingston WS Atkins Houston, TX (832) 476-3301 michael.livingston@atkinsamericas.com
Safety Instrumented Systems & Safety Critical Devices	Harry West Mary Kay O'Connor Process Safety Center, TAMU College Station, TX (979) 845-3489 hhwest@houston.rr.com	Angela Summers SIS-Tech Solutions Houston, TX (281) 922-8324 asummers@sis-tech.com
Case Histories & Lessons Learned	John F. Murphy Retired - US Chemical Safety Board & Dow Chemical Punta Gorda, FL (941) 624-0171 hamjfm@earthlink.net	Robert P Benedetti National Fire Protection Association Quincy, MA (617) 984-7433 bbenedetti@nfpa.org

MARK YOUR CALENDAR AND JOIN US

9TH PROCESS PLANT SAFETY SYMPOSIUM
HOUSTON HILTON & GEORGE R. BROWN CONVENTION CENTER
HOUSTON, TEXAS
APRIL 22-26, 2007

SYMPOSIUM CHAIR

Philip M. Myers - President
Advantage Risk Solutions, Inc.
P.O. Box 510, Sunbury, OH 43074
Phone: (740) 965-6304
Fax: (740) 965-9159
Email: pmyers@arisksolution.com

CALL FOR VOLUNTEERS AND IDEAS

Professionals will be needed to:

- Assist on the symposium organizing committee
- Serve as session chairs and co-chairs
- Author and present new developments and applications in risk and process safety

We will be developing a call for papers and welcome inputs for topics of wide interest to industry.

If you are an experienced professional, actively involved in the fields of risk and process safety, and are interested in assisting with the 9th Process Plant Safety Symposium (PPSS), please contact me.

We look forward to your participation in the 9th Process Plant Safety Symposium in 2007—plan to join us!

Philip Myers
2007 PPSS Chair

8th Process Plant Safety Symposium

Session 3 - Global Congress Joint Keynote Session

The Global Congress Keynote session recognizes AIChE's accomplishments in process safety over the past 40 years and sets the stage for the three Global Congress symposia as well as AIChE's future efforts in process safety.

Cochairs: Erdem A. Ural, PhD
James R. Thompson

Chair: Tim Overton

Session 32 - Loss Prevention: Past, Present, and Future

An invited paper will introduce the session by highlighting the 40-year history of the Loss Prevention Symposium (LPS) and by addressing how this forum has remained on the forefront of ever-changing process safety and loss prevention technologies and practices. Papers defining the "State-of-the-Art" and illuminating where this "Art" must go in the future are invited. In addition to emphasizing life-safety issues, the LPS emphasizes the business importance of accident prevention. Papers are encouraged that discuss the long-term consequences of industrial accidents, with special emphasis on continued corporate viability and financial health, on corporate loss prevention programs, and on the ever-changing regulatory roadmap.

CoChair: Walter L. Frank

Chair: David G. Clark

Session 53 - Improving Safety Culture

A strong Safety Culture is of critical importance to Safety over the long-term in any organization. This session invites papers demonstrating approaches to reducing incidents through management processes such as Operational Discipline, Behavior Sampling, Operator Training, effective operating procedures, interactive task management systems, etc.

CoChair: Dr. Stephanie C. Payne

Chair: Dr. M. Sam Mannan

Session 106 - Risk Assessment

Risk Assessment is key to evaluating and improving the safety and business risk of process systems. This session invites papers demonstrating the application of both qualitative and quantitative Risk Assessment techniques (such as HAZOP, Layer of Protection (LOPA), Fault Tree, and QRA), as well as Risk Screening Techniques.

CoChair: Pen Lewis

Chair: Albert I. Ness

Session 127 - Facility Siting Issues

Facility Siting remains important to Process Safety, and has many aspects. This session invites papers on novel approaches and applications to Facility Siting, and on important ancillary issues such as: controlling the location of temporary buildings/trailers, controlling access of non-essential personnel to process areas during startup/shutdown/emergency operations, and best practices for Fire Protection systems (testing/maintenance/reliability).

CoChair: Michael Livingston

Chair: Jack Chosnek

148 Facility Siting and the Bp Texas City Incident, What It Means for Our Industry
P. C. (Pat) Berwanger

157 Lessons Learned through Osha's Enforcement of the Facility Siting Provisions in the Psm Standard
Mark S. Dreux

158 Facility Siting Following a Merger of Two Large Oil Companies
Gregg A. Sapp, John A. Alderman

Break

159 Techniques for Siting New Buildings in Petrochemical Facilities
Donald E. Ketchum, Martin L. Goodrich, John N. Dyer, J. Kelly Thomas

174 Infrastructure Impacts during Hurricanes Katrina and Rita and Facility Siting Issues
David A. Moore

187 Facility Siting Database and Analysis
Randy Nyman, Alex M. See, James M. Hudson, Jon D. Chrostowski

Session 190 - Safety Instrumented Systems and Safety Critical Devices

The proper application of Safety Instrumented Systems (SIS) is obviously important for process plants. However, non-SIS Safety Critical Systems (e.g., Safety Critical Devices) also play an important role. This session invites papers in the areas of:· Safety Instrumented Systems - SIS design, SIL level determination, identification of IPLs, probability calculations, application of LOPA to SIS, etc.· Safety Critical Devices (SCDs) - Examples, opportunities to apply SCDs instead of SIS, management of SCDs, application of LOPA to SCDs, etc.

CoChair: Angela Summers

Chair: Harry West

211 Risk Abatement Provided by Safety Instrumented Systems May Cause Remote Hazards with Higher Risks
Adrian Sepeda

222 Failure Conundrum
Michela Gentile, Angela Summers

Session 213 - Case Histories and Lessons Learned

Reviews of Process Safety Incidents and near misses provide valuable learning opportunities. Papers dealing with incidents, near misses and the lessons learned are requested.

CoChair: Robert P. Benedetti

Chair: John F. Murphy

Session 215 - Control and Optimization in Refining I

This session offers an opportunity to hear about process control success stories within the refining and petrochemical industry. The preferred subject matter includes innovative solutions to process control issues along with the benefits of optimization programs. If you have a process control application that would be interesting for others to learn about, please submit an abstract for approval.

CoChair: James Harris

Chair: Tim Olsen

Session 230 - Joint Case Histories and Lessons Learned

Joint Case Histories and Lessons Learned

Keynote Address
Process Safety in the Global Chemical Industry

Gary Veurink
Executive Vice President of Manufacturing and Engineering
The Dow Chemical Company

About the speaker:

Gary Veurink is Corporate Vice President, Manufacturing and Engineering. Veurink became a member of the Office of the Chief Executive (OCE) in September 2004.

Veurink joined Dow's Michigan Division in 1972 in Research in the Electro Mechanical Research Lab. In subsequent years he has worked in both technical and managerial roles in a number of Dow's businesses, functions and locations. His business and technology experiences include ChlorAlkali, Hydrocarbons, Agricultural Chemicals, Plastics Films, Engineering Plastics and Performance Chemicals. His functional experiences, in addition to Research, include Manufacturing, Engineering & Maintenance, Environmental Management, Quality Management, Materials Management and Purchasing. He has served as the leader of four of Dow's global technology centers, concurrent with several of his business and functional leadership roles. He was appointed Global Manufacturing Director for Engineering Plastics in 1995. In 1998, he was named vice president, Global Purchasing. In 2000, he was named site director for Michigan Operations and Business Operations director for Performance Chemicals. In 2004, Veurink was named business operations leader and vice president of Manufacturing and Engineering for the Chemicals and Intermediates Portfolio of businesses and in late 2004 was named to his Corporate Vice President role. He is a member of the Board of Directors of The Dow Corning Corporation, of the Dorinco Reinsurance Company, and the National Association of Manufacturers. He is currently located in Midland, Michigan.

Veurink earned a Bachelor of Science degree in Chemical Engineering from the South Dakota School of Mines and Technology. He has done graduate studies in Chemical Engineering from the University of Michigan and is a graduate of the Advanced Management Programme at INSEAD, Fontainebleau, France. He has been named an Outstanding Alumnus by SDSM&T and granted an Honorary Doctorate degree in 2002. He serves on the institution's Academic Advisory Board.

Veurink has served on numerous community boards and leadership positions in the local communities, while working for Dow in Michigan, Ohio and Texas. He is currently President and on the Executive Council of Boy Scouts of America, Lake Huron Area Council, as well as on the Board of Trustees of the Michigan Chapter of The Nature Conservancy. Veurink is a member of AIChE.

He is married to wife Ruth and has three grown children.

Keynote Address
Process Safety in the Global Petroleum Industry

John Mogford
Senior Group Vice President
Safety and Operations
BP, UK

Additional information is not available

The Expanding Role of the Loss Prevention Professional – Past, Present, and Future

By Dennis C. Hendershot[1] and John F. Murphy[2], PE

Abstract

The forty years of the Loss Prevention Symposium has paralleled the careers of several of us in the loss prevention profession. During this time period, at the Dow Chemical Company, Rohm and Haas and many chemical companies, the role of the loss prevention professional (LPP) has expanded from a concern for fire protection engineering to a more proactive concern for process hazards and process safety management. The role of the LPP has changed primarily because of lessons learned from major chemical processing incidents that have occurred. The Loss Prevention Symposium has served as a forum for discussing these incidents and related process safety technology. For example, Flixborough (1974) demonstrated the concern for large quantities of flammable material in the process. Seveso (1976) demonstrated the necessity to review processes for reactive chemical potentials. Mexico City (1984) illustrated the hazard of large inventories of flammables, the consequences of inadequate separation distances and the need for adequate firewater supply. Bhopal (1984) clearly emphasized the need to evaluate processes for large toxic inventories that could be released and cause harm to large populations. Pasadena (1989) reemphasized the hazards of large inventories of flammables. In 2003, the U. S. Chemical Safety Board investigated several serious dust explosion incidents which have awakened industry to the hazards of combustible dusts. The March 2005 Texas City refinery explosion has highlighted concerns for facility siting and the hazards of start up operations. Because the root causes of all these incidents were management system deficiencies, the current role of the LPP now focuses on the evaluation of process safety management systems as well as the evaluation of process hazards. Recently the need for a positive safety culture to successfully implement process safety management systems has been recognized. This paper will explore the effects of major chemical incidents on the role of the LPP past, present and future.

1. Introduction

The role of the LPP has expanded greatly over the past forty years. The expanded role has been driven by the lessons learned from major chemical processing incidents as well as lessons learned by companies themselves from their own incidents. The authors have selected ten incidents or series of incidents to demonstrate the connection between the role of the LPP and the lessons learned from chemical processing incidents. Many other incidents could have been selected to illustrate the point but the authors feel these incidents are especially relevant. The Loss Prevention Symposium has been a forum for discussing the lessons learned from incidents

[1] Chilworth Technology, Inc., Plainsboro, NJ, dhendershot@chilworth.com
[2] Retired U. S. Chemical Safety and Hazard Investigation Board and the Dow Chemical Company, 2304 Kenya Lane, Punta Gorda, FL 33983, Tel: 941-624-0171, hamjfm@earthlink.net

and related loss prevention technology. The interaction of LPPs at the Loss Prevention Symposium has also contributed toward their expanded role.

This year, and the recent couple of years, process safety and loss prevention professionals have celebrated significant anniversaries of a number of important activities and organizations in the field.

- 2004 – Safety and Health Division of AIChE - 25th Anniversary
- 2005 – Center for Chemical Process Safety – 20th Anniversary
- 2005 – 50th Safety in Ammonia Plants and Related Facilities Symposium
- 2006 – 40th Loss Prevention Symposium
- 2006 – 30th Anniversary of the original Design Institute for Emergency Relief Systems (DIERS) project AIChE – industry alliance
- Coming up in 2007 – 25th Anniversary of publication of *Process Safety Progress*
- Coming up in 2008 – 100th Anniversary of the American Institute of Chemical Engineers

In recognition of these anniversaries, we welcome this opportunity to discuss some of the history of safety and loss prevention, and to think about future challenges and opportunities.

2. Some "Ancient" History

Clearly, safety has always been a key element of all engineering disciplines, including chemical engineering, from their earliest origins. As often cited, the Code of Hammurabi [1], king of Babylonia from approximately 1792-1750 BC, includes the following:

- If a builder build a house for some one, and does not construct it properly, and the house which he built fall in and kill its owner, then that builder shall be put to death. (229)
- If it kill the son of the owner the son of that builder shall be put to death. (230)
- If it kill a slave of the owner, then he shall pay slave for slave to the owner of the house. (231)
- If it ruin goods, he shall make compensation for all that has been ruined, and inasmuch as he did not construct properly this house which he built and it fell, he shall re-erect the house from his own means. (232)
- If a builder build a house for some one, even though he has not yet completed it; if then the walls seem toppling, the builder must make the walls solid from his own means. (233)

Clearly, society's expectations for safety performance have increased over the years. And, while some of the penalties and punishments for poor performance may not be as harsh today, they are still significant. The Code of Hammurabi also makes it clear that it has been an expectation of civilized societies for at least 4000 years that the designer and builder of a technological artifact, such as a house, or in today's world, a chemical plant or an oil refinery, is responsible for any harm or damage that might result from the failure of that artifact. There is no role defined in the Code of Hammurabi for oversight of the builder by a safety professional to remove any of this responsibility!

The chemical industry has a long history, and safety performance at a level consistent with the expectations of the society of the time has always been a requirement for those practicing chemical technology. For example, in The Canon's Yeoman's Tale in the *Canterbury Tales*, Geoffrey Chaucer (1340?-1400) describes alchemical processes, including accidents, and an early example of a multiple cause incident investigation following the explosion of a crucible [2]:

> "One said that it was due to fire-making,
> One said it was the blowing of the thing
> (There I was scared, for that was what I did);
> 'O straw! You silly fool!' the third one chid,
> 'It was not tempered as it ought to be.'
> 'Nay,' said the fourth, 'shut up and list to me;
> It was because our fire was not of beech,
> That's why, by all the wealth I hope to reach!'"

Note that there seems to be some fear of being blamed for the explosion by at least one of the operators in this process – something that continues to be a concern with incident investigation to this day, and a difficult barrier in establishing incident and near miss reporting systems.

In the early chemical industry, processes were relatively simple and societal expectations regarding safety were low by today's standards. There was no need for a separate "safety", "loss prevention", or "process safety" role in the design and operation of chemical processes. In fact, the discipline of "Chemical Engineering" itself did not begin to evolve from chemistry, industrial chemistry, and mechanical engineering until the late 1800s. The first 4 year Chemical Engineering program began at the Massachusetts Institute of Technology in 1888, the University of Pennsylvania began offering chemical engineering degrees in 1892, and Tulane University in 1894. As chemical technology evolved and increased dramatically in complexity, and, simultaneously, societal expectations for safety in all industrial activities increased, it became clear that there was a need for increasingly specialized expertise and knowledge in safety and loss prevention for the chemical industry. As with any change, the evolution of a separate safety and loss prevention discipline does have its downside. The process designer and operator can easily succumb to the notion that safety is not his responsibility – his job is to design or operate the plant, and the safety professional's job is to look over his shoulder and tell him what to do to make sure it is safe. Much of the work in creating an effective safety culture is to continually remind those designing and operating facilities that they have the primary responsibility for safety, and that the safety and loss prevention professional can provide resources to aid in fulfilling this responsibility, but they cannot remove it.

2. Evolution of the Role of the Loss Prevention Professional

The role of the LPP encompasses a variety of jobs ranging from a process safety generalist involved primarily in project reviews, process hazard analysis, training, incident investigation and auditing to jobs that specialize in specific technical disciplines such as quantitative risk assessment, vapor cloud dispersion and hazard modeling, reactive chemicals testing, and hazard research. When we discuss the role of the LPP in this paper, we include all of these jobs. The authors have identified five stages in the evolution of the role of the LPP. The role of the modern day LPP includes activities in all these stages. The beginning and growth of these stages were overlapping and are continuing today. The five stages are:

- Fire protection engineering
- Process engineering and hazard analysis
- Hazard modeling and quantitative risk assessment
- Process safety management
- Safety culture

2.1 Fire protection engineering

Through the mid 1960's, the role of the LPP was largely influenced by the interest of the insurance companies such as Industrial Risk Insurers and Factory Mutual, and by the interests of operating companies in reducing their potential exposure to large fire and explosion losses. Incidents up to this point in time had the LPP concerned primarily with fire protection, water supply, spacing between process units and business interruption issues. These issues are, of course, still important today.

This period of time saw the initial development of many consensus standards - for example, fire prevention and protection systems, protection of vessels from overpressure when exposed to fire, spacing guidelines for storage tanks and manufacturing facilities. Much of the work of the LPP was focused on the development and application of these consensus standards.

2.2 Process engineering and hazard analysis

In the late 1960's through the 1980's, an increasingly global economy and the economic benefits of large, world scale petrochemical plants resulted in construction of a new generation of plants which were larger than plants built in earlier times. Recognition of the material and energy conservation benefits of more integrated facilities also increased the complexity of many of these facilities. This increase in size also increased the potential magnitude of accidents, and the increased complexity made it more difficult to predict potential accident scenarios. The incidents which did occur taught the value of a more proactive approach to identifying and evaluating process hazards so they can be minimized and controlled in the process design. In this period, ICI first developed the Hazard and Operability Study (HAZOP) methodology [3], and actively promoted the use of this technique through papers at technical meetings and by teaching the method to other companies. Also in 1978 Trevor Kletz of ICI first described the concept of Inherently Safer Design – an approach to risk management which focuses on eliminating hazards or significantly reducing them rather than controlling hazards through "add on" safety devices or procedures [4]. During this time the LPP became more involved in reviewing projects during the various engineering phases from conceptual design to process flow sheets, piping and instrument diagrams, and plant layout and detailed engineering. Issues such as reactive chemicals, safe automation, and inherent safety began to take center stage.

2.3 Hazard Modeling and Quantitative Risk Assessment

During the 1980's and 1990's, with the rapid maturing of computer technology, the LPP began to use various modeling techniques as tools to quantify hazards and their effects and to develop protection against process hazards. These tools were developed internally in the chemical industry, and important tools were also borrowed from other industries and technologies. For example, tools such as fault tree analysis were developed in the aerospace and nuclear power industries and were applied to the chemical industry. Other tools such as failure modes and effects analysis had a long history in aerospace and defense applications and were readily adapted to chemical processing operations. Other tools such as vapor cloud dispersion models, fire and explosion models, and impact models to predict the effect of toxic chemicals, fires and explosions on people and property, were developed and enhanced for application to the chemical industry. Quantitative risk analysis (QRA) techniques were developed to combine all of these tools and models into a comprehensive estimate of chemical process risk – QRA requires hazard identification, incident frequency (likelihood) modeling tools such as fault tree analysis, consequence estimation, modeling of the impact of fires, explosions, and toxic vapors on people, and combining all of this information into one or more quantitative risk measures. Some early and important QRA applications included the Rijnmond Study in the Netherlands [5] and the Canvey Island Study in England [6]. In 1989, the CCPS published the first edition of the *Guidelines for Chemical Process Quantitative Risk Analysis* [7], which provided a complete methodology for conducting a QRA for a chemical process.

2.4 Process Safety Management

Beginning in the 1970's, and formalized by the Occupational Safety and Health Administration (OSHA) Process Safety Management regulation in 1992, the importance of managing process safety became apparent. Two major incidents in 1984 – the major fires and BLEVEs in Mexico City and the toxic gas release in Bhopal, India – highlighted the importance of community awareness, emergency planning, and emergency response as a critical component of any process safety management program. The LPP began to develop and audit management systems to control process hazards as well as to identify and evaluate hazards. Process safety management was formalized into the elements that we know today including:

- Process hazard analysis
- Mechanical integrity
- Management of change
- Incident investigation
- Auditing
- Emergency response
- Training

2.5 Safety Culture

During the 2000's, safety culture has become an issue that the LPP has begun to deal with. The National Aeronautics and Space Administration's investigations of the Challenger and Columbia shuttle accidents suggested that management systems alone are not enough to prevent major incidents. Process safety management is important, but a positive safety culture is also necessary to ensure that process safety management systems are properly implemented and that this implementation does not deteriorate over time. Safety culture involves the beliefs and attitudes of the organization and is strongly influenced by the actions of top management. The LPP is now developing tools to measure safety culture so that safety performance can be improved further.

3. Lessons Learned from Major Incidents

The follow table summarizes some of the major lessons learned from selected incidents and their effect on the role of the LPP.

Year	Incident Descriptions	Major Lessons Learned	Effect on the Role of the LPP
1950s and 1960s	A series of incidents, for example, 1957, Montreal, Canada, (BLEVE of 3 vessels) and 1966, Feyzin, France BLEVE (18 fatalities) [8]	• Importance of spacing, fire protection systems, fire fighting systems and procedures	• Major fire and explosion incidents in the 1950s and 1960s made loss prevention professionals in industry aware of the benefits of sharing information, lessons learned, and good design and operation practices. • Safety in Ammonia and Related Facilities Symposium (1957) • Loss Prevention Symposium (1966)
1974	Flixborough, England: Catastrophic vapor cloud explosion (VCE) resulting in 28 fatalities [8]	• Public control of major hazard installations • Siting of major hazard installations • Management system for major hazard installations • Limitation of inventory of flammables in the plant • Design and location of control rooms and buildings • Control of plant and process modifications • Importance of accident prevention in maintaining business viability	• More involvement in process design • Interest in management systems begins • Development of VCE models • Inherent safety concept development begins
1976	Seveso, Italy: Reactive chemical event results in toxic cloud affecting the community [8]	• Siting of major hazard installation • Hazards of ultra toxic substances • Hazards of reactive chemicals • Inherent safer design of chemical processes • Control and protection of reactors	• More involvement in process design • Reactive chemicals becomes an issue • Recognition of importance of long term environmental contamination as a significant consequence of an accident • Process control and safe automation become an interest
1984	Mexico City: Major fire and explosion resulting in over 500 fatalities [8]	• Siting of major hazard installations • Layout and protection of LPG storage • Fire fighting in BLEVE situations • Planning for emergencies	• Reemphasized the importance of fire protection, water supply and spacing • Involvement in emergency planning
1984	Bhopal: Toxic gas release resulting in over 2000 fatalities in the community [8]	• Public control of major hazard installations • Siting of major hazard installations • Management of major hazard installations • Effects of highly toxic substances • Reactive chemicals • Limitations of inventory in the plant • Maintenance of plant equipment and instrumentation • Control of plant and process modifications • Importance of accident prevention in maintaining business viability	• Acute toxic releases become an issue resulting in loss of containment reviews • Facility siting for toxics becomes an issue • Importance of management systems such as management of change and mechanical integrity becomes apparent • Reactive chemicals is a concern • Inherent safety is further emphasized
1989	Pasadena, Texas: Vapor cloud explosion resulting in 23 fatalities [8]	• Management of major hazard installations • Hazard assessment of major hazard installations • Location of control rooms • Isolation procedures for maintenance	• Major driver toward the OSHA PSM standard resulting in involvement in process safety management both regulatory and non regulatory

Year	Incident Descriptions	Major Lessons Learned	Effect on the Role of the LPP
1998	Longford, Australia: Vapor cloud explosion results in 2 fatalities, 8 injuries, and significant disruption of the economy because of loss of natural gas supplies for up to 10 days [8]	• Inadequate operator training for upset conditions and process understanding • Organizational factors and loss of engineering expertise • Management of change	• Recognition of the potential for wide spread economic damage from incidents • Awareness of organization, training, and cultural issues in PSM
1998 through 2002	A series of reactive chemical explosions, including Paterson, NJ (1998) [9], Allentown, PA (1999) [10], Augusta, GA (2000) [11], Toulouse, France (2001) [12], Pascagoula, MS (2002) [13]	• Inadequate process hazard analysis for reactive chemicals • Inadequate regulatory coverage for reactive chemicals • Inadequate understanding of reactive chemistry hazards	• Reactive chemicals management becomes a major issue • Formation of Reactive Chemical Alliance by CCPS, industry organizations, and regulatory agencies
2003	A series of dust explosions, including West Pharmaceuticals [14], CTA Acoustics [15], and Hayes Lemmerz [16]	• Hazards of combustible dusts, including metal dusts • Recognition of dust hazards • Ineffective investigation program • MSDS and hazard communication deficiencies • Management of change deficiencies • Hazard communication and material safety data sheets	• Awareness of the hazards of combustible dusts • Awareness of loss prevention concerns and process hazards in industries outside main stream chemical processing
2005	Texas City, Texas: Vapor cloud explosion resulting in 15 fatalities [17]	• Location of facilities occupied by people, including temporary buildings • Hazards of start up operations • Hazards of emergency vent systems	• Involvement in safety culture as a loss prevention issue

In summary, lessons learned from chemical processing incidents have expanded the role of the LPP from a primary focus on fire protection engineering to wider range of concerns including process hazard evaluation and engineering, hazard modeling, process safety management, and safety culture.

4. Role of the Loss Prevention Professional in AIChE

During this same forty years the role of the LPP in AIChE has greatly expanded because of chemical processing incidents. Bradford, et al. [18] and Davenport [19] describe the history of loss prevention in the AIChE in some detail, and we will just list a few important milestones:

1950s The first AIChE symposium focusing on what would become to be known as safety and loss prevention topics was held in Baltimore in 1957 because of problems in air separation plants. This was the beginning of the Safety in Ammonia and Related Facilities conferences that are still held today.

1960s A number of sizable fires and explosions in the oil and petrochemical industries, influenced Russell Miller of Monsanto and William Doyle of Factory Insurance Association (now HSB Industrial Risk Insurers) to arrange the first Loss Prevention Symposium at the AIChE National Meeting in Houston in 1967.

1976 At the 1976 AIChE meeting in Kansas City an ad hoc meeting was held that began the process which resulted in formation of the Design Institute for Emergency Relief Systems (DIERS).

1978 Gene DeHaven of the Dow Chemical Company formed the group that proposed the formation of the Safety and Health Division to AIChE, and developed the organization of the Division.

1985 Dr. Charles Forman and Harold Kemp organized a task group formed by AIChE at the request of chemical industry leaders that established the Center for Chemical Process Safety (CCPS), with seventeen initial sponsors, in response to the Bhopal incident.

1992 Under the leadership of Owen Kubias, Joe Louvar, and Dan Crowl the Safety and Chemical Engineering Education (SACHE) organization was formed under CCPS.

5. The Future

The future role of the LPP will continue to be affected by the lessons learned from major incidents, both in the oil and chemical industries, and increasingly in other technologies and industries. In the near term methods will be developed by the LPP to measure safety culture and to develop correlations between safety culture and safety performance. There are a number of specific areas where there are concerns and opportunities in the future. Some specific examples which we have identified include the following.

- The continued increase in availability of computing power and software tools makes sophisticated modeling tools widely available. This provides an opportunity for better understanding of the likelihood of incidents, consequences of incidents, and, from that information, the risk of incidents. Increased computing power makes it possible for evaluation of alternative approaches to risk management and optimal allocation of resources, both financial and technical. The increased availability of these tools makes it possible for them to be more broadly used for understanding and managing risk – it no longer takes a team of specialists many months and hundreds of thousands of dollars to conduct a risk analysis such as the Canvey Island and Rijnmond studies of the 1970s and 1980s.

  But there is also a danger from "user friendly" hazard and risk assessment tools. They can provide an "illusion of competence" for users who do not understand the models embedded in the software, do not know the assumptions in the models, do not recognize when they are not appropriate to apply, and do not recognize when results are not reasonable. They may use the tools outside of the bounds of their assumptions, generate results which do not give an accurate understanding of the risks, and make poor decisions on managing those risks.

- The LPP has a significant challenge in sharing his knowledge with the broad community of engineers and chemists who are actually operating facilities. Many of the "new" concerns in process safety management – for example the U. S. Chemical Safety and Hazard Investigation Board's investigations into reactive chemistry hazards and dust explosion hazards – are not likely to be considered "new" to a great majority of the attendees of the 2nd Annual Global Conference on Process Safety, where this paper is being presented. But an important cause of many of the incidents which lead to these investigations was the lack of awareness of the hazard at the facility where the incident occurred. And this is not limited to small companies or other organizations – it can also happen in a small plant which is part of a big company, or even a part of a large plant with extensive technical resources, but a part of that plant which does not receive a lot of attention from those technical resources.

  Part of the problem is that, despite some significant improvements in recent years, process safety is still not adequately covered in the education of most chemists and chemical engineers. It is all too common for a chemist or chemical engineer to complete his education and begin to work in industry without ever learning that finely divided solid materials which can burn can become explosive if they are suspended in air as a fine dust, or that a mixture of materials containing nitro groups, double bonds, or other reactive structures may react violently if heated. One challenge of the LPP is to convince the chemistry and chemical engineering educational community that safety education is a part of their responsibility to society.

- "Program overload" – plants have to deal with a huge variety of Three Letter Acronyms (TLA's) – OSHA Process Safety Management (PSM), EPA Risk Management Program (RMP), equivalent programs in other countries, "add on" programs in some state and local jurisdictions, Reliability Based Maintenance (RBM), Reliability Based Design (RBD), various International Standards Organization (ISO) certifications, Safety Instrumented System (SIS) standards, Security Vulnerability Assessment (SVA), Inherently Safer Technology (IST), Manufacturing Excellence, High Reliability Organization (HRO) programs, individual corporate programs, and others. This can become overwhelming and lead to poor performance in all. Many of these programs share similar elements – what we need is an overall framework that incorporates all of them into a single program meeting all needs. If this is not done, there will be a lot of repetitious and inefficient work, and that work is likely to not be done as well as it might be [20].

- Increased use of risk based safety programs to direct resources to the areas where they have the most benefit. With continuing pressure on financial and people resources, we want to devote those resources to managing risk in the parts of a plant where they have the most impact. We don't want to spend ¾ of our time working on something that only represents 5 % of the risk. Improved tools for rapidly prioritizing risk can be used to understand where the highest risks are, and devote the appropriate resources to managing those risks. On the other hand, there is a danger that risk analysis can become an exercise with the objective of "proving" that the plant is safe and that no additional actions need to be taken. In our opinion, that is a misapplication of risk analysis – qualitative or quantitative risk analysis should be used to understand and manage risks, and not to prove safety.

  Increased use of risk analysis through increased availability of user friendly computer tools makes misapplication and over reliance on quantitative risk analysis a concern. People using QRA may come to believe that the results are truly "objective," rather than a quantification of judgment of the analyst. This is not to devalue QRA – there is enormous benefit to quantifying one's judgment, and QRA allows that quantification of judgment to occur at a level where there is more data and it may be considered to be more credible. We have a lot more faith in our ability to judge the reliability of pumps, instruments, pipes, vessels, valves, and other equipment than in our ability to judge the reliability of a complex plant. Risk analysis tools allow us to understand the reliability of the complex plant by breaking it down into smaller components and understand how they interact. But, there are hundreds or thousands of judgments in every QRA – which reliability data to use for equipment, which model to use for fires, explosions, toxic material releases, which set of incidents to model to represent to risk of the plant –

because the actual number of possible things that can happen is virtually infinite but there will never be enough time to analyze all of them. In the past, QRA specialists have been well aware of this, but more widespread use of QRA by less expert practitioners may result in a loss of perspective on the meaning of the QRA results.

- Complacency – serious incidents truly are not common, and it is likely that most people will work a full career in the chemical process industry and never have any involvement in one. This can lead to complacency – "unlikely" turns into "impossible". When we read about accidents which occur at other facilities, it can be too easy to attribute it to problems at that company or in that organization and to believe that "it can't happen here". A good safety system can be its own worst enemy – if it performs well there will be no incidents. After some time, we forget why we do the things that we do which result in no incidents. There will be cost and other resource pressures to stop doing these things – why do we need to do something to protect ourselves against something which never happens? It is too easy to forget that the reason "it never happens" is because we are doing these things to prevent incidents, and to succumb to pressures to eliminate important safety activities.

  We can fight this common human tendency through increased sharing of incidents, to remind people that incidents can happen, and also by sharing of the lessons learned from the incidents. This should not be restricted to the chemical and related industries – there is a lot to be learned from incidents in completely unrelated technology – especially about management systems and safety culture. Also, we need increased reporting and sharing of challenges to process safety systems – "near misses" – which demonstrate why we have the systems in place and what their purpose is. In doing this, we should extrapolate the potential consequences of these challenges – what might have happened if the safety systems had not been in place. Keep records of the challenges – otherwise people will forget them and, when asked, they will not remember the challenges and not object to changes which compromise the safety systems because they do not remember challenges.

- The need to compete in a world economy, with different safety standards in place in some developing economies. For example, in the winter of 2005-2006, there was major media attention to incidents resulting in 16 fatalities in coal mines in West Virginia. This attention resulted in appropriate calls for improvements in mine safety and regulation, and a call for a "safety stand-down" of all mines in West Virginia, to inspect mines and correct safety problems, on February 1, 2006 by the state governor [21]. But, how much media attention was given to the 6027 coal mining fatalities (approximately 16 fatalities per day) in China [22] in 2004? Hopefully this is a transient condition, and people in developing economies will demand safety and environmental performance equivalent to those of developed countries in the future. That is certainly the experience in the United States and Western Europe, where a safety record which was tolerated as necessary for economic development in the late 19th and early 20th century would be intolerable today. The developed economies have learned how to improve safety and environmental performance over many years, and have an opportunity to share this knowledge with developing economies, allowing them to take advantage of this experience and move more rapidly to better safety and environmental performance. International companies in developing countries can drive improved process safety in those countries. For example, if these companies require compliance with reasonable process safety practices as a condition of doing business with local suppliers, compliance with such practices by suppliers will be made more economically attractive. However, if such companies, reasoning that they do not own or control the suppliers' facilities and that they are "just customers", choose local suppliers based only on cost, with no regard for the adoption of safe practices, then the non-compliance of such practices will be perpetuated.

- Recognition that "managers are human," even though we are sometimes inclined to wonder, particularly if we read "Dilbert" too much. Management system failures are, at their root, human failures. Research in human behavior and human factors also applies to managers and management systems, and we have an opportunity to improve performance by understanding what drives both a good and a bad safety culture. We have an opportunity to apply human factors research to management systems to improve performance. We must avoid the use of tools such as behavior based safety and human factors as just another management strategy to blame the guy at the bottom of the organization.

6. Summary

Incidents cost companies money both directly and indirectly. The LPP must continue to develop the business case for top management. The LPP must advocate that loss prevention is not only the right thing to do, but also strongly impacts the profitability and survival of the corporation.

The role of the LPP will continue to expand from the refinery and petrochemical industries to industries that handle hazardous chemicals such as the electronics industry and the other industries that are customers of the traditional chemical processing companies.

Future developments in technology and, unfortunately, future incidents will likely result in lessons learned that will further expand the role of the loss prevention professional into new areas while maintaining the important activities that he/she has today. In the case of future, and past, incidents, it is essential that we continue to study them, learn and share the lessons, and apply them to prevent future incidents. We have paid the price for the lesson by having the incident, and it is foolish not to learn from the experience. We owe that to the victims of past, and any future, incidents.

7. References

[1] http://www.yale.edu/lawweb/avalon/medieval/hammenu.htm, as translated by L. W. King, accessed February 2, 2006.

[2] http://classiclit.about.com/library/bl-etexts/gchaucer/bl-gchau-can-can.htm, accessed February 2, 2006.

[3] *A Guide to Hazard and Operability Studies*. Chemical Industries Association, London, 1977.

[4] Kletz, T. A., "What You Don't Have, Can't Leak." *Chemistry and Industry*. 6 May 1978, pp. 287-292.

[5] Rijnmond Public Authority. *Risk Analysis of Six Potentially Hazardous Industrial Objects in the Rijnmond Area, a Pilot Study*, 1982.

[6] Health and Safety Executive, *Canvey: An Investigation of Potential Hazards from Operations in the Canvey Island/Thurrock Area*. HM Stationary Office, London, 1978.

[7] CCPS. *Guidelines for Chemical Process Quantitative Risk Analysis*. American Institute of Chemical Engineers, New York, 1989.

[8] Mannan, M. S. (ed.), *Lees Loss Prevention in the Process Industries,* 3rd Edition, Volume 3, Butterworth-Heinemann, Oxford, UK, 2005.

[9] "Chemical Manufacturing Incident, Morton International, Inc.", U. S. Chemical Safety and Hazardous Investigation Board, Report No. 1998-06-I-NJ, issued August 16, 2000.

[10] "The Explosion at Concept Sciences: Hazards of Hydroxylamine", U. S. Chemical Safety and Hazardous Investigation Board, No. 1999-13-C-PA, March 2002.

[11] "Thermal Decomposition Incident: BP Amoco Polymers, Inc.", U. S. Chemical Safety and Hazardous Investigation Board, Report No. 2001-03-I-GA, June 2002.

[12] Creemers, A.F.L., R.J.A. Kersten, A.C. van der Steen, G. Opschoor. "The ammonium nitrate explosion in Toulouse, France – The incident and its consequences for industrial activities", *2nd International ESMG (European Safety Management Group) Symposium*. October 8-10, 2002, Nurnberg, Germany.

[13] "Explosion and Fire: First Chemical Corporation", U. S. Chemical Safety and Hazardous Investigation Board, Report No. 2003-01-I-MS, October 13, 2002.

[14] "Dust Explosion, West Pharmaceutical Services Inc", U. S. Chemical Safety and Hazard Investigation Board, Report No. 2003-07-I-NC, September 2004.

[15] "Combustible Dust Fire and Explosion, CTA Acoustics Inc." U. S. Chemical Safety and Hazard Investigation Board, Report No. 2003-09-I-KY, February 2005.

[16] "Aluminum Dust Explosion, Hayes Lemmerz International", U. S. Chemical Safety and Hazard Investigation Board, Report 2004-01-I-IN, September 2005.

[17] News Release October 27, 2005, CSB Preliminary Findings in BP Texas City Refinery Accident. www.csb.gov.

[18] Bradford, W. J., J. A. Davenport, and R. F. Schwab. "Safety, Health and Loss Prevention in the AIChE." *Process Safety Progress* **17** (2), 1998, pp. 83-85.

[19] Davenport, J., paper to be presented at the 40th Annual Loss Prevention Symposium, April 2006, Orlando, Florida.

[20] Dowell, A. M. "Regulations: Build a System or Add Layers?" *Process Safety Progress* **20** (4), 2001, pp. 247-252.

[21] "Mine deaths spur call for 'stand-down'", www.cnn.com, February 1, 2006.

[22] http://www.chinadaily.com.cn/english/doc/2005-01/17/content_409640.htm, "Coal mine accidents kill 6,027 in China", accessed February 1, 2006.

History of the Loss Prevention Symposia
Forty Years
1967 - 2006

John A. Davenport
4889 Wakema Road
West Point, Virginia 23181
john.davenport@worldnet.att.net

Abstract

In the early 1960s, a number of sizable explosions and fires occurred in the oil and petrochemical industries. A light hydrocarbons plant in Louisiana, suffered a vapor cloud explosion. This phenomenon, while it had occurred before, was largely unrecognized in industry. At about the same time, the process industries were moving toward the "jumbo" plants. Formerly, ammonia was produced in multiple plants of 100 to 300 t/d capacity with multiple parallel lines of reciprocating compressors. However, in the early 1960s ammonia plants of 750, 1000 and 1500 t/d capacities with centrifugal machines were rolling off the designers boards. Large, single train olefins plants, with capacities of 1.5 billion pounds per year, were also being designed and erected.

As a response to these trends, a number of people from industry and insurance put together a symposium to explore ways to improve loss prevention in the process industries. This symposium contained six sessions and was held at the AIChE National meeting in Houston in 1967. Attendance was excellent and a committee was formed to prepare annual symposia on the topic. These symposia are still held at AIChE National meetings with the 40th Loss Prevention Symposium being held this year. This paper will discuss some of the highlights of these symposia and some of the people that have made them so successful over the years

* * * *

1. Introduction

This is the Fortieth annual Loss Prevention Symposium sponsored by the American Institute of Chemical Engineers (AIChE) Loss Prevention Committee. The Loss Prevention Committee is designated Area 11a. This group is a part of the Safety and Health Division of the AIChE.

The first Loss Prevention Symposium was held in February 1967 in Houston. TX.

The world in 1967 was very different from today's world. For example, in the US there was no Occupational Safety and Health Administration or Environmental Protection Agency. OSHA was formed in 1970. And even though Rachel Carson's Silent Spring was published in 1962, the EPA was not started until 1970.

In the years leading up to 1967 there were a number of large oil, chemical and petrochemical industry losses. For example:

- In 1962 an explosion in an ethylene oxide day tank at the Olin Doe Run, Kentucky, plant led to one death and three injuries and an $59 million (2004 dollars) property damage loss.
- In 1963 an explosion and fire occurred in the Dow Chemical Plaquemine, Louisiana light hydrocarbon plant. This $29 million (2004 dollars) loss was the first time that the unconfined vapor cloud explosion was recognized as a real problem even though this phenomenon have been documented as far back as 1921.
- In 1965 an explosion in the Ethyl Corporation plant in Baton Rouge resulted in a $6.5 million (2004 dollars) loss. This explosion occurred from a release of ethyl chloride from a ruptured reactor recycle line and led to the collapse of the control room.

Also in the years just prior to 1967 the industry was moving toward the jumbo single train plants. To use ammonia production as an example, prior to this period the largest ammonia plants were producing 600 tons per day. These plants had multiple parallel banks of reciprocating compressors. The loss of one bank would result in the loss of only a fraction of the overall capacity. In the early '60s, M. W. Kellogg and others began designing plants with 1000 and 1500 tons per day capacity based upon large single train centrifugal compressors. Previously it cost $30,000-$40,000/ton to build an ammonia plant. The single-train plants could be built for $12,000-$15,000/ton at tripled or quadrupled capacity. These are 1967 dollars but the economic ramifications are clear. The specter of a fire, explosion or machinery breakdown in one of the "jumbo" plants sent shudders through the loss prevention professionals in industry and insurance.

2. The Genesis

In 1965 the Executive Board of the AIChE National Program Committee asked Dr Walter Howard of Monsanto to be Chairman of the Technical Program Committee (TPC) for the Spring 1965 National Meeting of AIChE at Houston. One of the sessions for that meeting was on safety. The titles of the papers and the authors have been lost to posterity.

The successful safety session provided germination of the thought about the need for programming on process safety. The National Safety Council had regular programs on personnel safety. The National Fire Protection Association was active on fire protection projects. Another group in the AIChE had been conducting successful symposia on Safety in Air and Ammonia Plants since 1957. However, very little programming was being presented on process safety, a relatively new topic at that time.

Dr Howard then presented to the Executive Board the idea for a process safety or loss prevention symposium. He asked his boss at Monsanto at the time, Russ Miller, to be an organizer and suggested that Bill Doyle of the Factory Insurance Association be a co-organizer. These two pioneers in Loss Prevention laid the foundation for the forty years of successful symposia. Dr Howard is credited with being the light that started the ball rolling.

3. The First Symposium – February 1967.

The first Loss Prevention Symposium was held in Houston, Texas in February 1967. This symposium had 23 papers. It is interesting to explore in detail these papers, the authors that presented them and the companies that supported the authors. Of the 23 papers, 16 were presented by authors employed by process industry companies. Companies included Monsanto

(3 papers), Dow (2 papers), Amoco (2 papers), Eastman Kodak, Ethyl Corporation, Esso Research, DuPont, Tennessee Eastman, Hoffman-LaRoche, Avisun, Merck, and Celanese. Three papers were presented by insurance company employees, two by equipment vendors, one by a testing laboratory, one by a government laboratory, and none by process safety consultants. Contrast these presenters with recent loss prevention symposia where very few papers are presented by process companies and many are presented by process safety consultants.

The paper subjects were also interesting. Six papers were case studies of incidents, seven were on process safety management, three were on reactive chemicals data, two were on building design and two were on equipment design. The Loss Prevention Symposia continue the tradition of presenting case histories as each symposium has a full session on these incidents.

A number of the papers were pioneering papers which had major impact on future loss prevention activities. Bill Bradford and Tom Culbertson of Esso Research and Engineering presented a paper on design of control houses. This was the first paper to espouse the designing of control houses for explosion resistance. The design methods contained in this paper were used for many years by the industry until more rigorous methods, based on testing, were developed.

Bill Boyle of Monsanto presented a paper titled Sizing of Relief Area for Polymerization Reactors. This was the first time that the phenomenon of two-phase flashing flow through relief vents was discussed. Boyle showed that sizing relief vents based on vapor-only flow resulted in undersized vents. It was nine years later that the Design Institute for Emergency Relief Systems was founded following the presentation of a paper by Dr Alan Duxbury of ICI at the 1976 Loss Prevention Symposium in Kansas City.

Four papers were presented that highlighted work that was being done in the area of chemical reactivity:

- Ronald Wankel of Tennessee Eastman presented a paper titled *Purpose & Function of Tennessee .Eastman's Reactive Chemical Program,*
- L. Silver of Merck presented a paper titled *Screening for Chemical Process Hazards,*
- Fred Lorentz presented a paper titled *Case Study of Incident Involving Acetylenic Alcohol*
- Bob Coffee and John Wheeler of Eastman Kodak presented a paper titled *Explosibility and Stabilization of Propargyl Bromide.*

These papers and others to follow in later years have culminated in the Reactivity Management Roundtable, an effort by the AIChE Center for Chemical Process Safety to raise awareness of chemical reactivity hazards and share appropriate process safety tools.

4. Comment, Question and Answer Periods

At the first symposium and in symposia until 1972, the question and answer period following a paper or a session were recorded and transcribed for inclusion in the published proceedings. The transcribing of the comments was a major effort which usually was done by the symposium chairperson's secretary. At all sessions the session chair would remind commenters to use the microphones and give their names and affiliations. In addition, the commenter had the right to ask that the tape recorders be turned off. One very memorable incident occurred in the first symposium following the two papers dealing with reactivity of acetylenic compounds. Bill

Doyle arose to indicate that he was aware of a recent explosion incident involving a tank car full of such a compound and wondered if the representative of that company cared to comment. The representative rose, identified himself and asked that the tape recorders be turned off. A hush fell over the attendees – surely some good information was about to be given. Once the recorders were turned off the company representative said "Off the record, I have no comment."

The comment period gave a number of people a chance to state in great detail their views on a subject. In some cases these comments were a "mini-paper." Chief among these verbose commenters were Dr Howard, Dr Trevor Kletz and Bill Doyle.

5. Visual Aids

Over the years one of the most challenging jobs of session chairs has been to get speakers to use effective visual aids. Until very recently the technology did not include Microsoft® PowerPoint®. Thirty five millimeter slides and overhead projection were the norm. Either of these two media can be effective if the presenter would abide by one major principal:

"Don't put too much information on a single slide."

Invariably presenters would show up with overhead transparencies made from a standard sheet of paper with the information typed by a standard typewriter . Thirty five millimeter slides would have graphs on them that contained so much information that the lines and legends were unreadable. There were several presenters over the years that did not bring notes to speak from, intending to speak from their projected slides or overheads. However, their slides were so bad that they could not read them themselves from the podium.

To aid speakers, the AIChE produced a pamphlet on developing visual aids. The 11a committee obtained and supplied to authors a pamphlet from Eastman Kodak on developing effective visual aids. Dr Howard wrote many letters to the committee protesting poor slides and on a few occasions rose in the symposium to comment on particularly poor slides used in a presentation. He was not only a founder of the symposia but also a guardian of paper and slide quality.

Today, it is much easier for authors not proficient in visual aids to produce excellent slides using Microsoft® PowerPoint® However, authors still come with slides that are unreadable from the rear of the meeting halls because they do not abide by the above simple principle. Today, Dr Howard does not attend symposia to berate authors and the Loss Prevention Committee about poor slides.

6. Preprints and Proceedings

The presentations at the Loss Prevention Symposia are, of course, very important. However, in order for the technology presented to have lasting value, the written papers are the most important. For the first thirteen years the AIChE published the papers in Loss Prevention series booklets that took about nine months to produce. Part of the delay was caused by the time it took to transcribe the question and comment tapes.

For the next number of years, there was no publication of the papers. During these years and the preceding thirteen, authors were requested to bring 250 to 300 copies of their papers to the symposium for hand-outs. The preprints were placed in the back of the meeting room for the attendees to take. These preprints were the only copies of the papers once the AIChE stopped publishing the Loss Prevention series.

Later, the Loss Prevention Committee began requiring authors to send a copy of their paper to to the Symposium Chairman several months before the symposium. The Chairman would assemble the received papers into a book which was printed for sale to symposium attendees. Of course there were always authors that did not make the publication deadline. These were asked to bring the 250 to 300 copies for hand-out at the symposium. Thus, during these years, it was necessary to obtain the proceedings book and grab any preprints in the back of the meeting room to have a full set of papers. Later, the AIChE agreed to take over the publication of the proceedings book.

In 1998, the AIChE and the Center for Chemical Process Safety (CCPS) published all of the Loss Prevention Symposia papers from 1967 to 1997 in compact disk format. Randy Freeman, then of Monsanto, spearheaded this effort. Most people who had been attending the symposia over the years had dispersed their copies of the papers into personal or company libraries. Fortunately, some members of the Loss Prevention Committee, this author included, had kept all of the papers in one lot. These were used to populate the AIChE produced CDs. Now, for a nominal fee, everyone had all of the valuable loss prevention information in a searchable, easily retrievable format. The CCPS/AIChE issued an updated CD in 2004 that covers all papers from 1967 through 2003.

From the start of the symposia, certain chosen papers were published in *Plant Operation Progress*, *Plant Safety Progress* and *Chemical Engineering Progress*. So, in some cases, there are two versions of some papers – one as handed out at the symposia and one as published in the AIChE publications.

7. Membership in the Loss Prevention Committee

The number of members in the Loss Prevention Committee is limited by the by-laws to 20 active Members plus Emeritus Members and International Members. Membership is by a majority vote of the committee members. Over the years we have identified 87 people who have been members of the group. A list of these people is contained in the Appendix. 1. Note the companies that are represented. Just about every major chemical, petrochemical and petroleum producer in the country is or has been represented.

It is interesting to note that Dr Walter Howard, the virtual founder of the group, has never been a member of the Loss Prevention Committee. He has been a frequent contributor and a more frequent critic.

8. DIERS and the Loss Prevention Symposia

At the April 1976 Loss Prevention Symposium in Kansas City, Dr. Alan Duxbury of ICI presented a paper outlining the weaknesses in our knowledge of chemical reactor venting design. Dr. Walter B. Howard arranged for a meeting that was held after the regular Loss Prevention Symposium in the basement of the of the meeting hotel. About 50 interested people attended the meeting and the first seeds of the Design Institute for Emergency Relief Systems (DIERS) were planted. Mr. Harold Fisher of Union Carbide was an early founder of DIERS.

DIERS was a consortium of 29 companies formed under the auspices of AIChE to develop methods for the design of emergency relief systems to handle run-away reactions. Of particular interest were the prediction of when two-phase flow venting would occur and the applicability of various sizing methods for two-phase vapor-liquid flashing flow. DIERS spent approximately \$1.6 million to investigate two-phase vapor-liquid onset / disengagement dynamics and the hydrodynamics of emergency relief systems. The primary contractor on the original DIERS work was Dr. Hans Fauske.

The DIERS research program came to a conclusion in 1985 with the publication by AIChE of:

- SAFIRE Computer program and documentation
- A design manual "Emergency Relief System Design Using DIERS Technology"
- Technical Information in 19 manuals, 1950 pages

The work of DIERS was presented at the Loss Prevention Symposium in 1985 in Houston.

Today DIERS is the DIERS Users Group which is made up of 126 companies. This group meets twice per year to cooperatively assimilate, implement, maintain and upgrade the DIERS methodology. This group also sponsors sessions at AIChE National Meetings.

9. The William H. Doyle Award

In 1984 following the death of Bill Doyle, the Loss Prevention Symposium Committee created the William H. Doyle Award. This annual award is presented to the person presenting the most outstanding paper at the previous year's Loss Prevention Symposium. The criteria for the award are:

- A paper that is clearly and well presented with good effective slides
- A paper has wide applicability to the Chemical Industry Loss Prevention efforts
- A paper that adds substantial knowledge to the field of Loss Prevention
- A paper that engages and stimulates the intellects of the audience
- A paper that would have caused Bill Doyle to rise to his feet, ask a penetrating question, make a decisive comment and lead the applause.

The first recipient of the award was Dr. James McQuaid of the United Kingdom's Health and Safety Executive. Annual awards are still presented at each Loss Prevention Symposium. A list of all recipients to date is in Appendix 2.

10. Conclusion

The AIChE Loss Prevention Symposia has for the last 40 years presented papers with the very latest technical and programatical information on loss prevention in the chemical and petroleum industries. Many papers have been the germination of important loss prevention concepts and programs. The 1985 paper by Trevor Kletz introduced the concept of inherent safety. Papers on risk assessment, consequence analysis, management of change, auditing, chemical reactivity screening and other subjects formed the basis for discussion leading to the various process safety management programs in use today.

The improvement of visual aids production such as PowerPoint® and electronic publishing of papers on CD have made the symposia more interesting and the information more accessible. It can only be speculated what will impact future symposia. The AIChE Loss Prevention Symposia Committee is already planning for future meetings.

Appendix 1.

Members of the AIChE Loss Prevention Symposia Committee 1967 – 2005

The following is a list of people who have served on the AIChE Loss Prevention Symposia Committee since its inception in 1967. The list was compiled from fragmentary committee meeting minutes, committee rosters and the failing memories of committee members. If there are omissions or errors, we apologize and welcome comments.

Name		Company/Affiliation (at the time of joining the LPS 11A committee)	1st Year
Adcock	C. T.	Dow Chemical Co	1975
Albrecht	A. R.	Dow Chemical Company	1968
Arendt	J. Steve	JBF Associates, Inc.	1987
Auzine	M. J.		1969
Ayers	A. W.		1968
Badger	Fred W.	Hartford Steam Boiler Inspection and Insurance Co.	1968
Benedetti	Robert E.	National Fire Protection Association	1987
Birtwistle	John	Monsanto Company	1987
Bosworth	David A.	Mobil Research & Development Corp.	1987
Bradford	William J.	Olin Corp.	1968
Brock	Norman H.	Air Products & Chemicals, Inc.	1975
Burklin	C. Raymond	Brown & Root	1975
Chatrathi	Kris	Fike Corporation	1993
Clark	David G.	DuPont Company	2001
Corona, Jr	Albert A.	Mobile Research & Development Corp.	1977
Crowl	Daniel A.	Wayne State University & Michigan Tech	1989
Davenport	John A.	Industrial Risk Insurers	1973
DeHaven	Eugene S.	Dow Chemical Co.	1978
Denevers	Noel H.	University of Utah	1968
Doyle	William H.	Factory Insurance Association/Consultant	1967
Dunbobbin	Brian R.	Air Products & Chemicals, Inc.	2003
Duxbury	H. Alan	Imperial Chemical Industries, PLC	1983
Febo	Henry L.	Factory Mutual Research Corp.	1991
Frank	Walter L.	EQE International	2000
Freeman	Randy	ABS Group Incorporated	2000
Gibson	Thomas O.	Dow Chemical Company	1989
Gillian	J. E.		1973
Going	John E.	Fike Corp.	2005
Grabowski	G. J.	Fenwal Inc.	1968
Griffin	Michael L.	Procter & Gamble	1995
Grossel	Stanley S.	Hoffmann-LaRoche; Process Safety & Design, Inc.	1969
Grounds	Cheryl A.	Baker Engineering & Risk Consultants, Inc.	2005
Hanauska	Chris	Hughes Associates Inc.	2001
Hendershot	Dennis C.	Rohm & Haas Company	1997
Hill	Russel G.	Emeritus Member	1984
Howard	Walter	Monsanto	1982

Name		Company/Affiliation (at the time of joining the LPS 11A committee)	1st Year
Johnson	Robert W.	Battelle Memorial Institute	1998
Kirby	David	Union Carbide Corporation	1987
Kletz	Trevor A.	ICI, Ltd.	1978
Kubias	F. Owen	SCM/Glidden Coatings and Resins	1973
LaCoursiere	Jean Paul	Universite de Sherbrooke	2005
Leonard	Paul	Hartford Steam Boiler Inspection and Insurance Co.	1981
LeVine	Richard. Y.	Olin Mathieson Chemical Corp.	1968
Long	Lisa	U. S. CSB	2005
Louvar	Joseph F.	BASF Corporation	1987
Ludwig	Earnest E.	Ludwig Consulting, Engr., Inc.	1979
Mannan	M. Sam	Texas A&M University	1998
Mashuga	Chad V.	BASF Corporation	2003
McKinley	Clyde	Air Products & Chemicals	1971
Mill	Robert C.	Exxon Chemical	1981
Miller	Russell L.	Monsanto/Consultant	1967
Munson	Richard E.	DuPont Company	1979
Murphy	John	RMT Inc.	1998
Natale	Joseph R.	Mobil Technology Company	1998
Noronha	John A.	Eastman Kodak Company	1979
Ormsby	Robert W.	Air Products & Chemicals, Inc.	1981
Ostrowski	Scott W.	Exxon Chemical Co.	1998
Page	Gary A.	American Cyanamid Co.	1991
Peterson	Parker	Fenwal Inc.	1971
Phillips	Gary C.	Dow Chemical Company	1999
Powers	G. J.	Carnegie-Mellon University	1973
Preston	Malcolm	ICI	1995
Puckett	R. A.		1969
Quarton	Evan F. R.	Johnson & Higgins	1987
Rogerson	John E.	Procter & Gample Co.	1977
Sachere	R. M.		1970
Sanders	Roy E.	PPG Industries	1979
Santos	Anthony	Factory Mutual Research Corp.	1979
Scheier	Ephraim A.	OxyChem	1998
Schreiber	Sanford	Allied Chemical Corporation	1968
Schwab	Richard F.	Allied Signal Inc.	1975
Senecal	Joseph A.	Fenwal Safety Systems, Inc.	1992
Sethuraman	R.	Indian Petrochemicals Corp. Ltd.	1987
Swift, PhD	Ian	American Cyanamid; Fike; Himont, U.S.A. Inc.	1979
Sykes	Wendell G.	Arthur D. Little Company	1968
Thornberg	William A.	HSB Industrial Risk Insurers	2000
Ural	Erdem A.	Fenwal Safety Systems, Inc./Loss Prevention Science & Tech.	1998
Ventrone	Theodore A.	American Cyanamid Co./Consultant	1968
Walker	J. J.	Union Carbide Corporation	1970
Weldon	George E.	Factory Mutual Research Corp.	1968
Willey	Ronald J.	Northeastern University	2005

Name		Company/Affiliation (at the time of joining the LPS 11A committee)	1st Year
Williams	Mary M.	U.S. DOT/Coast Guard	1979
Windsor	Donald G.	DuPont Company	1971
Wixom	Everet D.	Exxon Chemical Co.	1991
Wood	William S.	Sun Oil Company/Consultant	1979
Worthington	David R. E.	DNV Technica	1995

Appendix 2
William H. Doyle Award Recipients 1987 - 2005

Symp. Year	Presenter	Title
1984	James McQuaid	Trials in Dispersion of Heavy Gas Clouds
1985	Trevor A. Kletz	Inherently Safer Plants
1986	Laurence G. Britton	Thermal Stability for Ethylene at Elevated Pressures
	Ian Swift	The Performance of Low Pressure Explosion Vents
1987	Robert A. Mancini	The Use (and Misuse) of Bonding for Control of Static Ignition
1988	Dean K. Wilson	Failure Mode Management: A Loss Prevention Philosophy for Programmable Logic Controllers
1989	Laurence G. Britton	Combustion Hazards of Silane and Monosilicon Chlorides
1990	Daniel A. Crowl	Using Thermodynamic Availability to Determine the Energy of Explosion
1991	Arthur. M. Dowell	Flashback from Waste Gas Incinerators into Air Supply Piping
1992	James.E.S. Venart	To BLEVE or Not to BLEVE: Anatomy of a Boiling Liquid Expanding Vapor Explosion
1993	Norman E. Scheffler	Vapor Suppression of Chemicals using Foams
1994	T. C. Hofelich	The Determination of Compatibility via Thermal Analysis and Mathematical Modeling
1995	Ronald G. Eierman	Improving Inherent Safety with Sealless Pumps
1996	Francesco Tamanini	Modeling of Panel Inertia Effects in Vented Dust Explosions

Symp. Year	Presenter	Title
1997	Norbert. Jaeger	Determination, Prevention, and Mitigation of Potential Hazards Due to the Handling of Powders during Transportation, Charging, Discharging, and Storage
1998	Dennis C. Hendershot	Peroxide Drum Explosion and Fire
1999	Peggy Westfall-Lake	Human Factors: Preventing Catastrophic Human Error in 24-Hour Operations
	Chad. V. Mashuga	Flammability Zone Prediction – Using Calculated Adiabatic Flame Temperatures
2000	Angela. Summers	Using Instrumented Systems for Overpressure Protection
	Peter Lodal	Case History: Steam Line Rupture at Tennessee Eastman Division 11/3/98
2001	Robert Johnson & Thomas Oakey	Combustion Safeguards Test Intervals - Risk Study and Industry Survey
2002	Dennis C. Hendershot	Connections: How Pipe Failure Resulted in Resizing Vessel Emergency Relief Systems
2003	Erdem A. Ural	Airplane Fuel Tank Explosions
2004	Peter N. Lodal	Distant Replay: What Can Reinvestigation of a 40-year Old Incident Tell You?
2005	Erdem A. Ural	Dust Explosion Venting Through Ducts

Looking from the Past to the Future: Is Loss Prevention affected by Globalisation?

F. Alfert
Inburex Consulting GmbH
August-Thyssen - Str. 1-59067
Hamm, Germany

The globalization process challenges not only existing industrial and governmental structures. The globalization feeds an accelerating process of industrial concentration, rationalization and moving of entire production facilities to third world countries. For economical reasons production of goods is concentrated in larger, "mono product" facilities, plants are designed to deliver 100 % of designed production capacity 365 x 24 a year. The paper presented describes a rather interesting finding that the formation of common markets and the step to a free global market with all the changes and challenges involved create new and sometimes unexpected risks and challenges to the Loss Prevention.

Globalization has increased the trans-boundary nature of both risk and vulnerability, decision making in safety matters is not subject to a single global authority, but to different institutions and authorities being or feeling responsible for risk framing, assessment and management. Inconsistencies, misunderstandings and communication problems are immanent, Guidelines and Standards addressing the same problem but offering deviating results and solutions increase the confusion.

Contributing to the design of Guidelines, Standards and risk assessment procedures from a national platform (Germany, VDI) through the European Standardization Process (CEN Working Groups) to international working groups (ISSA, NFPA) over the last 25 years has given the insight and contributed to develop an understanding that the globalization process is a challenge to the Loss Prevention and that industrial installations in practice may need a more differentiated risk assessment with networking actors and new evaluation and protection techniques that consider not only the present protection aspects but also aspects such as loss of production and its impact on others. Risk benefit evaluations and integrated risk disaster management may be other aspects.

The Current Status of PSM Diligence

Carolyn W. Merritt
Chair and CEO of the US Chemical Safety Board
2175 K St NW Suite 650
Washington, DC 20037

The influence of the Process Safety Management Regulations on chemical and petroleum manufacturing and processing industries has been profound. In the early years there was a genuine effort of the covered industries to comply with the provisions that most good managers recognized as good engineering practice. There was a lot of effort put into the Process Hazard Analysis and into developing and implementing mechanical integrity and management of change and all the other provisions. In the 10 years since its implementation, the question can be asked whether the same diligence has maintained the programs in these critical industries. This presentation will present some examples of accidents where the diligence to maintain mechanical integrity, management of change and perform and complete process hazard analysis may indicate that the rigor of the rule may be deteriorating. The paper will look at what may be contributing to that loss of rigor.

The impact that a good corporate culture for safety can have on the maintenance of PSM compliance or the impact of a weak culture will be explored. The presentation will offer some ideas that might give the audience measures to use to identify if a facility could be at risk of the effects of a sliding PSM diligence due to specific cultural issues.

Evolution of the Corporate KPC HSE Management System

Fadhel A. Al-Ali
Manager Health & Safety
HSE Department
Kuwait Petroleum Corporation
P.O.Box 26565, Safat
Pincode 13126, Kuwait
Phone: 00965 - 499 4112
f.ali@kpc.com.kw

Abstract

Kuwait Petroleum Corporation (KPC) is committed to balancing the professional and optimal exploitation of Kuwait's hydrocarbon resources, the proactive management of the health and safety of our employees, contractors and the community, and the protection of the environment. We believe that good HSE performance is an essential and integral part of efficient and profitable business.

Regrettably during 2000 – 2002, there were two major accidents in the KPC Oil Sector, which resulted in major explosions and fires. We also had two other safety incidents during maintenance. These four incidents resulted in the tragic loss of life. The two major explosions resulted in asset and business losses in the hundreds of millions of dollars (US). The challenge was how to learn from these tragic accidents and to dedicate the resources needed to avoid them in future.

Toward that end, in 2002, KPC adopted a mission statement that included the goal of becoming the regional leader in HSE Performance. This necessitated the implementation of an improved HSE management system (MS). This paper presents important aspects of that journey.

With the implementation of the HSE MS at KPC and at its subsidiaries it is hoped that a drastic reduction in the number and in the severity of injuries and incidents will occur as KPC continuously pursues world-class HSE Performance.

Table of Contents

Acronyms

KPC	Kuwait Petroleum Corporation
KOC	Kuwait Oil Company
KNPC	Kuwait National Petroleum Company
KUFPEC	Kuwait Foreign Petroleum Exploration Company
KPI	Kuwait Petroleum International
PIC	Petrochemical Industries Company
KAFCO	Kuwait Aviation Fuelling Company
KOTC	Kuwait Oil Tanker Company
HSE MS	Health Safety Environment Management System
BPD	Barrels per day
VLPC	Very Large Product Carrier
EAMD	Executive Assistant of Managing Director

Definitions

Vision - a positive picture of where a Corporation or Company sees the business going. A vision is positive, aims to create value, engages and motivates all stakeholders who share the Vision. It specifies the destination, but not the route.

Mission – the means and route to achieve the 'Vision'. The Mission specifies the key elements of the strategies used and is involved with objective setting. Strategy content must be both feasible and achievable.

Stakeholders – are shareholders or owners, customers or clients, employees and contractors, community and interested parties.

Policy – is a corporate position statement that directs current and future business decisions and activities to carry out core values and behaviors held critical to the corporation and its businesses.

Protocol – provides the framework for implementing the corporate commitment to HSE and defines the managing process for developing, approving and maintaining HSE policies, standards, guidelines and goals.

Standards – a document containing at least one mandatory requirement, and may also contain non-mandatory advice, instructions, and / or education.

Guidelines – a document containing only non-mandatory optional advice, instruction, and/or education.

Signpost – a document that guides an interested person to see how two differently structured systems match each other

Responsible Groups – a group that is responsible, accountable for implementing the requirement provided by the document.

Contact Active Groups – is a task force networked group within organization who are responsible for conducting the implementation of requirement provided in the document.

'Irrelevance' deviation – a deviation from requirement which is unrelated to the concerned working environment or process or operation.

'Gap' Deviation – a deviation from requirement which needs to be planned in the future

'non-compliance' deviation – a deviation from requirement which is not being met or planned to be complied with.

Requirement for audit plans – the marking of a requirement which must be placed in considered audit program.

Evolution of the Corporate KPC HSE Management System

1. Kuwait Petroleum Corporation (KPC)

Kuwait Petroleum Corporation is the state-owned entity responsible for Kuwait's hydrocarbon interests throughout the world. As part of the global energy industry, KPC helps to supply the world with its vital oil and gas needs by exploring, producing, refining, transporting and marketing these precious natural resources both in the home country and internationally. Kuwait Petroleum Corporation is universally recognized as one of today's top ten oil energy conglomerates, and as a leader in providing safe, clean energy to the global markets.

Kuwait Petroleum Corporation (KPC) directs petroleum exploration and production, refining and marketing, petrochemicals and transportation towards a holistic and sustainable outcome. KPC commitment is to balance the professional and optimal exploitation of Kuwait's hydrocarbon resources with proactive management of the health and safety of our employees, contractors and community and care for the environment. We believe that good HSE performance is an integral part of efficient and profitable business.

KPC was first established in 1980 in order to bring together all state-owned elements of the Kuwait oil sector under one corporate umbrella. Today, KPC oversees a fully-integrated industry with operations on six continents. We bring hydrocarbon energy from our own domestic reservoirs and our upstream interests abroad direct to the consumer through a series of specialized subsidiary operating companies.

The Corporation is managed by the Board of Directors which in turn report to the Supreme Petroleum Council. The chairman of the Board is Kuwait's Minister of Energy.

The Vice Chairman, Mr. Hani Hussain, serves currently as the Executive President; other Board Members include Managing Directors for Transportation; Kuwait Project; Petrochemical Industries; International Marketing; Refining and Local Marketing; in addition to Petroleum Services and Exploration and Production as well as the Under secretary of the Ministry of Energy (Oil) and other non-executive directors.

2. KPC Affiliates

Kuwait Oil Company (KOC) was formed in 1934 as a joint venture between BP (U.K.) and Gulf Oil (U.S.A.) for crude oil exploration and development. Despite catastrophic damage resulting from the Iraqi invasion of 1990, production today has recovered to 2 million barrels per day, a testimony to KOC's strength and pride.

Kuwait Oil Tanker Company (KOTC) was founded in 1957 and, at present, has one of the Arab world's largest fleets for marine transportation of fuel - with a combined lifting capacity of more than 4 million dead-weight tones. Pioneers in the field, KOTC was the first company in Europe to use the Very Large Product Carrier (VLPC).

Kuwait National Petroleum Company (KNPC) is responsible for refining and gas liquefaction activities in Kuwait, with a three-refinery complex of over 800,000 bpd capacity. KNPC supplies the fuel for Kuwait's power and desalination plants, as well as operating a chain of filling stations across the nation.

Kuwait Petroleum International Limited (KPI) directs refining and marketing throughout Europe and operates a wholesale, retail and aviation fuel sales network in Asia. KPI, which introduced the first unleaded fuel and advanced diesel for passenger vehicles in Europe, now markets more than 225,000 barrels per day in 12 countries through 5,500 stations in the Q8 network.

Kuwait Foreign Petroleum Exploration Company (KUFPEC) manages the Corporation's foreign upstream activities with interests in oil and gas fields in eight countries worldwide. The largest are the South China Sea's Yacheng gas field, and in the Jannah complex of fields in Yemen.

Kuwait Aviation Fuelling Company (KAFCO), which serves Kuwait International Airport, underwent rapid growth in the 1970s with the "jet revolution" in air transportation. Today, KAFCO has a storage capacity of more than 30 million liters, and the facilities to service over 1,500 aircraft per month.

Petrochemical Industries Company (PIC) is a leader in the fertilizers and petrochemicals sectors. The first company to produce fertilizers in Kuwait using natural gas, PIC now exports its products fertilizer, feedstocks, and ammonia to more than 50 countries world-wide.

3. KPC Strategic Directions

During mid 2002 KPC worked on the new strategic directions for the next 20 years. In that plan, KPC created a new mission and vision statement challenging KPC to become the regional leader in HSE Performance. Apparently KPC wanted to continuously enhance their HSE statistics and prove to the world that when they talk business, they put HSE on top.

The KPC's strategic directions on the HSE front are:

- Comply with local HSE Standards and Regulations
- Pursue application of the best and most appropriate international HSE Standards in order for KPC to become a regional leader in HSE performance.

The action plan drafted to achieve the strategic directions on HSE was:

- Develop the HSE policy, vision and mission.
- Develop Corporate HSE Standards and Guidelines in line with local and international standards.
- Assist KPC Subsidiaries to develop HSE procedures to implement KPC HSE Standards.
- Implement and comply with KPC HSE Standards and Local Regulations
- Assess (by a third party) the subsidiaries' systems against KPC HSE Standards and Guidelines.

4. The Need for the Development of an HSE MS

The four major accidents in 2000 - 2002 resulted in the tragic and unnecessary loss of twelve lives and resulted in asset and business losses in the hundreds of million of dollars. These losses implied that our HSE systems were not effective enough. So our highest priority then was to improve our process safety performance record. Our challenge was how to learn from these tragic accidents and to ensure that we dedicate all the resources needed to avoid them in future.

Improving HSE standards and performance did become a KPC priority. Logically, to become one of the best, you have to associate with the best as goes the old adage. To become a regional leader in HSE, KPC knew the smartest and wisest thing to do was to seek the expertise of a world leader in HSE. Thus KPC sought the help of experts in the field to help evaluate and improve its HSE Management Systems. Thus a partnership of improvement was born between KPC and DuPont.

5. Agreeing on the Consultancy

KPC did a pre-project analysis of various Consultants capabilities, HSE Records analysis, past history of the Consultants in terms of process safety and personnel safety records and the consultants' focus areas/expertise in the HSE fields. We have received very positive feedback from those who have chosen them as partners to improve their philosophy and systems dedicated to the protection of people and environment. DuPont was also very glad to cooperate with Kuwait Petroleum Corporation, one of the global leaders in the field of energy, in our drive to meet the long-term objectives of becoming a leader in protection of people, asset and the environment.

The criteria for selecting DuPont were its relatively unblemished HSE history and its current performance. DuPont is able to deliver consultancy services based not only on theory but also the practice of operating businesses at various locations. DuPont did demonstrate that its success at reducing incidents and injuries can be applied across the wide range of activities in the oil sector.

6. KPC Corporate HSE Management System

6.1 The Agreement

An agreement was signed in January 2003 between KPC and the consultancy to develop a corporate HSE Management System (MS) and the required standards and guidelines. DuPont provided the consultancy services for the Project, which is based on the proven successful system developed by DuPont. Signing this agreement is a clear manifestation of KPC's earnest effort to elevate its corporate HSE MS to its optimal level. Improvement in HSE has become a pressing priority for the oil sector and something KPC is strongly committed to. The first step in this 'direction and commitment' was this signing of the agreement.

This project is part of an on-going process to drive improvement in occupational health, safety and environmental protection at KPC and its affiliated companies.

6.2 Objective of the Project

The purpose of this project was that DuPont to provide consultancy services for KPC for the design and development of a Corporate Health, Safety and Environmental Management System (HSEMS), and necessary standards, guidelines to meet KPC vision. This is based on the successful system developed by DuPont and applied at all of its facilities and locations around the world. A DuPont team consisting of HSE senior consultants worked closely with KPC to plan and execute the project.

6.3 Focus Areas

The agreement with DuPont did cover three main areas.

- The first is establishing comprehensive HSE management systems in coordination with HSE officials in the Oil Sector. The systems to cover important elements such as Hazard and Risk Management, Contractor Safety Training, Mechanical Integrity and Quality Assurance, in addition to other elements of HSE MS.
- The second area is evaluating the operational status of all our facilities and schedule regular maintenance programs.
- The third is ensuring the proper training for all oil sector operators on how to recognize hazards in the work place, how to prevent the hazards, and safe evacuation in the event of emergencies.

6.4 Elements, Policies and Standards of the KPC HSE MS – Model Structure

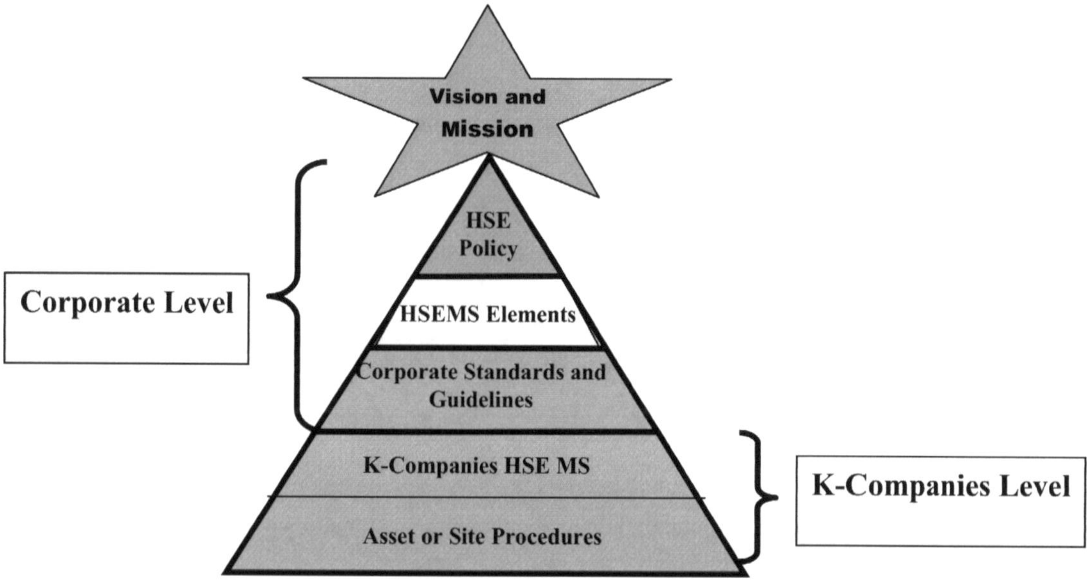

KPC Corporate HSE Management System Model

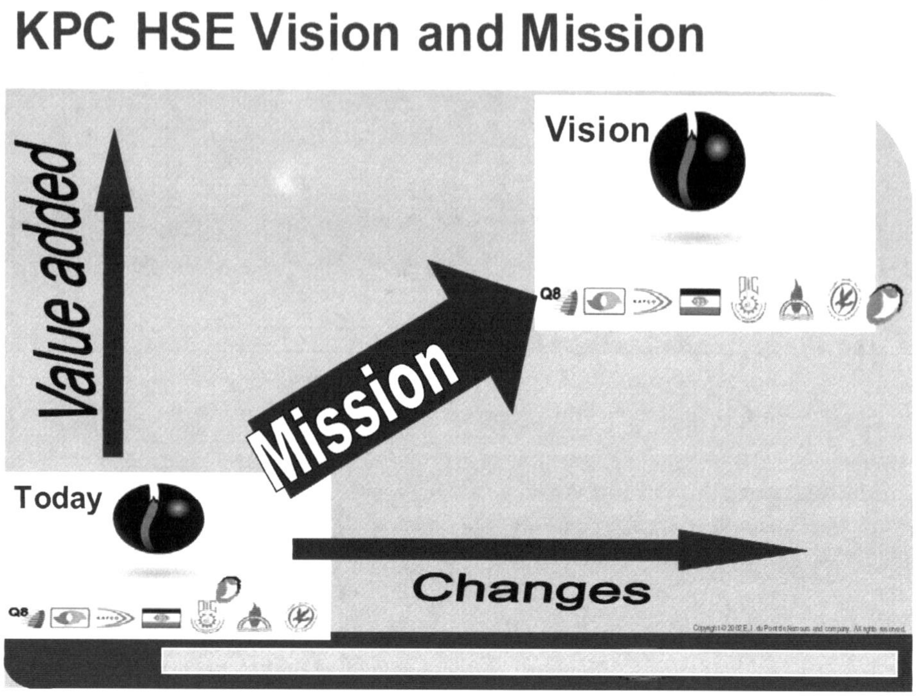

6.5 KPC HSE MS Policy Documents

KPC HSE MS Policy documents include the following:

1. HSE Vision and Mission
2. HSE Protocol
3. HSE Commitment
4. HSE Guideline

These four policy documents forms the basis for the development of KPC HSE MS Standards, which takes into account the essential elements of KPC HSE MS – 16 Elements.

There are 11 packages of HSE Standards, whose descriptions are given below.

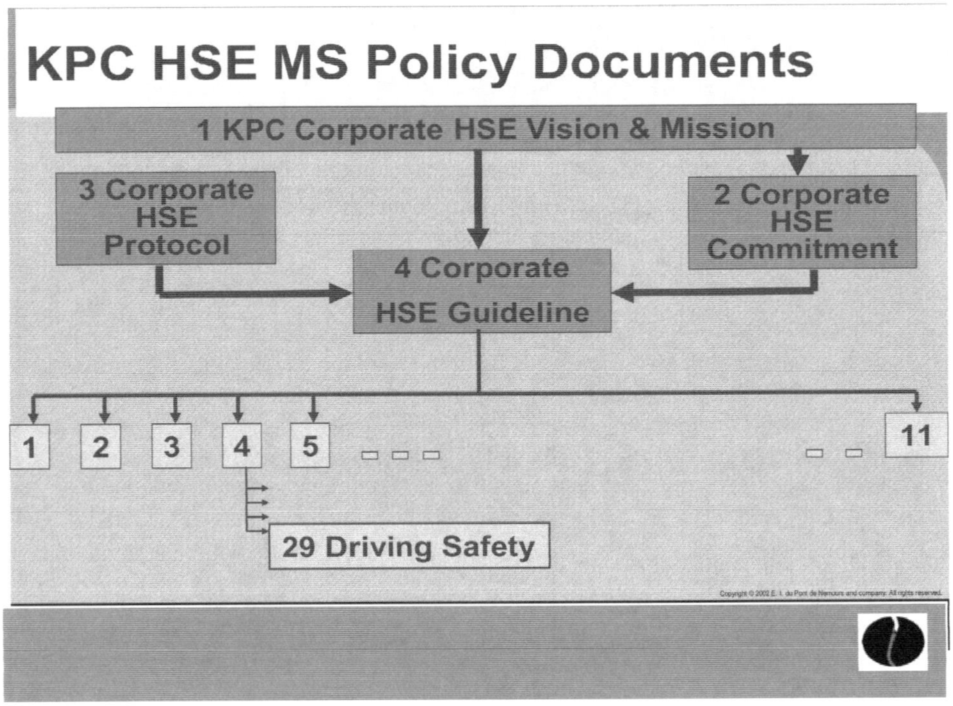

6.6 The Elements of the KPC HSEMS are:

1. Management Commitment and Responsibilities
2. Integrated HSE Organization
3. Challenging Goals and Objectives
4. High Performance Standards
5. Training and Performance
6. Effective Communication and Motivation
7. Behaviour and System Audit
8. Incident Investigation, Reporting and Analysis
9. Management of Change
10. Contractor Management
11. Mechanical Integrity and Quality Assurance

12. Hazard Identification and Risk Management
13. Emergency Response and Contingency Plan
14. Environmental Management
15. Occupational Health Management
16. Legal and other Requirements

The 59 standards and guidelines fall under the following 11 main categories.

- Policy - 4 Standards
- Process Safety Management – 7 standards
- Environmental Management – 9 standards
- Safety General – 10 standards
- Fire protection – 7 standards
- Occupational health – 7 standards
- Project HSE Requirements – 1 standards
- Distribution and Emergency Response – 6 standards
- Auditing - 1 standard
- Protective Equipment – 4 standards
- Incident Management – 3 standards

6.7 The Documents System

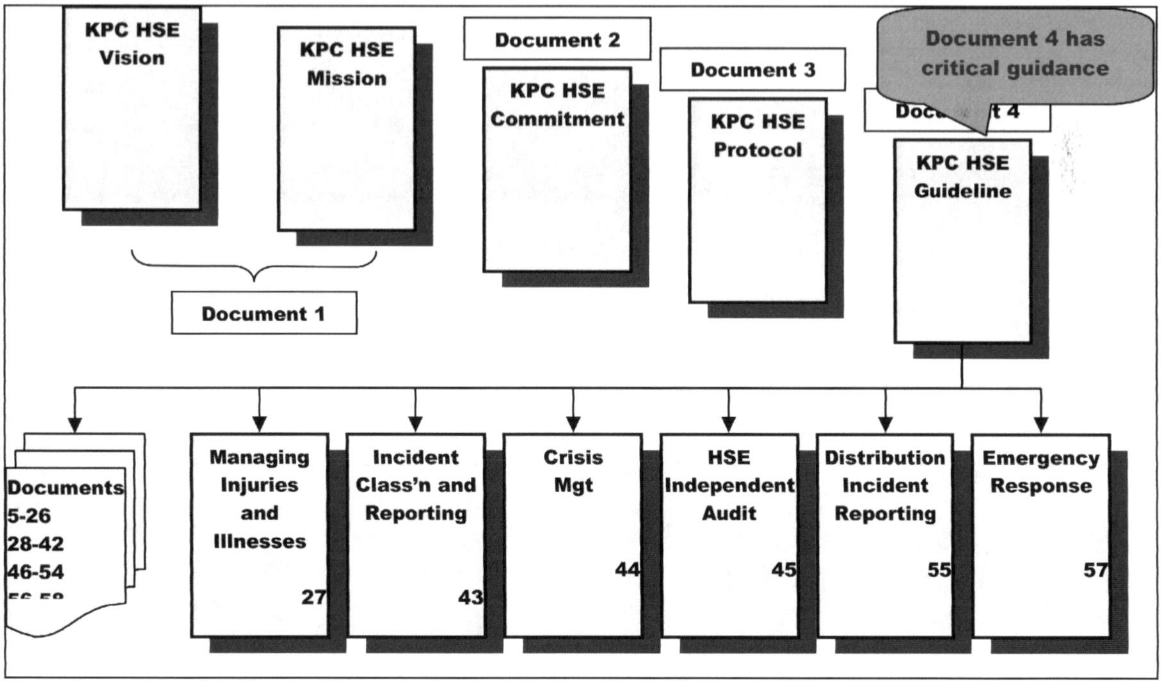

As noted from the above illustration, out of the four policy documents the document 4 – KPC HSE Guideline has critical guidance. i.e. all the standards are mainly evolved and is decisively based on this document contents, which contains the guidelines.

The following documents are specific documents and the Subsidiaries must adopt the contents in toto as they are common to all and require the involvement of all together, be it reporting or emergency response or auditing.

- Managing Injuries and Illnesses
- Incident Classification and Reporting
- Crisis Management
- HSE Independent Audit
- Distribution Incident Reporting
- Emergency Response.

6.8 HSE MS Project Duration

The work was envisioned to be carried out over a period of approximately two years. The project was substantially completed over a period of fourteen months pending a comprehensive audit of the results to be achieved. The timeline for the Project indicates that the following deliverables would be developed as per the time frame noted for each of the deliverables.

Deliverables/Phases	**June 2003**				**Oct. 2003**									**July 2003**
	1	**2**	**3**	**4**	**5**	**6**	**7**	**8**	**9**	**10**	**11**	**12**	**13**	**14**
Performance Audit and Initial Setup														
HSE Manual Development														
Training														
Third Party Audit	Will be conducted after the declaration of implementation of KPC HSE MS by KPC & Subsidiaries.													

6.9 The KPC HSE MS Project Timeline

- KPC HSE MS Project development signed – January 2003
- Gap Analysis in KPC & Subsidiaries – June 2003
- Steering and Implementation Committees established – October 2003
- HSE Policies, Standards and guidelines delivered - October 2003 to June 2004
- Final approval of the KPC HSE MS Policies and standards – July 2004
- Review of the KPC HSE MS with local experts – October 2004
- Implementation in progress in Subsidiaries – from July 2004
- KPC HSE MS application in the KPC HO – from January 2005

6.10 KPC Role

Through its collaboration with a world leader like DuPont, KPC strives to oversee and coordinate the development of HSE policies, standards and systems in all activities. It will have an important role in monitoring and auditing HSE performance in KPC & Subsidiaries.

6.11 The HSE future at KPC

Kuwait Petroleum Corporation cares about the health and safety of its employees and about the protection of the environment in the countries wherever it operates. KPC has been investing significant resources for protecting health and environment, reaching a position which is recognized worldwide. Now, through the support of DuPont, a global leader in safety consultancy, we aim at developing a culture - at every level of our corporation - enabling us to reach superior levels of excellence in safety and environmental protection, a culture which would contribute in the meanwhile also to the optimization of our economic returns and productivity as well as of the image of Kuwait Petroleum Corporation. Safety, quality and economic efficiency are interdependent and safety is the foundation for quality and productivity, as every modern and responsible company has to recognize today.

7. KPC HSE MS Working Groups

7.1 The Project Steering Committee

Purpose
The Project Steering Committee approves and authorizes the completed Policies, Guidelines and Standards Document submitted by the Implementation Committee.

Authority
The Steering Committee is formed and governed by an official approval of H.E the Minister of Oil that sets the roles and responsibilities, and use of the Steering Committee.

Function
The Committee review the overall progress of the project and direction, Resolve Strategic conflicts, approve funding and resources and approves and authorizes the use of the completed Policies, Guidelines and Standards Document approved and submitted by the Implementation Committee.

The Steering Committee will be responsible to evaluate the financial impact of the approved Policies, standards and Guidelines Document and design recommendations.

The approval period should be within 2 weeks and not later than one month of the date of submission of the approved document by the Implementation Committee.

Constitution

The Committee shall consist of a core of 12 members including the Chairman. The nominated members are:

H.E Minister of Oil (Chairman)
KPC Chief Executive Officer (Vice Chairman)
KPC Project Leader
KPC EAMD – HSE Department
KOC EAMD
KNPC EAMD
PIC EAMD
KAFCO EAMD
KOTC EAMD
DuPont Engagement Director
DuPont Project Manager
DuPont Senior Consultant

The core members may include one or more technically qualified member, selected by the chairman, who is identified with specific interests in the following expertise; Finance, Risk Management, Planning and Human resources.

Meetings

Meetings of the parent group shall be held approximately one time each two-month at the call of KPC HSE EAMD with the advance approval of KPC CEO who shall also approve the agenda. All members or their representative shall be present at all meetings.

Meetings shall be open to other KPC, Affiliates or DuPont personnel except as determined otherwise by the Chairman or designee. Meetings shall be conducted and records of the proceedings kept as required.

7.2 The Project Implementation Committee

Purpose

The Project Implementation Committee advises, approve content of Policies, guidelines and Standards Document and submit it to the Steering Committee members for final authorization.

Authority

The Implementation Committee is formed and governed by an official approval of all members of the steering Committee, which sets the roles and responsibilities, and use of the Implementation Committee.

Function

The Committee advises the Steering Committee regarding the progress of the project versus plan, manage Knowledge transfer from DuPont, approve content of developed Policies, standards and Guidelines Document, and formulate recommendation to the steering Committee for Final approval to authorize their use.

The Implementation Committee will be responsible to formulate the exact mechanism for the approval of the content of developed Policies, standards and Guidelines Document. The approval period should be within 2 weeks and not later than one month of the date of submission of the Draft document by DuPont.

Structure

The Committee shall consist of a core of 14 members including the Chairman. Members and the Chairman are selected based on their authorities knowledgeable in Health, safety Environment and risk management.

The nominated members are:

KPC Project Leader (Chairman)
KPC EAMD-Department and decision Support (Vice Chairman)
KPC Manager, Health and Safety
KPC Coordinator Health and Safety
KPC Coordinator Oil spill
KPC Coordinator Environment
KOC Manager Health, Safety and Environment
KNPC Manager Safety, Health and Environment
PIC Manager Safety, Health and Environment
KAFCO Manager Loss and Prevention
KOTC Manager Quality, Safety, Health and Environment
DuPont Project Manager
DuPont Senior Consultant
DuPont Project Assistant

The core members may include one or more technically qualified member, selected by the chairman, who is identified with specific interests in the following expertise; Training, Planning, Human resources and Info technology.

Meetings

Meetings of the parent group shall be held approximately one time a month at the call of KPC and DuPont Project Leaders with the advance approval of KPC HSE EAMD, who shall also approve the agenda. All members or their representative shall be present at all meetings.

Meetings shall be open to other KPC, Affiliates or DuPont personnel except as determined otherwise by the Chairman or designee. Meetings shall be conducted and records of the proceedings kept as required.

8. The Document Authorization Process

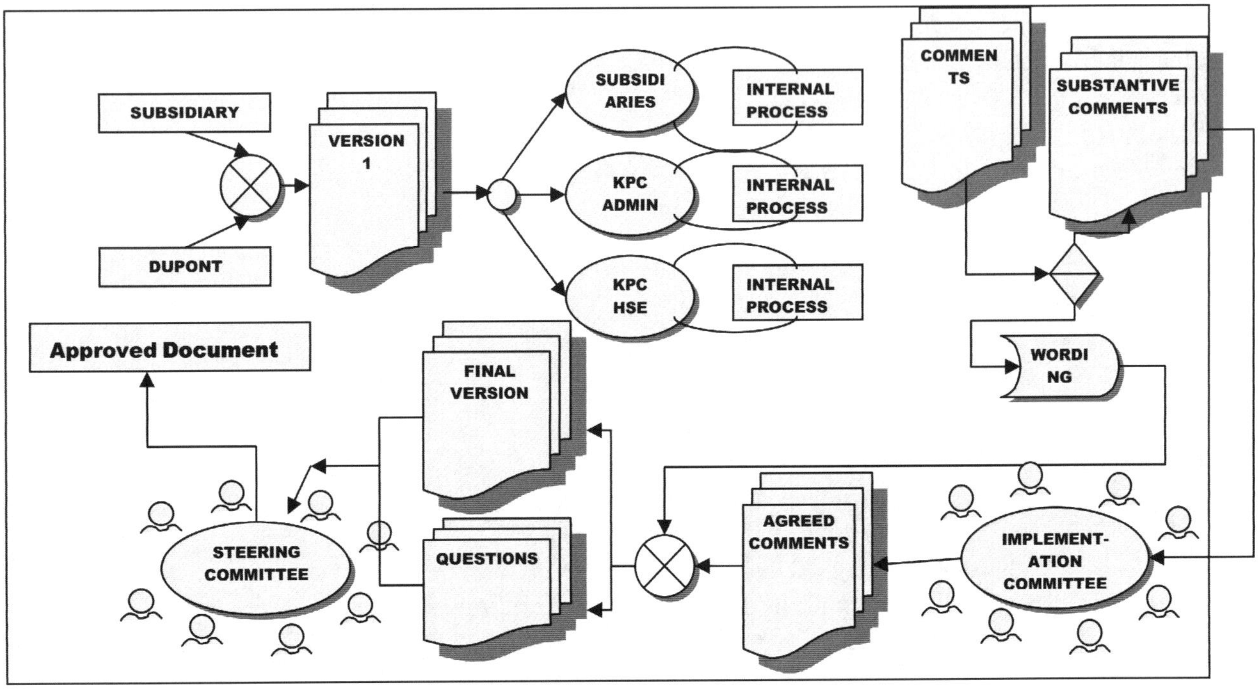

The version 1 of the document is prepared by embarking on the HSE MS developed by DuPont and the best international practices and standards being practiced. This version takes into account the identified gaps of the Subsidiaries of KPC operating with the whole range of oil sector business, viz., Exploration, Drilling, Refining, Petrochemicals, Marine Transportation and international operations.

The review of the above draft version is done comprehensively by KPC Corporate HSE department, and the HSE departments of Subsidiaries. The comments are obtained, reviewed and consolidated. The non-substantive comments are justified for their non-inclusion in the draft and remarked accordingly.

The substantive comments and important 'wording' replacement or providing alternate wording is carried forward to be discussed in the Implementation Committee. Though the key wording replacement or alternate wording is carried forward to the Implementation Committee, they are ultimately approved for changes in the Steering Committee only. The agreed comments from the list of substantive comments are taken up in the Steering Committee for discussion and final decision for inclusion or non-inclusion accordingly. The final version is prepared after duly incorporating the steering committee agreed comments, by literally adopting the changes or replacements.

9. The Interphasing Process

The implementation of the new HSEMS System throughout the group will be the responsibility of each of the affiliates. Many of them have already established a solid basis for safe and environmentally friendly running of their operations. The work with DuPont will support their efforts and help build a world-class HSEMS working environment for all our employees.

The development and deployment of HSE Management Systems in each KPC subsidiary requires a flexible approach to best meet the varying requirements of subsidiaries within KPC, resulting from the diversity of industries, regulatory environments, countries and cultures represented.

The framework for the individual subsidiaries HSE management is decided by each subsidiary, to reflect its specific needs and operational circumstances. However, all such systems will meet the requirements of the KPC HSE Management System, as they apply to their operations.

Each subsidiary will demonstrate the links between the elements of their HSE Management System and the KPC HSE Management System, during the inter-phasing process.

In addition to the requirements of the KPC HSE Policy, Standards, Guidelines and the KPC HSE Management System Elements Guideline, HSE Management Systems at all levels will comply with a range of external requirements - that is, requirements which are controlled by other bodies, outside KPC.

Generally these requirements include:

- Legislation (acts, regulations and laws);
- Statutory Codes of Practice;
- Statutory licences and permits;
- Standards (e.g. Country Standards, DIN, ISO, etc.) which are referenced in legislation; and
- Industry Codes of Practice (as applicable).

KPC Standards will be met where local legislation is less stringent

9.1 System Development and Alignment

Some of the key introspective factors need to be considered during the system development and alignment process. A few of them are:

- What difference is there between what Subsidiaries have in place and what they need to satisfy the requirements of the KPC HSE MS?

- Is there a framework, processes and procedures already in place in the Subsidiaries to meet the requirements of the KPC HSE MS? At what stages of development?

    - Not present or fragmented
    - Core documents in place
    - Development phase
    - System in place

- Are people in the Subsidiaries aware of their roles and responsibilities with respect to the alignment process
- Where HSE Management Systems are in place is there an active Plan, Perform, Measure, Improve – mechanism in place and working effectively?

10. The Signposting Process

10.1 Document Development Process

The DuPont system of document development is based on the following model.

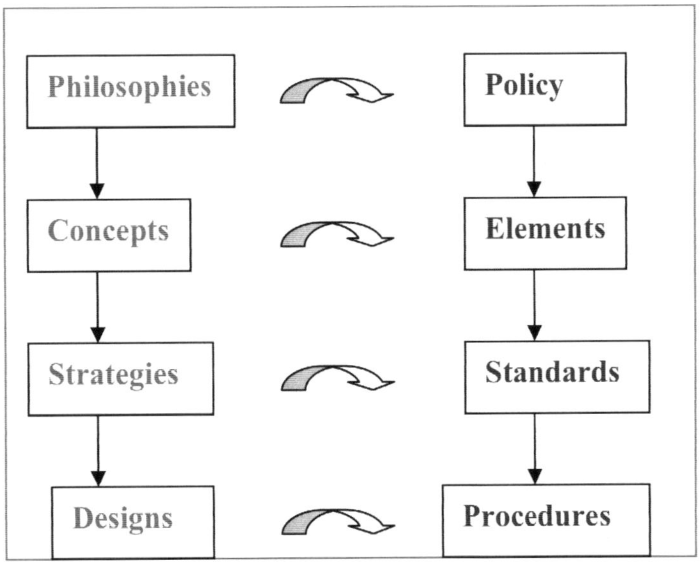

Fig. 1 - THE DOCUMENT DEVELOPMENT PROCESS

As seen from the above figure the DuPont system of document development process is as follows:

> The philosophies evolve into policies, which takes into account the following stake holders: people, community, owner and customers. The concepts evolve into the HSE MS elements, the strategies into standards, and the design into procedures.

Once the standards have been developed, then the process to develop asset or site procedures or Head office procedures is guided by the nature of the facility or organization. For e.g. the Affiliates with the operating facilities need to develop the asset or site procedures based on the standards being developed at the corporate level. At the same time, the Corporate Head office or similar Head office Buildings in the Subsidiary needs to develop the office procedures.

It is worth mentioning here that the corporate standards and policies are broad specifications and the detailed plan or program or procedures must be developed for the asset or office building.

10.2 Signposting

Before the signposting process the two main development or status of the Subsidiaries have to be agreed upon. They are:

- For the Subsidiaries with their own systems.
  Their effort would be concentrated on recognizing the KPC system in some way. Create a signpost document. Create deviations or addendums.

- The second is for the Subsidiaries without their own systems
  Develop HSE system based on the KPC System and develop the procedures.
  The signposting exercise guides a person to see how two differently structured systems match each other i.e. the essential or mandatory requirements of a parent system are checked for coverage in the developed documents or procedures at the plant or HO levels.

In the signposting exercise, initially the signposting database documents are developed. This signposting database documents provide the list of mandatory requirements in the parent document or the Corporate Standard, the responsible groups and the active groups and the mutually accepted deviations. The deviations must be mutually accepted by the corporate representative and the affiliates representative.

10.3 Implications for Implementation

For KPC the following steps were highlighted:

- o Act on the existing gaps
- o Prepare HO procedures
- o Create audit protocols and checklists
- o Create audit schedule by competence
- o Conduct negotiations with Subsidiaries
- o Plan audit.

For the Subsidiaries the following was highlighted:
- Either carry on with current implementation, await KPC Schedule, based on Subsidiary implementation and KPC Schedule plan signpost and develop deviations
- Or, write procedures from KPC Standards; based on Subsidiary implementation and KPC schedule plan signpost and develop deviations.

10.4 The HSE MS Implementation Groups

Responsible Groups
The signposting documents provide the number of mandatory requirements to be covered or cross-checked; the group responsible from the audited company who will be accountable to deliver/implement each of the particular requirements provided by the document.

Contact Active Groups
The task force network team responsible from the audited company who will actually deliver/implement the requirements provided by the signposting document.

Deviations and Audit Plans
A deviation from the mandatory requirements which is unrelated to the concerned working environment or process or operation is termed as 'irrelevant' deviation.

A deviation from the mandatory requirements which is required to be planned and completed somewhere in the future is considered a 'Gap' Deviation.

A deviation from the mandatory requirements which is not being met or planned to be met in future for compliance is called a 'non-compliance' deviation.

The accounted or taken-into-account mandatory requirements which must be placed or considered for the audit protocol or program is call 'actual' requirements.

Once the 'actual' mandatory requirements are finalized the audit protocol can be fabricated based on these 'actual' requirements.

The final compliance audits are conducted based on the audit protocols and checklists and is unique for every company, based on the exercise that is explained as above - i.e. based on the 'actual' requirements for each company.

10.5 The Document Alignment Process:

The steps in the document alignment process are as follows:

1. The KPC Parent HSE MS document and the Subsidiary's corresponding/relevant document review within KPC & Subsidiary's representatives

2. The signpost database documents is developed which produces the number of mandatory requirements that Subsidiary needs to adopt in their corresponding document.

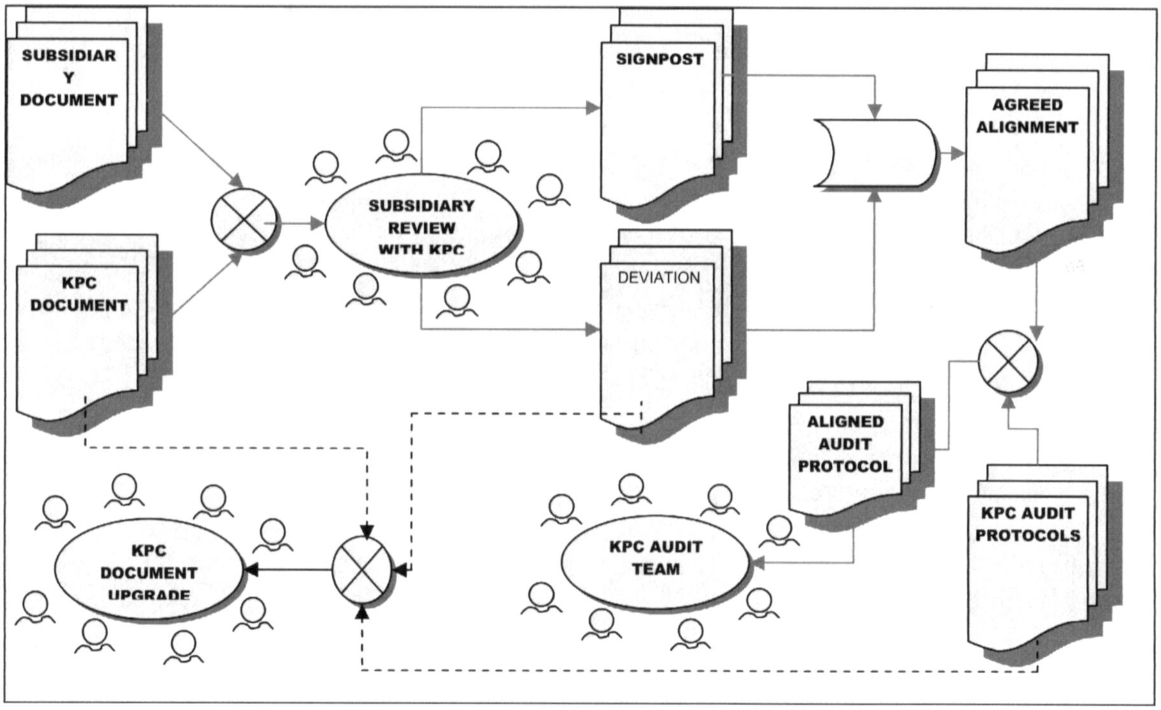

3. The agreed alignment signpost document and deviations are listed.

4. The deviations may fall into four main categories, viz.,
   - To comply in full
   - To comply with deviations
   - Already existing
   - Not applicable

5. The agreed alignment process and the KPC Audit Protocol are jointly studied and the Subsidiary audit protocol is developed.

6. The aligned protocol is provided to the KPC Audit Team for the audit of the KPC HSE MS in that particular Subsidiary.

7. The deviations developed during the signposting process are fed to the KPC Document upgrade committee/team for further modification or updating of the documents.

11. Conclusion

Now that the KPC HSE MS evolved and implementation roadmap laid, the burden is on the Subsidiaries to effect the KPC HSE MS implementation in their respective Subsidiaries.

The HSE MS Contents elaboration and the comprehensive analysis of the HSE MS Standards and Guidelines have been done between the Subsidiaries' concerned HSE personnel, the consultancy and the KPC HSE Department.

The Subsidiaries are periodically (fixed quarterly for Oil Sector Subsidiaries) reporting their progress of KPC HSE MS implementation in their respective Subsidiaries through a spreadsheet with the explanations and figures, where deemed necessary.

We are pleased to inform that progress in the HSE MS Implementation Process is being achieved as per the plan envisioned by the respective Subsidiaries. Numerous HSE programs, Activities and Training are being affected as part of the HSE improvement in the Oil Sector.

The HSE Statistics are improving, the field HSE Culture changing towards a better future. However, we feel it is premature to announce the change for better due to the adoption of the HSE MS in terms of numbers or statistics, at this juncture.

Our main aim was to present in a nutshell, the roadmap to the evolution of a HSE Management System, the deliberations between the Consultant, Corporate and the Subsidiaries.

We hope to achieve a world-class performance with our HSE MS in a reasonable time frame, if not in the near future.

Safety Culture in the CCPS Risk-based Process Safety Model

W. L. Frank, P.E.
Organizational Performance Assurance Center
ABSG Consulting Inc.
5301 Limestone Road, Ste. 210
Wilmington, Delaware 19806
(302) 239-0496
wfrank@absconsulting.com

Prepared for Presentation at
American Institute of Chemical Engineers
2006 Spring National Meeting
8th Process Plant Safety Symposium
Orlando, Florida
April 24-26, 2006

Safety Culture in the CCPS Risk-based Process Safety Model

W. L. Frank, P.E.
Organizational Performance Assurance Center
ABSG Consulting Inc.
5301 Limestone Road, Ste. 210
Wilmington, Delaware 19806
(302) 239-0496
wfrank@absconsulting.com

Abstract

The American Institute of Chemical Engineers (AIChE) Center for Chemical Process Safety (CCPS) is developing a guideline describing a new risk-based approach that an organization can use in defining and implementing a process safety management (PSM) system specific to its unique needs. Recognizing the importance of safety culture as a determinant of safety performance, CCPS has included safety culture as one element in the new risk-based process safety (RBPS) model. This paper will describe both the new safety culture element and its integration in the overall RBPS model.

1. INTRODUCTION

CCPS is developing a new book in its *Guidelines* series, *Risk-based Process Safety*, that is intended to address the dilemma increasingly faced by the chemical process industries – the need to respond to expectations by society, regulators, and management for continual improvement in process safety results, even while competitive pressures are tending to decrease the resources available for implementing and maintaining process safety systems.

The *RBPS Guidelines* presents ideas that can help organizations achieve better results in spite of having fewer resources. To do so, the RBPS system focuses on improving PSM effectiveness, which is defined according to the following function:

PSM Effectiveness = function [PSM performance + PSM efficiency]

Key to this search for increased effectiveness is the application of a more discriminating approach to ensure that the rigor of PSM element design and implementation is "proportional to need." In applying the principles contained in the *RBPS Guidelines*, organizations may follow either or both of two paths to increased effectiveness: (1) doing less PSM work, based on risk-based judgments or (2) doing the same PSM activities, but more efficiently. The intent and content of the new *RBPS Guidelines* were previewed in greater detail at the 20th Annual CCPS International Conference in 2005. [1,2]

Table 1 lists the 20 elements proposed for the RBPS model. As shown in the table, the elements are grouped under four main focal points or themes:

- Commit to Process Safety
- Understand Hazards and Evaluate Risk
- Manage Risk
- Learn from Experience

As the table indicates, **Safety Culture** is listed as the first element under **Commit to Process Safety**. This paper seeks to explain why CCPS has assigned this pre-eminent role to this particular element.

Table 1: CCPS RBPS Elements

A. Commit to Process Safety	**C. Manage Risk**
• Safety Culture	• Operating Procedures
• Standards, Codes, Regulations, and Laws	• Safe Work Practices
• Process Safety Competence	• Asset Integrity and Reliability
• Workforce Involvement	• Contractor Management
• Stakeholder Outreach	• Training and Performance
	• Change Management
	• Operational Readiness
	• Conduct of Operations
	• Emergency Management
B. Understand Hazards and Evaluate Risk	**D. Learn from Experience**
• Process Knowledge Management	• Incident Investigation
• Hazard Identification and Risk Analysis	• Measurement and Metrics
	• Auditing
	• Management Review and Continuous Improvement

For each element, the *RBPS Guidelines* identifies **Key Principles** associated with the implementation of the element. **Essential Features** required to support each key principle are identified, and **Possible Work Activities** to achieve the essential feature are proposed. Figure 1 illustrates this hierarchy. This paper also discusses the key principles and essential features for the *safety culture* element.

- **Element**
  - **Key Principles**
    - **Essential Features**
      - **Possible Work Activities**

Figure 1: RBPS Element Hierarchy

2. WHY IS SAFETY CULTURE IMPORTANT?

The OSHA PSM standard (29 CFR 1910.119) went into effect on May 26, 1992. The standard mandated the implementation of chemical accident prevention program requirements that were patterned after existing practices followed by industry leaders and endorsed by numerous industry technical organizations and trade associations. The later-promulgated EPA risk management program rule (40 CFR Part 68) required a similar chemical accident prevention program and stipulated full compliance by June 21, 1999.

In the nearly 14 years since the advent of the PSM standard, many organizations have devoted considerable time, effort, and expense to developing and implementing chemical accident prevention programs. Nevertheless, serious near misses, incidents, and program audits all too frequently reveal that some facilities are experiencing levels of PSM performance that are either falling or have plateaued below original expectations.

When seeking to understand the root cause of poor performance in ostensibly complete PSM systems, experience has shown that we must not overlook the prevailing beliefs, attitudes, and behaviors within the organization with respect to safety (i.e., the organization's safety culture). PSM programs can be highly sophisticated, well documented, and proudly showcased, yet not produce the intended results because the organizational culture does not fully value, support, and establish an imperative for PSM excellence.

Investigations of catastrophic incidents such as the Longford gas plant explosion[3] and the Piper Alpha oil platform disaster[4] have identified common safety culture weaknesses that can often be found in other serious incidents within the chemical and energy industries. Well-studied and documented parallels exist with the Challenger[5] and Columbia[6] space shuttle disasters.

The need for mature safety cultures has been described for high risk/high performance systems such as nuclear aircraft carriers, submarines, and air traffic control systems.[7] Relevant parallels exist within the chemical and hydrocarbon process industries, especially as our facilities (1) become larger, (2) are more tightly integrated, (3) operate at more severe conditions of temperature and pressure, and (4) become more cognizant of the risks to which they expose surrounding facilities and populations.

As noted above, an RBPS system addresses four main process safety issues: (1) committing to process safety, (2) understanding hazards and evaluating risk, (3) managing residual risk, and (4) learning from experience. By including safety culture as a new element in its RBPS model, CCPS is acknowledging that the safety culture of an organization is a principle determinant of how it will approach each of these issues. For example:

- Commit to process safety - A weak safety culture may assign little credence to the importance of process safety, while a strong safety culture may assign a moral imperative to the organization's responsibilities for safe operations.

- Understand hazards and evaluate risk - A weak culture may have a poorly developed sense of its process safety vulnerabilities, while a strong culture is likely to have an appropriate preoccupation with potential failures that drives it to seek a profound understanding of risk and the means to control it. [7]

- Manage residual risk - A weak culture, perhaps with an inadequate appreciation of (or "in denial" with respect to) its risk exposure may devote minimal resources to controlling residual risk while, in contrast, a stronger culture will seek to provide resources in a fashion proportional to the perceived level of risk it seeks to control.

- Learn from experience - Finally, strong cultures place an emphasis on learning from past experience and improving performance, while weaker cultures often overlook the "weak signals" of safety problems and accept the gradual deterioration caused by "normalizing" increasingly poor performance.

CCPS regards an organization's safety culture as an important input in helping determine the degree of success of the RBPS program. A sound safety culture should maximize both the diligence with which the organization implements each of the RBPS elements and the consequent effectiveness of element implementation. In turn, the health of the safety culture can often be inferred from the quality and results of RBPS element implementation.

These concepts become all the more important in an organization seeking to increase the effectiveness of its PSM efforts through application of the RBPS principles. As organizations seek to trim the scope of PSM efforts based on risk judgments, or seek to find more efficient ways of implementing PSM activities, it is imperative that the organization has high value for, and diligently implements, the remaining, streamlined PSM efforts.

3. KEY PRINCIPLES AND ESSENTIAL FEATURES OF THE SAFETY CULTURE ELEMENT

Three key principles have been identified for the *safety culture* element:

- Maintain a dependable practice
- Develop and implement a mature safety culture
- Monitor and steward the culture

As shown in Table 2, 12 essential features have been associated with these key principles. The desired state for each, as it might exist in a mature safety culture, is described in the following sections of this paper. Potential work activities illustrating how these essential features could be achieved are detailed in the *RBPS Guidelines*.

Table 2: Key Principles and Essential Features of the Safety Culture Element

Maintain a Dependable Practice
Espouse safety as a core value
Provide strong leadership
Establish and enforce high standards of performance
Document the safety culture emphasis/approach
Develop and Implement a Mature Safety Culture
Maintain a sense of vulnerability
Empower individuals to successfully fulfill their safety responsibilities
Provide deference to expertise
Ensure open and effective communications
Establish a questioning/learning environment
Foster mutual trust
Provide timely response to safety issues and concerns
Monitor and Steward the Culture
Provide continuous monitoring of performance

3.1 Maintain a Dependable Practice

In the context of the *RBPS Guidelines*, "maintaining a dependable practice" means that we want the practice to be implemented consistently over time. For some elements, this might entail a written procedure as an essential feature of the element. Unfortunately, however, maintaining a dependable safety culture practice cannot be addressed so prescriptively or mechanistically. With respect to the *safety culture* element, the *RBPS Guidelines* proposes the following four essential features.

Espouse safety as a core value. A clear vision of the importance of safety to the individual and the organization has been established, documented, and shared with all in the organization. The value placed by the organization on safety is reinforced on a frequent basis; consequently, a deeply ingrained sense of that value exists at all levels of the organization. In exceptionally strong cultures this value may be elevated to an ethical imperative. Each individual has an awareness of his/her responsibilities to self, co-workers, company, and society with respect to their safety performance. There is a strong individual and group intolerance for violations of norms of safety performance.

Provide strong leadership.[8] Managers have been educated in the basic concepts and dynamics of culture and are aware of their roles and responsibilities for fostering a sound safety culture. Visible, active, and consistent support for safety programs and objectives exists at all levels of management within the organization. Visionary and inspiring managers are committed to doing what is right, and their values are demonstrated through their communications, actions, priorities, and provision of resources. The concept of "safety as a line responsibility" cascades responsibilities down through all levels of the organization.

Establish and enforce high standards of performance. High standards of safety performance are established and reinforced, for both organizations and individuals. Where changing circumstances warrant, standards are carefully modified, but "normalization of deviance" (i.e., gradual erosion of performance standards due to increased tolerance of nonconformances) is not acceptable.[9] The organization understands that human errors can be prompted by management system failures; however, there is zero tolerance for willful violations of safety standards, rules, or procedures.

Document the safety culture emphasis/approach. Safety culture cannot be designed or manufactured; rather it is the output of thousands of individual management and employee actions (and inactions) that create the foundation for future individual attitudes and behaviors. Nonetheless, it is important for an organization to record, for its own institutional memory, the salient features of any significant complex activity that the company wants to be consistently effective over the lifetime of the organization. Thus, a company should consider documenting the rubrics of its safety culture. This may involve simply recording basic process safety tenants, such as in an EHS policy or mission statement. Or, a company may choose to formalize its culture-nurturing efforts by documenting some specific culture evaluation, monitoring, and learning activities - expected to be carried out by someone or some group on a periodic basis - in the form of a simple written management practice.

3.2 Develop and Implement a Mature Safety Culture

The attitudes and behaviors that an organization accepts as valid and subsequently incorporates into its culture are those that have been demonstrated to successfully deal with the internal and external challenges faced by the organization.[10] The following seven essential features are proposed to help the organization deal with the safety challenges it faces:

Maintain a sense of vulnerability. A high awareness of process hazards and their potential consequences exists within all in the organization. There is constant vigilance for indications of system weaknesses that might foreshadow more significant safety events. Complacency that might be stimulated by past performance and a good safety record is avoided. Where uncertainty about the safety of a particular course of action exists, the burden of proof is placed on determining that the activity or condition is safe before it is permitted to proceed, rather than requiring that staff prove that it is unsafe in order to prevent it.

Empower individuals to successfully fulfill their safety responsibilities. Clear delegation of, and accountability for, safety-related responsibilities has been established. Correspondingly, staff members are provided requisite authority and resources to allow success in their assigned roles. Personnel accept their individual responsibilities for safety. Management expects and encourages the sharing of safety concerns by all members of the organization.

Provide deference to expertise. A high value is placed upon the training and development of individuals and groups. The authority for key safety decisions naturally migrates to the proper people based upon their knowledge and expertise, rather than their rank or position. Competent authorities have a substantive role in the deliberation of safety issues. The organization maintains the "critical mass" of expertise required for safe operations.

Ensure open and effective communications. Healthy communications channels exist both vertically and horizontally within the organization. Vertical communications go both ways; managers listen as well as speak. Horizontal communications ensure that all have the information needed to satisfy their safety responsibilities. Communications channels are monitored for their effectiveness, and necessary "repairs" are implemented. Redundant and/or nontraditional communications channels exist where necessary to (1) ensure adequate communications and (2) provide safe means for those seeking anonymity to share their concerns. There is a strong emphasis on promptly observing and reporting nonstandard conditions to permit the timely detection of "weak signals" that might presage safety issues.

Establish a questioning/learning environment. There is an organizational imperative for enhancing risk awareness and understanding as a means to continuous improvement in safety performance. This is implemented through various means, including (1) appropriate and timely hazard/risk assessments, (2) prompt and thorough incident investigations, and (3) looking beyond the site or company for applicable learnings. The organization recognizes that catastrophic events are typically complex in their causation; consequently, overly simple solutions are avoided when addressing safety issues. For example, the response to a corrosion-caused release does not stop with just replacing the pipe. A questioning/learning organization would ask questions like: Was the corrosion rate excessive? If so why? Does this suggest a process problem that must be addressed?

Foster mutual trust. Employees trust managers to "do the right thing" in support of safety. Managers trust employees to shoulder their share of responsibility for safety performance and to report potential problems and concerns. Peers trust the motivations and behaviors of peers. Employees have confidence that a just system exists where honest errors can be reported without fear of reprisals. Organizational performance, communications, and behaviors are such that the community can trust the facility, and the facility can be confident of a "continued license to operate."

Provide timely response to safety issues and concerns. The organization recognizes that there is often only a brief period between the recognition of a problem and the potential demonstration of the significance of the problem.[7] Priority is placed on the timely response to learnings from incident investigations, audits, process hazard analyses, etc. Mismatches between practices and procedures (or standards) are resolved in a timely manner to prevent normalization of deviance. Emphasis is placed on the timely reporting and resolution of employee concerns.

3.3 Monitor and Steward the Culture

Any management system requires a "feedback loop" to determine whether desired objectives are being achieved. Many of the potential work activities described in the *RBPS Guidelines* for each of these essential features suggest relevant leading indicators (i.e., inputs in the effort to establish and maintain a sound safety culture) that could be monitored. And, as noted previously, the degree of success achieved in implementing certain other RBPS elements can reflect directly on the health of the safety culture (i.e., serve as lagging indicators).

Provide continuous monitoring of performance. There is a healthy curiosity, perhaps anxiety, with respect to "How are we doing?" Where appropriate, relevant, clear metrics exist addressing both leading and lagging indicators. Metrics are tracked, trended, and responded to. There is a high "sensitivity to operations" (i.e., close, frequent contact to monitor performance and conditions within process operations, management systems, and interpersonal issues that could have a bearing on safety performance).[7]

4. STRIVING FOR RBPS EFFICIENCY

In striving to achieve increased PSM efficiency, as described at the beginning of this paper, the *RBPS Guidelines* proposes a new paradigm that (subject to the satisfaction of any applicable regulatory requirements) PSM systems, practices, and procedures should be the simplest that they can be while still being fit-for-purpose. The following three factors are proposed for consideration when determining the "rigor" required in the management system:

- Perception of complexity, hazard, and risk
- Demand for the system results and the resources required to deliver them
- Current organization/facility safety culture

An appropriate management system strategy will seek an optimum "solution" based on the above issues and implement this solution using a blend of prescriptive and performance-based tactics. Any successful attempt at maintenance, correction, or improvement of a management system framework must also involve the consideration of these factors. [2]

It should not be construed that a lower hazard facility can accept a weak safety culture. However, high levels of operational risk, and circumstances where the rate or magnitude of challenges to the safety management system are high, necessitate particularly strong safety cultures to help ensure that the organization accepts, understands, and controls operational risk in a continuously improving fashion.

In the broadest sense, holding the first two factors constant, a stronger safety culture suggests the potential for lessening the rigor (i.e., the prescriptive nature) of the management system. For example, a culture that (1) places a high value on group and individual awareness of the hazards inherent in the process, (2) emphasizes the importance of training individuals to a high level of competence, and (3) stresses the importance of being alert for, and promptly responding to, off-normal situations, suggests at least the potential for reducing the detail required in written operating procedures.

Each element-specific chapter in the *RBPS Guidelines* more explicitly addresses the consideration of safety culture in determining the rigor required of the particular PSM element.

5. SUMMARY

Regardless of whether one is seeking to establish a new safety management system, repair an existing underperforming system, or fine-tune a basically sound system to achieve higher performance, it is the actions or inactions of the individual working within the system that can ultimately be the limiting performance factor. Creating and sustaining a sound safety culture can be a decisive factor in determining the performance of the individual and the system.

This paper has, thus far, avoided providing a precise definition of the term *safety culture*, apart from an oblique reference to the "prevailing beliefs, attitudes, and behaviors within the organization with respect to safety." Many definitions, some quite scholarly, have been proposed for *safety culture*. But, perhaps one of the most utilitarian definitions of *safety culture* is one proposed by a member of the CCPS RBPS subcommittee: "Safety culture is how we behave when no one is watching."

If we can have the confidence that all individuals at every level in our organizations share an imperative for sound safety performance and can be trusted to behave in a fashion supportive of such performance (regardless of whether they "are being watched"), then our search for an efficient risk-based approach to PSM implementation becomes a good deal more straightforward and less fraught with risks of its own.

6. ACKNOWLEDGEMENTS

The 12 key features of a strong organizational safety culture included in the *RBPS Guidelines* and described here are excerpted from the ABS Consulting continuing education course *Creating and Sustaining a Sound Safety Culture*.

The works of Professors Weick, Sutcliffe, and Schein were of particular value in suggesting/stimulating many of the concepts that were distilled into these 12 key features.[7,8]

I would also like to note the work of the CCPS Safety Culture Area of Interest Committee. Those interested in additional CCPS guidance on safety culture may wish to consult the Committee's work product on the CCPS Web site: "Building Process Safety Culture: Tools to Enhance Process Safety Performance." [11] This product was inspired by earlier, insightful work by Dave Jones (Chevron) and Shakeel Kadri (Air Products) who drew parallels between the safety culture learnings from the Columbia accident investigation and safety culture issues within the chemical and energy industries. [12]

7. REFERENCES

1. McCavit, Jack, "Risk-based Process Safety," 20th Annual CCPS International Conference, Atlanta, Georgia, April 11 through 13, 2005.
2. Arendt, Steve, "Continuously Improving PSM Effectiveness - A Practical Roadmap," 20th Annual CCPS International Conference, Atlanta, Georgia, April 11 through 13, 2005.
3. Hopkins, Andrew, *Lessons from Longford*, CCH Australia Limited, 2000.
4. UK Department of Energy, *The Public Inquiry Into the Piper Alpha Disaster* (2 vol), London: HMSO, ISBN 010113102, 1990.

5. *Report of the Presidential Commission on the Space Shuttle Challenger Accident* (includes NASA report on the implementation of recommendations) http://history.nasa.gov/rogersrep/51lcover.htm.
6. *Columbia Accident Investigation Board, Report Volume 1*, Government Printing Office, 2003. http://caib.nasa.gov/news/report/volume1/default.html.
7. Weick, Karl E. and Sutcliffe, Kathleen M., *Managing the Unexpected*, Jossey-Bass, 2001.
8. Frank, W. L., "Essential Elements of a Sound Safety Culture," AIChE, Process Plant Safety Symposium, Atlanta, GA, April 2005.
9. Vaughan, Diane, *The Challenger Launch Decision*, University of Chicago Press, 1996.
10. Schein, E. H., *Organizational Culture and Leadership*, 3rd Ed., San Francisco: Jossey- Bass, 2004.
11. CCPS, "Building Process Safety Culture: Tools to Enhance Process Safety Performance," 2005. http://www.aiche.org/CCPS/PSCulture.aspx,
12. Jones, D. and Kadri, S., "Nurturing a Strong Process Safety Culture," 20th Annual CCPS International Conference, AIChE Spring Meeting, Atlanta, GA, April 2005.

Safety Climate and Decision Making

Stephanie C. Payne, Mindy E. Bergman, Jaime B. Henning, and Carolyn J. Stufft
Department of Psychology, Texas A&M University

Prepared for Presentation at
American Institute of Chemical Engineers
2006 Spring National Meeting
Process Plant Safety Symposium
Orlando, FL
April 24-26, 2006

Abstract

This paper describes a pilot study examining the influence of individual difference characteristics and safety climate on safety-related decision making. Safety climate is defined as employee perceptions of the policies, practices, and procedures concerning safety at their organization. Safety climate can be understood by examining employee perceptions of the types of safety behaviors that are rewarded, supported, and expected. We conducted a survey-based experiment in which we manipulated the level of risk (high, uncertain, and low) for two factors: people and the organization, described in two hypothetical scenarios. Participants indicated how likely they would be to choose one of two responses to reflect how they would proceed if they were in this situation. They also responded to a series of questions designed to measure safety climate for their organization as well as individual propensities to take risks and/or be productive. This methodology can reveal the relative influence of individual differences and safety climate on safety-related decision making.

Safety Climate and Decision Making

Employees have to make decisions in the workplace every day. Whether or not they will make the right decision depends on factors internal and external to the individual employee. Internal factors include such individual differences as job knowledge, skill level, beliefs, and propensities, whereas external factors include environmental conditions, the quality of the equipment and materials, and the climate of the organization. The *purpose of this study* is to gain further insight into the relative influence of individual difference characteristics and environmental characteristics on employee decision making regarding issues of safety and productivity in the organization. To do this, we manipulated information conveyed in hypothetical scenarios and asked respondents to report how they would respond to each scenario. We also asked participants to rate a battery of their own personal characteristics and to assess their organization's safety climate.

Individual Differences

In this study we examine a wide range of individual difference characteristics hypothesized to relate to employee decision making including risk propensity, regulatory focus at work, the Big Five personality variables, and efficiency orientation. We will briefly describe each one and how we expect it to influence safety-related decision making.

First, we examined *risk propensity,* or the extent to which individuals are inclined to take risks, seek adventure, and engage in risky behaviors such as hang-gliding or bungee-jumping. Because individuals who have a greater risk propensity are more likely to make rash decisions, to gamble on lesser odds, and to act with less caution, we expect risk propensity to be negatively related to safe decision making.

Second, we measured *regulatory focus at work* (Chen & Wallace, 2006). Regulatory focus theory describes the cognitive approaches individuals use to strategize about enacting behaviors to meet goals (Forster, Higgins, & Bianco 2003; Higgins, 1997). According to the theory, there are two approaches to goal pursuit: (1) *promotion focus* and (2) *prevention focus*. Individuals with a promotion focus approach tasks by strategically seeking out ways to accomplish the task, whereas individuals with a strong prevention focus strategically avoid things that deter successful task execution (Higgins, Roney, Crowe, & Hymes, 1994). That is, a promotion focus is related to a desire to avoid errors of omission, whereas a prevention focus is related to a desire to avoid errors of commission (Crowe & Higgins, 1997). Researchers have shown that employees with a strong promotion focus engage in strategies that maximize productivity, whereas employees with a strong prevention focus engage in strategies that maximize quality or safety (Chen & Wallace, 2006; Crowe & Higgins, 1997). We hypothesized that promotion focus would be negatively related to safe decision making and that prevention focus would be positively related to safe decision making.

Third, we assessed personality dimensions. Years of research have led to the identification of five robust personality traits that have been documented across sex, race, culture, and generation; they are frequently referred to as the Big Five (Costa & McCrae, 1992; Digman, 1990; Goldberg, 1990). The Big Five consists of agreeableness, conscientiousness, extroversion, neuroticism, and openness to experience.

Agreeableness is a personality characteristic that encompasses cooperation, empathy, selflessness, and identification with others. Agreeable individuals are motivated to be altruistic toward others (Graziano & Eisenberg, 1997). In an organizational setting, this might foster feelings of responsibility to the people within the organization, which would lead to a higher degree of safe decision making. On the other hand, highly agreeable employees are also likely to have higher levels of organizational commitment or stronger affect-based loyalty to the organization, making them concerned about the productivity of the organization. This might lead these employees to focus more on productivity than safety, which could result in more risky decision making. Given competing theories for the relationship between agreeableness and safe decision making, we do not offer a formal hypothesis about this relationship.

Extraversion describes the extent to which one seeks out and enjoys companionship and is confident in social situations. Highly extraverted individuals are characterized as bold, gregarious, sociable, assertive, lively, enthusiastic, optimistic, and talkative (Watson & Clark, 1997). Excitement seeking is a facet of extraversion which might be related to risk taking tendencies. Thus, extraversion is expected to negatively relate to safe decision making.

Conscientiousness is a broad trait consisting of both dependability and drive to achieve (Mount & Barrick, 1995). Highly conscientious individuals tend to be careful, thorough, achievement-oriented, persevering, and dutiful hard workers (Hogan & Ones, 1997). One of the facets of conscientiousness is cautiousness (Costa & McCrae, 1992), which has been measured by the extent to which individuals avoid mistakes (Goldberg, 1999). Conscientious individuals follow the rules; thus they are highly aware of what they should do in a given situation. We expected conscientiousness to be positively related to safe decision making.

Neuroticism is the opposite of emotional stability. Individuals with high levels of *neuroticism* are more likely to be anxious, depressed, angry, embarrassed, emotional, worried, and insecure (McCrae & Costa, 1985). They also tend to have high levels of negative affect, leading to recurring negative emotions and experiences. Because of their high level of anxiety, individuals with a high level of neuroticism may be more cautious and less inclined to make hasty decisions. Thus, neuroticism is expected to be positively related to safe decision making.

Openness to experience has been defined as "tolerance for the unfamiliar, interest in ideas and problems, and appreciation of experiences involving actions, fantasy, values, feelings, and aesthetics" (Tesch & Cameron, 1987, p. 617). Openness to experience has been associated with being imaginative, cultured, curious, original, broad-minded, intelligent, and artistically sensitive (Barrick & Mount, 1991). We did not speculate regarding the relationship between openness to experience and safe decision making.

The final individual difference characteristic we assessed was *efficiency orientation,* which is the tendency to approach a task with the goal of obtaining the most out of the resources used or using the least amount of resources possible (Payne & Philo, 2002). Individuals high on efficiency orientation are concerned about maximizing productivity through efficient processes. Also, preferring to use the least amount of resources emphasizes productivity over safety. Given the inverse relationship between productivity and safety, we expected efficiency orientation to be negatively related to safe decision making.

Environmental Characteristics

We also assessed a number of environmental characteristics that are likely to contribute to individual employees' decision making processes. These included safety climate, assigned safety priority, and perceptions of risk to people and to the organization.

Safety climate refers to shared employee perceptions of the policies, practices, and procedures concerning safety. Consistent with this definition, Zohar (2000) offered a more parsimonious definition of safety climate as *the relative priority of safety in an organization*. Organizations often have competing operational goals, and one goal that often competes with safety is productivity. Thus, the level of safety climate reflects the priority given to safety versus productivity. Typically climate is assessed by administering a battery of items to all employees within a given work group or organization and aggregating their responses. When climate questions are administered to only one employee of an organization (as in this study), the construct assessed is referred to as psychological rather than organizational climate. Safety climate is expected to positively relate to safe decision making.

Another environmental characteristic we assessed was *assigned safety priority*. This is the level of priority the employee's immediate supervisor assigns to safety within the employee's organization (Zohar, 2002). Assigned safety priority is expected to be positively related to safe decision making.

A final environmental characteristic that we examined is perceived *risk to people and to the organization.* With any given industrial situation, there are risks to the people (employees and the surrounding community) as well as to the organization (process, equipment, etc.). In order to determine the relative influence of each of these factors, we trichotomized these two variables into high, uncertain, or low and experimentally manipulated them so that we could empirically examine the impact of each on safety-related decision making.

Method

Sample

Attendees of a safety symposium were recruited to participate in a survey study on safety climate. Although the symposium attracted attendees from industry, consulting, and academia, only those attendees who worked for industry were eligible to participate. This is because only industry employees would be embedded in an organizational setting in which the organization's safety climate could influence the decisions that they made about the organization's processes, safety, and productivity. Thirty-four usable surveys were returned from 30 males and 4 females. The majority of participants (91.2%) were Caucasian, and the average age was 49 years. Twenty-four respondents were employed in chemical process manufacturing, six were in oil and gas production, and four reported "other." Twenty-two respondents identified themselves as plant management, eight were process supervisors, and four had other roles within industry.

Measures

We assessed risk propensity with a nine-item scale from the International Personality Item Pool (Goldberg, 1999). Promotion and prevention regulatory focus at work were measured with twelve items (six items each) developed by Chen and Wallace (2006). The Big Five personality constructs were measured with the 41-item Big Five Inventory (John, Donahue, & Ketle, 1991). We measured efficiency orientation with Payne and Philo's (2002) ten-item scale. Finally, safety climate was assessed with sixteen items from Zohar and Luria (2005), while assigned safety priority was captured with the same five items that Zohar (2002) used.

Design

We used a 3x3 factorial design manipulating the level of risk described for the people and for the organization within the hypothetical scenarios. Risk was described as either high, uncertain, or low for each. Participants were randomly assigned to one of nine conditions. Each participant read two scenarios (described below), both from the same condition. The number of participants in each condition appears in parentheses in the corresponding cell of Tables 1 and 2.

Procedure

Participants anonymously completed an eight-page survey consisting of the measures described earlier, and responded to two hypothetical scenarios. Scenarios were written to emulate real-life safety decisions made on the job. The first scenario (Scenario 1) described a situation in which the respondent heard a report from an employee that turnaround intensive maintenance in a chemical process plant has been field stripped. Resuming the process could generate lower quality output and damage equipment and/or release chemicals into the air. The possible harm to individuals at the plant and in the local community and the risk of endangering the organization's materials were varied across conditions. The scenario ended with the following statement: "Given what you know about the plant and its materials, estimates that you've made in the past, and incidents in other organizations, you estimate that there is a $\underline{10^{-2}}$ chance of injury to people off-site and there is a less than $\underline{10^{-8}}$ chance of damage to the plant or process." The underlined numbers were manipulated according to the condition assigned (and were not underlined in the scenarios given to the participants). The numbers in this example represent condition 3 (people high, organization low). Participants then indicated their likelihood of choosing one of two response options on an 8-point scale (1 = most risky, 8 = most conservative/safe). The options were (a) proceed as planned with the scheduled inspection or (b) require incomplete maintenance to be redone.

The second scenario described a styrene leak from a tank in the tank farm which could explode. The effects of styrene, the results of an air quality check, and information about the materials log were described. Risk levels to the local community and to the nearby chemical process were varied across conditions. The scenario ended with the following statement: "Given previous styrene releases from processes and tanks, the location of the tank farm relative to the plant and to the community, and the current wind patterns, the potential danger to the plant and processes is not known and there is a $\underline{10^{-3}}$ chance of injury, illness, or other harm to the community." The underlined numbers were manipulated according to the condition assigned. The numbers in this example represent condition 6 (organization uncertain, people high). Again, participants then indicated their likelihood of choosing one of two response options on an 8-point scale (1 = most risky, 8 = most conservative/safe). The options were (a) conduct containment and clean-up activities or (b) counsel city authorities to institute a mandatory evacuation.

Upon completion of the survey, respondents had the opportunity to include their name in a raffle drawing for a set of industry handbooks.

Results

Preliminary results are presented in Tables 1 through 3. The descriptive statistics for the decisions to the hypothetical scenarios displayed in Table 3 are particularly interesting. It seems that the participants made more conservative decisions for scenario 1 (*M* = 6.82; field stripping turnaround maintenance) than scenario 2 (*M* = 3.81; styrene leak in the tank farm). There also was considerably more variance in responding to scenario 2 (*SD* = 2.18) than scenario 1 (*SD* = 1.67).

Table 1 depicts the mean responses on an 8-point scale (1= most risky, 8 = most safe) to scenario 1 (field stripping turnaround maintenance) by experimental condition. Unfortunately, due to some missing data, we did not obtain an equal number of participants in each experimental condition (see numbers in parentheses), and extreme caution is warranted when examining these data given the small number of respondents in each cell. An analysis of variance did not reveal any significant differences between these eight means ($F_{(8,22)} = 1.012$, $p > 05$). This was more likely a function of low power and a small sample size. Based on these data, it seems that respondents in conditions 1 and 9, 2 and 3, 4 and 8, and 6 and 7 responded similarly to scenario 1.

Table 2 depicts the mean responses on an 8-point scale (1= most risky, 8 = most safe) to scenario 2 (styrene leak in the tank farm) by experimental condition. Again, due to missing data, we did not obtain an equal number of participants in each experimental condition (see numbers in parentheses), and extreme caution is warranted when examining these data given the small number of respondents in each cell. An analysis of variance did not reveal any significant differences between these eight means ($F_{(8,22)} = 0.402$, $p > 05$). Again, this was more likely a function of low power and a small sample size. Based on these data, it seems that respondents in conditions 1, 5, 6, and 7 responded similarly to scenario 2 as did those in conditions 2, 3, and 4. Participants in condition 8 and 9 responded uniquely.

Table 3 depicts the correlations between the individual difference variables, environmental factors, and participants' responses to the scenarios. General trends are reported, but conclusions cannot be drawn based on this small sample. Contrary to expectations, only two of the individual difference variables (promotion focus and extraversion) related significantly to safety-related decision making. This is most likely a function of low power due to a very small sample size. Contrary to expectation, promotion focus was positively related to safe decision making in scenario 1. Also contrary to expectation, extraversion was positively related to safe decision making in scenario 2. These results suggest that participants with a strong promotion focus and a high level of extraversion were more inclined to make safe decisions.

Discussion

The purpose of this study was to gain further insight into the relative influence of individual difference characteristics and environmental characteristics on employee decision making regarding issues of safety and productivity in the organization. To do this, we asked participants to rate a battery of their own personal characteristics and to assess their organization's safety climate. Then, we manipulated information conveyed in hypothetical scenarios and asked respondents to report how they would respond to each scenario.

Participants varied considerably in their responses to the two hypothetical scenarios, making more conservative decisions in scenario 1. Experimental condition did not significantly influence participants' decisions to either scenario. The only consistent trend was for respondents in conditions 2 (people uncertain, organization low) and 3 (people high, organization low) to respond similarly.

We also examined the influence of a number of individual difference characteristics on safe decision making. Only promotion focus and extraversion were significantly related to the respondents' decisions. Contrary to expectation, individuals with a strong promotion focus (they approach tasks by strategically seeking out ways to accomplish the task) made more safe decisions. Also, individuals with a high level of extraversion (they seek out and enjoy companionship and are confident in social situations) were more likely to make more safe decisions. Perhaps because of their sociable nature, extraverted individuals are more likely to be concerned about risks to others, driving them to make cautious decisions when others' safety is of concern.

Contrary to expectation, the participants' perception of the safety climate in their current organization did not relate significantly to their decision making. It is unclear if this was a function of a small sample size, the artificiality of the hypothetical paper-based scenarios, or a lack of a relationship between safety climate and decision making. Given previous research on safety climate (e.g., Neal & Griffin, 2004; Zohar, 2000; Zohar, 2002; Zohar & Luria, 2005), we believe it is more likely the former reasons than the latter.

We are currently administering these surveys to a larger sample of chemical engineering students and will share these results at the conference.

Table 1

Safety decision making for Scenario 1 by experimental condition

	People		
Organization	**Low**	**Uncertain**	**High**
Low	1. People Low Organization Low 7.500 (4)	2. People Uncertain Organization Low 6.400 (5)	3. People High Organization Low 6.200 (5)
Uncertain	4. People Low Organization Uncertain 8.000 (3)	5. People Uncertain Organization Uncertain 5.000 (3)	6. People High Organization Uncertain 6.900 (5)
High	7. People Low Organization High 7.000 (1)	8. People Uncertain Organization High 8.000 (2)	9. People High, Organization High 7.333 (3)

Note. Mean responses, sample size in parentheses. 1 = most risky, 8 = most safe/conservative.

Table 2

Safety decision making for Scenario 2 by experimental condition

	People		
Organization	**Low**	**Uncertain**	**High**
Low	1. People Low Organization Low 3.250 (4)	2. People Uncertain Organization Low 4.400 (5)	3. People High Organization Low 4.200 (5)
Uncertain	4. People Low Organization Uncertain 4.333 (3)	5. People Uncertain Organization Uncertain 3.000 (3)	6. People High Organization Uncertain 3.600 (5)
High	7. People Low Organization High 3.000 (1)	8. People Uncertain Organization High 2.000 (2)	9. People High, Organization High 5.000 (3)

Note. Mean responses, sample size in parentheses. 1 = most risky, 8 = most safe/conservative.

Table 3

Correlations Between Experimental Condition, Individual Difference Constructs, Environmental Factors, and Participant Decisions to Scenarios

	M	*SD*	1	2	3	4	5	6	7	8	9	10	11	12	13
1. Condition	4.41	2.56													
2. Risk Propensity	2.14	.48	-.07	.64											
3. Prevention Focus	4.53	.32	-.20	-.20	.69										
4. Promotion Focus	3.59	.61	-.08	.01	.14	.74									
5. Openness to Exp	3.67	.44	-.19	-.03	.01	.07	.64								
6. Conscientiousness	4.07	.53	-.24	-.38*	.37*	.25	.22	.84							
7. Extraversion	3.13	.60	.04	-.27	.13	.21	.27	-.08	.79						
8. Agreeableness	3.87	.41	-.02	-.57*	.16	.09	-.02	.57*	-.06	.68					
9. Neuroticism	2.44	.60	.17	.13	-.19	-.08	-.06	-.53*	-.13	-.19	.81				
10. Efficiency Orientation	4.29	.40	-.15	.11	.19	.33	.01	.26	.06	.18	.02	.87			
11. Safety Climate	3.82	.67	-.14	-.47*	.05	-.03	.09	.15	.37*	.40*	-.43*	.22	.95		
12. Assign Safety Priority	1.67	.70	.24	.40*	-.45*	-.09	-.02	-.20	-.12	-.29	.19	-.25	-.34	.79	
13. Decision for Scen 1[a]	6.82	1.67	.10	-.06	.04	.39*	.19	.10	.14	-.25	-.12	-.14	.15	.15	
14. Decision for Scen 2[a]	3.81	2.18	-.03	-.01	.01	-.08	-.16	-.30	.40*	-.29	.17	-.09	.00	-.25	.05

Note. $N = 30\text{-}31$, $^{*}p < .05$. [a] 1 = most risky, 8 = most safe/conservative.

References

Barrick, M. R., & Mount, M. K. (1991). The big five personality dimensions and job performance: A meta-analysis. *Personnel Psychology, 44,* 1-26.

Chen, G., & Wallace, C. (2006). Regulatory focus in the workplace: Development and validation of a work specific measure. Manuscript submitted for publication.

Costa, P. T., Jr., & McCrae, R. R. (1992). *Revised NEO Personality Inventory (NEO-PI-R) and NEO Five-Factor Inventory (NEO-FFI) professional manual.* Odessa, FL: Psychological Assessment Resources.

Crowe, E., Higgins, E. T. (1997). Regulatory focus and strategic inclinations: Promotion and prevention in decision-making. *Organizational Behavior and Human Decision Processes, 69*, 117-132.

Digman, J. M. (1990). Personality structure: Emergence of the five-factor model. *Annual Review of Psychology, 57,* 195-214.

Goldberg, L. R. (1990). An alternative "description of personality": The Big-Five factor structure. *Journal of Personality and Social Psychology, 59*, 1216-1229.

Forster, J., Higgins, E. T., & Bianco, A T. (2003). Speed/accuracy decisions in task performance: Built in trade-off or separate strategic concerns. *Organizational Behavior and Human Decision Processes, 90*, 148-164.

Goldberg, L. R. (1999). A broad-bandwidth, public-domain, personality inventory measuring the lower-level facets of several five-factor models. In I. Mervielde, I. Deary, F. DeFruyt, & F. Ostendorf (Eds.), *Personality Psychology in Europe* (Vol. 7). The Netherlands: Tilberg University Press.

Graziano, W. G., & Eisenberg, N. H. (1997). Agreeableness: A dimension of personality. In R. Hogan, J. Johnson, & S. Briggs (Eds.), *Handbook of personality psychology* (pp. 795-824). San Diego, CA: Academic Press.

Higgins, E. T. (1997). Beyond pleasure and pain. *American Psychologist, 52*, 1280-1300.

Higgins, E. T., Roney, C., Crowe, E. & Hymes, C. (1994). Ideal versus ought predilections for approach and avoidance: Distinct self-regulatory systems. *Journal of Personality and Social Psychology, 66*, 276-286.

Hogan, J., & Ones, D. S. (1997). Conscientiousness and integrity at work. In R. Hogan, J. Johnson, & S. Briggs (Eds.), *Handbook of personality psychology* (pp. 849-870). San Diego, CA: Academic Press.

John, O., Donahue, E. M., & Kentle, R.L. (1991*). The "Big Five" Inventory- Versions 4[a] and 54.* Berkeley: University of California , Institute of Personality and Social Research.

McCrae, R. R., & Costa, P. T., Jr. (1985). Updating Norman's "adequate taxonomy": Intelligence and personality dimensions in natural language and in questionnaires. *Journal of Personality and Social Psychology, 49,* 710-721.

Mount, M. K., & Barrick, M. R. (1995). The big five personality dimensions: Implications for research and practice in human resources management. *Research in Personnel and Human Resources Management, 13,* 153-200.

Neal, A., & Griffin, M. A. (2004). Safety climate and safety at work. In J. Barling & M. R. Frone (Eds.), *The psychology of work place safety* (pp. 15-34). Washington, DC: American Psychological Association.

Payne, S. C., & Philo, J. (2002). *Identifying those who prefer to do more with less.* Paper presented at the 17th annual meeting of the Society for Industrial and Organizational Psychology Conference, Toronto, Canada.

Tesch, S. A., & Cameron, K. A. (1987). Openness to experience and development of adult identity. *Journal of Personality, 55* (4), 615-630.

Watson, D., & Clark, L. A. (1997). Extraversion and its positive emotional core. In R. Hogan, J. Johnson, & S. Briggs (Eds.), *Handbook of personality psychology* (pp. 767-793). San Diego, CA: Academic Press.

Zohar, D. (2000). A group-level model of safety climate: Testing the effect of group climate on microaccidents in manufacturing jobs. *Journal of Applied Psychology, 85*, 587-596.

Zohar, D. (2002). The effects of leadership dimensions, safety climate, and assigned priorities on minor injuries in work groups. *Journal of Organizational Behavior, 23,* 75-92.

Zohar, D., & Luria, G. (2005). A multilevel model of safety climate: Cross-level relationships between organization and group-level climates. *Journal of Applied Psychology, 90,* 616-628.

Maintaining a Healthy Safety Culture

James W. Bayer
Sr. Vice President, Manufacturing, Lyondell Chemical Company

Prepared for Presentation at
American Institute of Chemical Engineers
2006 Spring National Meeting
8th Process Plant Safety Symposium
Orlando, Florida
April 24-26, 2006

UNPUBLISHED

Maintaining a Healthy Safety Culture

James W. Bayer
Sr. Vice President, Manufacturing, Lyondell Chemical Company

The Chemical Industry of today is very different from thirty years ago. Technology has enabled safer designs; regulators have established Process Safety Management standards; and best practices in the area of safety are routinely shared between companies. If you were to visit a facility of any of the major producers today, you would find very similar safety procedures and practices. From hazardous work permitting and process change management, to behavioral safety observations and incident reporting; the basic safety systems are there. Why is it then, when a tragic event occurs it always appears to be the unlikely confluence of multiple failures? Having safer designs, proper procedures, and multiple layers of protection is not enough. Safety must be seen and expressed as a cultural value.

Operational Excellence

I learned this early in my career. In July of 1990 I was working for ARCO Chemical Company, one of the predecessor companies of Lyondell. On the night of July 5th a tragic accident occurred at our Channelview Texas plant. A wastewater tank that was being commissioned after some maintenance exploded and killed 17 workers. The entire company was in shock. The Channelview plant was considered the flagship of the company. It was run by some of the best and brightest people in the industry. It was beyond belief that something like this could happen. But it did, and when the investigation was completed it appeared once again to be the unlikely confluence of multiple failures.

In our industry we control risk through engineering design, operating procedures, and training. Collectively this gives us multiple layers of protection from the risk. There is a human element in every layer so no one layer is perfect enough to rely on by itself. There are gaps in each layer and, when these gaps line up, the risk is not mitigated and an accident occurs. That is why it always appears to be an unlikely combination of bad luck.

The Channelview tragedy fundamentally changed the company. We realized that it was not enough to have some of the best and brightest people in the industry operating our facilities. We needed to have common systems in place at all of our sites to ensure the multiple layers of protection and a culture that would continuously work to minimize the gaps in each layer. At Lyondell we call this Operational Excellence.

Operational Excellence is a broadly institutionalized and disciplined framework which achieves reliable performance and drives continuous improvement in support our strategic focus areas of:

Safety and Health
Environment
Quality
Reliability
Efficiency
Community
People.

For each one of these areas we have developed a set of beliefs and principles; standards and procedures; key measurements of performance; and an auditing function to support continuous improvement. For example, this is a summary of Operational Excellence in the area of Safety and Health:

Belief

All injuries are preventable.

Principles

- We are committed to preventing all injuries and work-related illnesses.
- Management is responsible for providing resources and leadership to ensure accountability for safe performance.
- Everyone is responsible for working safely and for preventing injuries and exposures to self and others.
- Working safely is a condition of employment.
- We eliminate at-risk behaviors and unsafe conditions.
- We communicate openly with employees and contractors about hazards in the workplace.
- We investigate and respond to concerns associated with the occupational health status of our employees and contractors.
- We audit our safety processes and facilities and promptly address deficiencies to ensure compliance and continuous improvement.

Requirements

Management Systems exist that:

- Provide effective leadership to achieve continuous improvement in health, safety and process safety performance.

- Protect and promote the health and safety of employees and contractors.
- Ensure that process hazards are identified, assessed, and when appropriate, the risk mitigated.
- Ensure that all equipment required for safe operation is in service and functioning as required.
- Ensure that the current risk level of the facility is maintained or enhanced when changes are made.
- Ensure that operating instructions which promote safe, consistent operations are maintained.
- Ensure that documentation containing key process safety information is maintained.
- Identify, assess and evaluate potential hazards and risks associated with new and existing processes and products and distribution of these products.
- Protect the security of the company's human, financial, intellectual property, and physical assets.
- Ensure facilities are prepared for emergencies.
- Ensure that incidents are reported, investigated, communicated and proper follow up occurs.
- Ensure that assessments are periodically conducted to determine conformance to requirements and identify opportunities for improvement

We have established standards and procedures in order to meet the requirements outlined above. Since we operate on five continents, the standards are broadly written for the most part to allow for local customs. In some areas such as Process Hazards Analysis, the standard is very prescriptive and detailed. The standards are written and maintained by standing committees of plant managers and subject matter experts, and all updates are balloted by all the sites.

Key Process Indicators (KPI's) are published and tracked. For example the KPI's for this focus area would include: Total Recordable Injury Rate; Behavioral Observation number and quality; Open and overdue action items from incident investigations, audits, PHA's; Safety Critical Variable Deviations; Preventive Maintenance Compliance; Safety Work Order Status; Overdue Inspections; etc. The plant's standard compliance is audited by a team of volunteers from other sites. This is done on a staggered basis covering all of the standards on a three-year cycle.

Cultural Values

It may surprise you that we don't consider safety a priority. We consider safety a value. Priorities can change with conditions, values do not. We expect the same behaviors and actions whether we are in the peak or the trough of the business cycle. This is important because just as we found you can't just rely on the quality of your people, you also can't just rely on standards and procedures. Values are important to any culture. They are what aligns the entire organization from the CEO on down.

Early on in the development of Operational Excellence I was at a committee meeting, and our CEO posed this question to the group: "How do I know that in the middle of the night when a technician is faced with a new unique situation requiring action he will make the right decision?" It was a good question and, being a group of engineers, we attacked it logically- good procedures, good training, proper tools and resources at his disposal. All correct, but we were missing the point. While you need all of those things, they alone are not enough. You can't write a procedure or train for every possible scenario. When you are faced with a problem, you run the possible courses of action through your value system and make a choice based on the results. The answer to his question was that he and the technician must share the same value system. Stating your values, beliefs, and principles may appear to be like mom and dad and apple pie, but it is the critical glue of a healthy safety culture.

Entropy

A final point to make about cultures is that they are living and breathing things. At any given time, a company may have good people, great management systems and shared values, but can lose it without proper care and maintenance. There is a phenomenon known as "normalization of deviance" that is always active in the culture. This is where certain behaviors or actions alien to the culture are tolerated and, after enough exposure, become accepted as the new "normal." It's very easy to see if you are looking for it, but by its very nature, if you aren't, it will be invisible. Take this example of a near miss:

A guard, two technicians and a truck driver are involved in loading a tank truck. The truck driver arrives at the gate and gives the guard his paperwork. The guard looks at it, weighs the empty truck and tells the driver where to go and be loaded. He drives to the loading area and shows the first technician the paperwork. The technician begins a preloading check list and directs the driver to the loading spot. Another technician completes the preloading checklist, loads the truck, completes a post-loading checklist and gives the driver the paperwork and he drives back up to the guard gate. The guard weighs the loaded truck and gives the finished paperwork to the truck driver. As the driver walks to his truck he looks at the weight and notices it is different from usual. He walks back to the guard and that is when they both realize he has been loaded with the wrong product. Not only that, he is at the wrong plant. How can this happen? There were multiple checks by multiple people. You can see the system had several layers of protection. Or did it? In reality, the guard was always the one who made sure everything was correct. The technicians never bothered to check the same items the guard checked because in the last ten years he had never missed anything. That was until he took vacation and his replacement missed the name of the company as well as the product. Granted, both the product name and company were similar to the correct ones, he still missed it and all the others did not check it. There really was only one layer of protection at work here because over time, not double-checking the guard's work became the "new norm."

In this example no one got hurt; there are many more where the outcome was tragic. The loss of the Challenger and Columbia are examples of normalization of deviance. The growing acceptance of o-ring blow-through led to the failure of the Challenger. With the Columbia it was the growing acceptance of insulation coming off of the external fuel tank during liftoff. Normalization of deviance is a subtle thing and can only be controlled through awareness. That is why audits and near miss reporting are critical to the health of the culture.

Summary

Maintaining a healthy safety culture requires effort. We have found that it requires the right people, strong management systems, shared values, and the awareness that it must be actively managed or it will drift over time.

53d Impact of Engineering Ethics on Safety Culture

Charles E. Harris Jr., Philosophy Department, Texas A&M University, College Station, TX 77843-5697

Engineering codes state that engineers must hold paramount the safety, health, and welfare of the public. Engineers therefore pledge to protect the public from unacceptable risk. Implementation of this pledge however, can be challenging because of different criteria for risk estimation applied by engineers, the public, and the government. This interactive presentation will discuss issues of engineering ethics, their involvement in the assessment and management of acceptable risk, and their affect on the safety culture of the organization.

Do You Know Your Engineered Systems?

By

Kumar R. Bhimavarapu* and William W. Doerr, Ph.D.

Risk, Reliability & Failure Prevention, Research, FM Global

1151 Boston Providence Turnpike, Norwood, MA 02062

Paper Prepared for Presentation at the Session

T 3001 Risk Assessment, Process Plant Safety Symposium

April 23-27, 2006

2006 Spring National Meeting, Orlando, Florida

American Institute of Chemical Engineers

---

* *To whom the correspondence should be sent : Kumar.Bhimavarapu@FMGlobal.com, Phone: (781) 255-4919*

Do You Know Your Engineered Systems?

*Kumar R. Bhimavarapu** and William W. Doerr, Ph.D.*
Risk, Reliability & Failure Prevention, Research, FM Global

Abstract

The process industry has been adopting different scenario-based methods for risk assessment. As a function of the requirements and available resources, scenario-based approaches vary from qualitative to quantitative in nature. Layer of Protection Analysis (LOPA), and Safety Integrity Level Analysis (SILA) are examples of hands-on methods that use scenario-based approaches. This paper takes a closer look at scenario-based approaches, analyzes the underlying principles, and explains the basic concepts in terms of engineered systems. Engineered systems are defined as the systems necessary to operate process plants productively and safely; thus, all equipment and facilities (i.e., the hardware associated) as well as procedures, and their combinations constitute engineered systems. Every engineered system will have one or more functional intents, and an engineered system is considered to have failed if it fails to fulfill its functional intent. Losses occur only when engineered systems fail. Engineered systems fail because of the gaps between the functional and design intents, and due to deficiencies. All engineered systems are susceptible to failure, but the failure modes and the failure likelihoods vary significantly. The failure modes of interest change with the loss scenarios. Scenario paths leading to the loss events are comprised of scenario events, which are failures and successes of engineered systems, in general. Deficient engineered systems fail with higher likelihoods, and the effect of deficient engineered systems on risk is discussed. As an example, a simplified approach that is built on engineered systems is provided to evaluate the fire and explosion related risk. Finally, the efficacy of engineered systems based approaches to perform risk assessments for a variety of requirements is emphasized.

INTRODUCTION

There are many feasible approaches for risk assessment, and they vary in sophistication. Each approach has different uses and potential benefits and has different resource requirements, for both development and implementation. Scenario-based evaluations are important means of risk assessment.

A scenario path consists of a series of scenario events that lead to a loss event[1] (see Figure 1).

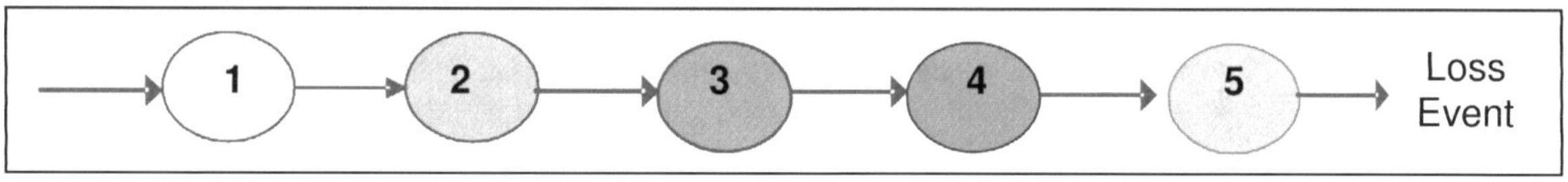

Figure 1: Scenario Path, a Series of Scenario Events

In scenario-based evaluations, the sequential list of scenario events is identified and then the likelihood and consequences associated with the loss event are evaluated to assess the risk and the necessary risk reduction.

* *To whom the correspondence should be sent: Kumar.Bhimavarapu@FMGlobal.com, Phone: (781) 255-4919*

[1] *Loss events could involve losses to property, personnel, or environment; and their consequential impact. The examples of loss events provided in this paper are associated with property damage.*

This paper discusses scenario events in terms of their basic components to promote effective use of scenario-based evaluations. The concept of engineered systems is introduced to explain the underlying principles of scenario events. Further, engineered systems based approaches are proposed to facilitate quantitative risk assessment to address a variety of requirements in a consistent fashion.

SCENARIO-BASED EVALUATIONS

The scenario-based analysis could be quantitative or qualitative in approach. As a function of the need and the available resources, analyses may involve single or multiple scenario paths to evaluate loss event/s. Event tree analyses use multiple scenario paths to evaluate the multiple outcomes of an undesirable event. Semi-quantitative scenario-based techniques are used to derive most of the benefits of the quantitative techniques, and to minimize the resources necessary for a rigorous quantification of risk.

Methods such as safety integrity level (SIL) analysis (*Ref: ISA 84, 2004*) and Layer of Protection Analysis (LOPA) (*Ref: LOPA, 2001*) use scenario-based approaches to account for the safety systems in place and their integrities. LOPA, promoted as a semi-quantitative technique, uses a single scenario path of concern from an event tree. LOPA uses the safety protection layer concept. Figure 2 shows the typical way the protection layers for a process are perceived.

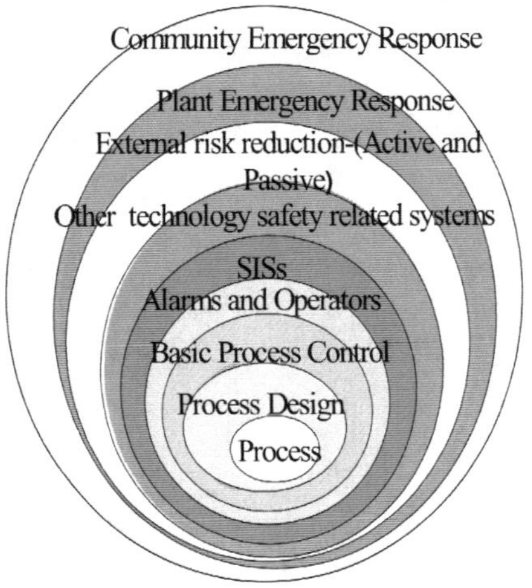

Figure 2: Protection Layers

The first (innermost) level of protection is built into the process and is considered as part of the inherent nature of the process; the final (outermost) level of protection is the emergency response action from the community. The prevention or mitigation of loss events depends on the availability and successful performance of the various safety protection layers shown in Figure 2. In case of failure of one protection layer, the protection layer at the next level will still be available to take the process to a safety state. As the number of protection layers and their integrities increase, the safety of the process increases. LOPA characterizes the functioning of the available protection layers and their order of functioning to generate scenario events and a scenario path.

Scenario and Scenario Events

Different logics are used to build the scenarios to meet different requirements. When a process parameter deviation or an upset in the process is of concern, the scenario is developed towards a loss event. This is an inductive approach followed in event tree type analyses, which starts with the identification of an initiating event. When a well-defined loss event is of concern, an evaluation is done backwards to identify the scenario/s that can result in this loss event. This is a deductive approach followed in fault tree types of analyses. When process safety engineers find deficiencies during safety audits, they have to generate the applicable scenarios to understand the impact of the deficiencies. Such evaluation needs both deductive and inductive approaches because the deficiency can impact any one of the scenario events on the scenario path. Explicitly or implicitly, all the scenario-based approaches include a common task, i.e., the identification of the "engineered systems" whose successes or failures impact the scenario outcome and the risk.

ENGINEERED SYSTEMS

Engineered systems[2] are simply the systems established to operate the plants productively and safely. Thus, all engineered equipment and facilities (that comprise hardware) as well as procedures, and their combinations constitute engineered systems. Generally, scenario events are successes or failures of engineered systems in place. Examples of engineered systems are process control systems, pressure relief systems, safety instrumented systems, pressure vessel, fire suppression systems, dikes, emergency response, fire fighting, hot work permit system, operator response, and mechanical integrity controls. Engineered systems are designed/established and maintained/operated appropriately for a productive and safe operation.

Some engineered systems specifically meant for productivity may not appear to be of interest to the process safety engineer. However, engineered systems not intended for safety could affect the risk since they could fail and place safety systems on demand. For example, a process temperature control whose primary function is productivity can fail, initiate a scenario, impose the demand on the safety systems, and lead to a loss event. As a result, while performing scenario-based analyses, process safety engineers evaluate all the relevant and applicable engineered systems - not only those specifically serve safety functions such as a sprinkler system or a relief system.

Failure Modes and Scenario Events

Engineered systems can fail in different modes, and each failure mode can result in one type of scenario event. Thus, each engineered system may be associated with more than one type of scenario event. Functional failure is the basic failure mode of interest to a process safety engineer. When the functional failure occurs, the engineered system will not perform / fulfill its functional intent. If the engineered system is an equipment containing hazardous material, the material will be released if a mechanical integrity (MI) failure occurs. Even though there are many variations in functional and MI failure modes, the chosen resolution is adequate for the current discussion because the objective is just to review the logic related to the failure modes of engineered systems.

A compressor can have a functional failure[3] and fail to compress the process medium to meet the functional intent. In addition, a compressor can have a mechanical integrity failure, which can result in a release from the compressor. Additional examples are provided in Table 1 for engineered systems and their failure modes that result in scenario events.

The process safety engineer should be cognizant of all engineered systems that can affect the risk. Every equipment or facility (hardware) and a procedure can be considered as an

---

[2]*Note: In this paper, we used the term engineered system in an encompassing fashion to include any system that an engineer may come across in a process plant. The objective is to develop a consistent and uniform approach to address a variety of risk assessment requirements.*

[3]*Note: Even though the failure of a piece of equipment like a pump or a compressor can be referred to as a breakdown, the term functional failure is used to keep the terminology uniform. The terminology can be changed to suit convenience and established practices if any. In addition, what constitutes a failure (or success) of an engineered system can be defined by the engineers performing the analyses.*

engineered system with specific functional intent. When the functional intents are clear, the causes for the failure of engineered systems and the scenario events can be identified easily.

Table 1: Engineered Systems and Scenario Events, Examples

Engineered System	Failure Mode	Scenario event (Failure of Engineered System)
Cooling water system	Functional failure	Failure of cooling water system
Sprinkler protection system	Functional failure	Failure of sprinkler protection system
Transformer	Functional failure	Failure of transformer
Operating procedure*	Functional failure	Failure of operating procedure Note: Failure of operating procedure is generally a failure in performing a predefined step, and many times manifests it self as an 'initiating event' of the scenario.
Operator response*	Functional failure	Failure of operator response Note: Operator response is generally needed when a preventive or corrective action is to be taken because some undesirable event has already occurred and has started the scenario.
Process vessel	Mechanical integrity (MI) failure	Release from the process vessel Failure of mechanical integrity (of process equipment) can result in an uncontrolled release.

* Note that it is convenient to designate operator response and operating procedure as engineered systems rather than an operator. Engineered systems should be defined appropriately to facilitate generation of clear-cut scenario event/s on association with specific failure mode/s. While performing quantitative risk assessment, all such scenario events can be assigned with likelihoods.

Causes for the Failures of Engineered Systems

Each engineered system can have one or more functional intents. From a loss analysis perspective, an engineered system is considered to have failed if it fails to fulfill its functional intent. Engineered systems are designed/established to meet the functional intent/s to the extent possible. Engineered systems fail for two different reasons as outlined below.

Gaps Between Functional and Design Intents

The design intents are generally based on established standards, guidelines and practices, and there can be a gap between the functional intent and the design intent. For example, the functional intent of a sprinkler system is to extinguish and or control the fire. Sprinkler systems are designed to provide specific flow rates, drop configuration etc., to address a fire of certain size and nature. However, it may not be feasible to design sprinkler systems to address every conceivable type and size of fire. Economics and technology impose the gap between the functional intent and the design intent. Thus, a sprinkler system can fail to fulfill its functional intent even when it performs as per the design intent. The same is true with any engineered system. In other words, engineered systems can fail even when there are no deficiencies and they perform as per the design intent. Such failures could be due to excessive demands placed on them, i.e., due to operation outside of the design envelope. The likelihoods of such failures depend on the likelihood of the excessive demands, which is a function of the size of the gap between the design intent and the functional intent. The gap could be greater if the implemented design intent is inferior to the intended design intent. As a function of the nature of the engineered systems, these gaps do vary.

Deficiencies

In the process industry, deviations from established standards, guidelines, and practices are generally considered as deficiencies, and deficiencies are always associated with engineered systems. Since engineered systems include hardware and procedures, deficiencies could be associated with hardware and/or procedural systems. In case of hardware, deficiencies could be associated with design and /or operation/inspection/testing/maintenance. In case of procedural systems, deficiencies can be associated with establishing (design) and/or implementation (operation) of the procedures. The failure modes change as a function of deficiencies.

Process safety engineers routinely ask the question, "How does a deficiency increase the risk?". The deficiencies affect the fulfillment of the design intent and thus it becomes more difficult for an engineered system to fulfill its functional intent. For example, if not tested and maintained properly, a sprinkler protection system can fail to fulfill its design intent when there is a demand; as a result, it would fail to fulfill its functional intent as well.

It is not necessary that deficiencies are noticeable all the time. Many incidents occur because of unnoticed/unnoticeable deficiencies. For example, many hardware items deteriorate over a period and such deterioration may not be obvious all the time.

IMPACT OF DEFICIENCIES ON ENGINEERED SYSTEMS

The likelihood of failure of an engineered system (i.e., the likelihood of a scenario event) increases with a deficiency. Severe the deficiency, higher is the likelihood of the scenario event. If any one of the scenario events is associated with a deficient engineered system, the likelihood of the scenario or the loss event will also be higher. Accordingly, the risk[4] increases as an outcome of the deficiency. Figure 3 depicts how the scenario event associated with a deficient engineered system constitutes the scenario path along with other scenario events.

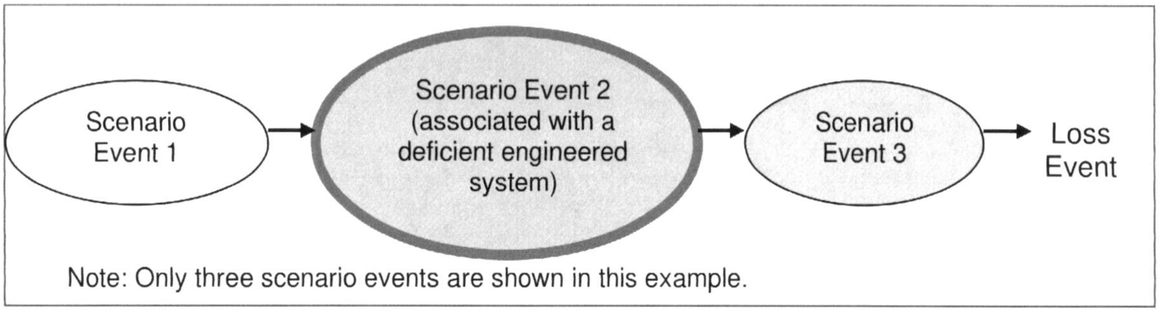

Figure 3: Impact of a Deficiency On the Scenario Path

Loss prevention recommendations address deficiencies in order to reduce the risk. Evaluation of the impact of a deficiency associated with an engineered system needs a decent understanding about the engineered system, its components, and their functioning and failure modes.

---

[4] *Risk is a function of the likelihood and the consequence associated with the loss event/s.*

Components of Engineered Systems

The engineered systems of interest in scenario-based analyses are generally macro systems. Macro systems may comprise many components, which we may designate as subordinate systems, for convenience. Success or failure of engineered systems depends on the success or failure of the subordinate systems. Obviously, when applicable, administrative controls and training are also subordinate systems of an engineered system since deficiencies associated with them can lead to the failure of the engineered system. Subordinate systems can be at different hierarchical levels and so are the deficiencies. As a function of the subordinate system that is deficient, the failure likelihood of the engineered system changes.

For example, a sprinkler protection system consists of a number of subordinate systems such as firewater storage, firewater pumping system, firewater headers, control valves, and sprinklers. Further, firewater pumping system can have subordinate systems such as main pumps, auxiliary pumps, associated instrumentation etc. If there is a deficiency associated with the maintenance of the motor of the pump, sprinkler protection may fail. In order to estimate the impact of a deficiency on the likelihood of failure of the engineered system of concern, the engineer should be knowledgeable about the subordinate engineered systems and their effect on the engineered systems. This becomes more important when the failure likelihood of the engineered system has to be evaluated quantitatively.

From Deficiency to Scenario Event

In order to understand the increase in the risk due to a deficiency, the first step is to identify the engineered system that would fail to fulfill its functional intent due to the deficiency. Through a simple-but-objective examination, the engineer can identify how a deficiency would result in or contribute to the failure of a specific engineered system. See a few examples provided in Table 2 for the identification of the scenario event from a deficiency.

Table 2: Deficiencies of Engineered Systems, and Scenario Events

Deficiency	Failure of the Engineered System	Engineered System (Failure Mode)	Scenario Event	Remarks
Inadequate number of sprinklers	Inadequate number of sprinklers can result in failure of sprinkler protection	Sprinkler protection (Functional failure)	Failure of sprinkler protection	
Noncompliance with DIERS†[5] methodology for relief- system design	Noncompliance with DIERS† methodology can result in failure of relief system	Relief system (Functional failure)	Failure of relief system	

[5] *DIERS: Design Institute for Emergency Relief Systems.*

Table 2: Deficiencies of Engineered Systems, and Scenario Events (Continued)

Deficiency	Failure of the Engineered System	Engineered System (Failure Mode)	Scenario Event	Remarks
Improper inspection of storage tank	Improper inspection can lead to mechanical integrity failure and in turn release of flammable materials	Storage tank (MI failure)	Release from the tank	Here, the outcome of interest is 'loss of containment'. Non-availability of the tank could also be a concern.
Failure of release controls	'Release control' is a broad engineered system. Releases can occur due to failures of a variety of engineered systems	Release control (Functional)	A release, generic	Subordinate systems include MI failure of process equipment, failure of an operator etc.

Sometimes however, there could be a need to generate appropriate terminology to define the engineered systems. Engineered systems can be defined based on the functional intent they are serving. For example, if the process safety engineer finds a deficiency related to electrical area classification specifications, the engineered system of concern can be defined as 'electrical ignition control', which is synonymous with its functional intent. The scenario event of concern is functional failure of 'electrical ignition control', i.e., an electrical ignition. The last engineered system in Table 2, i.e., release controls, is another such example of an engineered system that is defined based on functional intent. As mentioned earlier, only by appropriately defining the engineered systems and the failure modes, scenario events can be assigned with likelihoods. Likelihoods are needed for any quantitative risk assessment.

Engineered systems susceptible to failure may not be obvious all the time. For example, if a warehouse is storing pallets above a recommended height, what engineered system is going to fail from a scenario-based perspective? What is the scenario event of interest in this case? In order to evaluate such issues, it is necessary to understand the reasons for failure of engineered systems (see the Section 'Failures of Engineered Systems'). In this example, the engineered system of interest is 'sprinkler protection system'. In case of a fire, because of the excessive combustible load, the sprinkler protection system will not be able to fulfill its functional intent. Even though the risk improvement can be effected by reducing the pallet height or by upgrading the sprinkler protection system, from the risk evaluation perspective, the engineered system susceptible to failure is the sprinkler protection system. Due to the excessive combustible load, the failure likelihood of the sprinkler protection system will be higher.

Impact of Deficiency on a Scenario

Scenario path development is necessary to evaluate the impact of a deficiency on the risk. Once the scenario event associated with a deficient engineered system is identified, the remaining scenario path can be developed easily. The development of the scenario path starts with the identification of the event that places the demand on the deficient engineered system.

Table 3 presents the events that place the demand on the first two deficient engineered systems shown in Table 2. The remaining scenario events on the scenario path can be identified considering the loss event, the nature of process, layout, other engineered systems in place, etc. The engineered systems in place may change from place to place, and accordingly the scenario events may change.

Table 3: Identification of Demand Events

Deficient Engineered System	Event that Places the Demand on the Deficient Engineered System	Remarks
Sprinkler protection	Fire *	Every potential fire that has to be mitigated by the sprinkler protection (as per the functional intent) should be considered.
Relief system	Over-pressurization	Every potential over-pressurization event that places the demand on the relief system (as per the functional intent) should be considered.
* Note: The events fire and over-pressurization are the outcomes of the failures of a few or more engineered systems. If the likelihoods of fire or over-pressurization are known, then there is no need to include all those individual engineered systems on the scenario path.		

Categorization of Engineered Systems

A variety of approaches can be adopted for scenario-based analyses using the concept of engineered systems as the basis. In fact, engineered systems based approaches can be used to simplify the scenario building process. As a function of the scenario being analyzed, engineered systems can be grouped based on their functional intent. For example, for fire and explosion scenario analysis, engineered systems can be segregated based on four types of scenario events: initiating event (generally related to basic process operation), prevention event (generally related to safety systems), ignition event or ignition (related to ignition control systems), and mitigation event (related to mitigation systems) (see Figure 4). This type of classification facilitates consistency in the analysis and reduces the need for the discretion of the engineer.

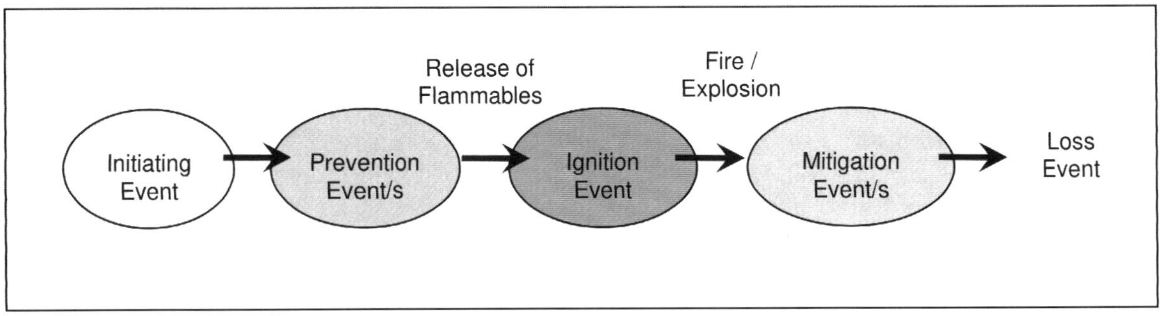

Figure 4: Types of Scenario Events

See the illustrative scenario paths shown in Figures 5 and 6 for the example cases showed in Table 3. The scenario path of interest shown in Figure 5 comprises success of fire fighting after the failure of sprinkler protection. One may build the scenario for the failure of fire fighting in which case the loss event would be different. In the case of the second scenario associated with the failure of the relief system (see Figure 6), the failure of containment is considered as certain outcome. The same is the case with the outcome explosion in case of an ignition. No mitigation event is considered on this path.

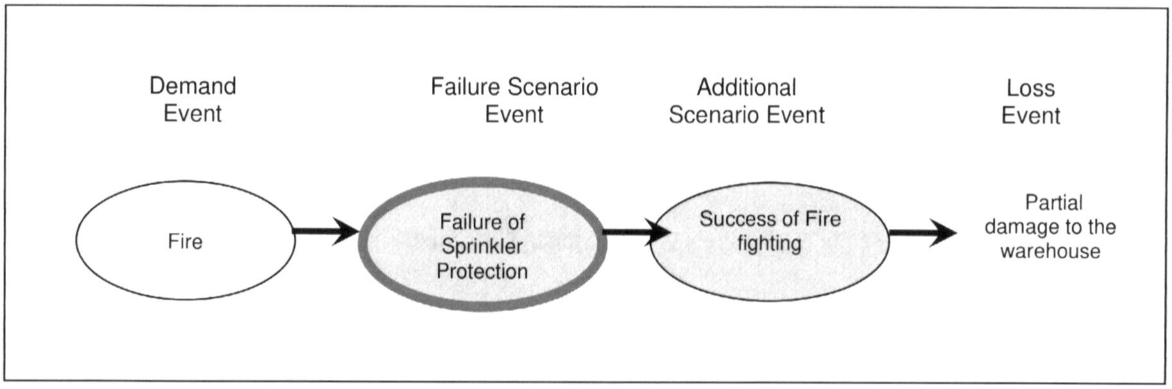

Figure 5: From Fire to Loss event, Scenario Path

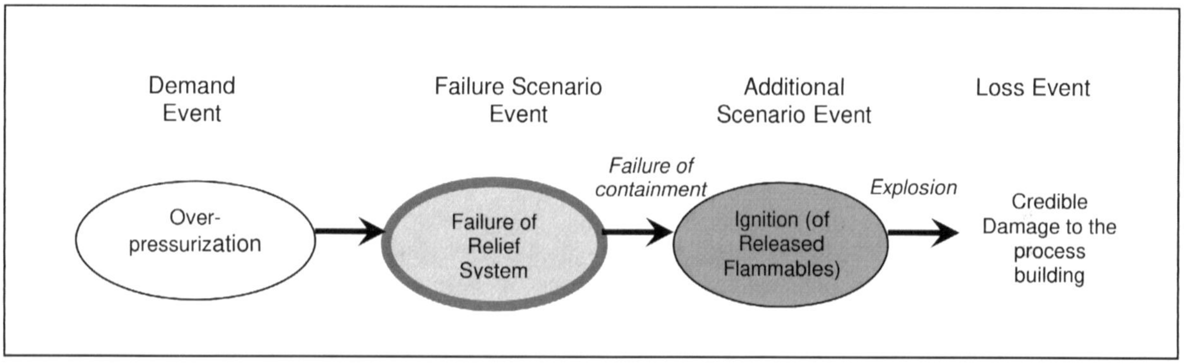

Figure 6: From Over-pressurization to Loss Event, Scenario Path

Evaluation of Likelihoods

Evaluation of the increase in the likelihood of the loss event due to a deficiency needs the failure likelihoods of the engineered system with and without the deficiency. If a deficiency is associated with a subordinate system, the engineer should assess the impact of the subordinate system on the success or failure of the engineered system.

In semi-quantitative approaches, the quantification of the likelihood associated with the scenario and loss events[6] is done approximately or even in terms of orders of magnitude. An example using the order of magnitude scenario likelihoods before and after removal of a deficiency is shown in Figure 7.

Engineers can develop their own likelihood evaluation methodologies at the desired and feasible resolutions. Such methodologies would have to consider the nature of the engineered system (hardware, procedure, or a combination), the criticality of the subordinate system, and the availability for recovery with other subordinate systems. If time and resources permit, fault tree analyses can be used to evaluate the impact of the deficiency of a subordinate system on the engineered system, more precisely. However, quantitative fault tree analyses may not be warranted for the resolution needed for semi-quantitative approaches.

---

[6] *A loss event is the event associated with specific loss value.*

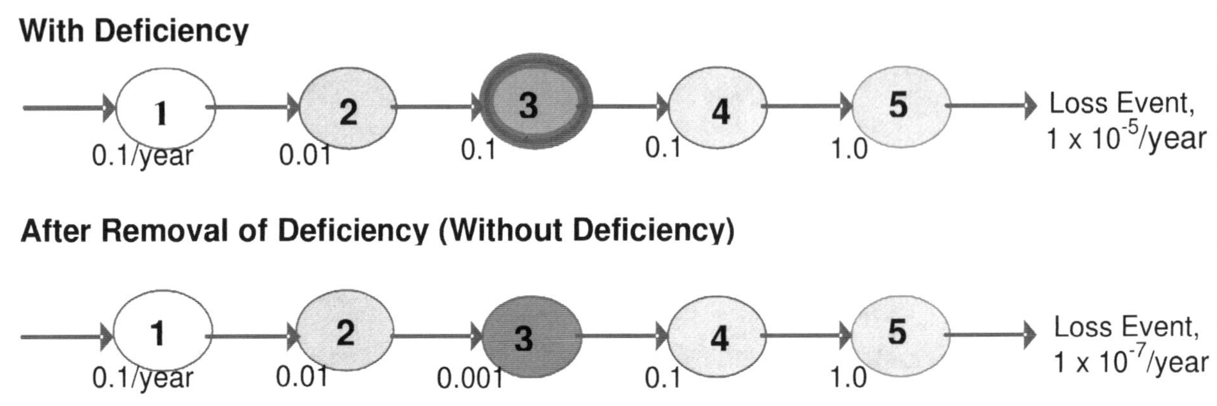

Note: 1. The deficient engineered system is associated with scenario event number 3.
2. The likelihoods of scenario events and loss event are shown below the event.

Figure 7: Likelihoods with and without deficiency

CONCLUSIONS

Proper identification of engineered systems and their functional and design intents is essential for a consistent and effective risk analysis process. For example, some scenario-based techniques define process design as a protection layer. The underlying factor is that a higher design pressure can improve the integrity of the process vessel. However, defining the process design as a parameter is not conducive to accounting of the impact on risk. Better definition of systems with a clear functional intent facilitates easier definition of scenario events. Process design is not an engineered system, but a process vessel is. The functional intent of the process vessel is to hold the contents and to prevent loss of containment. If the vessel can be designed very robust, the gap between the design intent and the functional intent will be reduced. A robustly designed process vessel can be subjected to greater demands but still may not fail. The failure likelihood of such engineered system will be low. If there is a deficiency in the process design of the vessel, its failure likelihood will increase.

Engineered systems based approaches facilitate easy identification of scenario events, scenarios, allotment of likelihoods for the scenario events and the estimation of likelihood of a scenario or loss event. In turn, scenario-based evaluation can be adopted to perform risk assessment for a variety of circumstances. Particularly, these approaches will be useful when deficiencies are to be evaluated. Obviously, engineered systems based approaches facilitate evaluation of risk reduction associated with any improvement, and thus can help better utilization of resources.

The process safety engineer should be cognizant that the hardware (equipment or facilities), procedures, and their combinations constitute engineered systems with specific functional and design intents and specific failure modes. Any loss event can be described as a combination of failures and successes of a few engineered systems. Only those engineered systems that fail to increase the risk and that succeed to mitigate the risk would be of interest for the process safety engineer. Engineered systems based approaches help to improve the utility and the efficacy of scenario-based analyses. Therefore, the question is, "Do You Know Your Engineered Systems?"

References:

LOPA, 2001: *Layer of Protection Analysis: Simplified Process Risk Assessment*, Center for Chemical Process Safety, AIChE, 2001.

ISA 84, 2004: *ANSI/ISA 84.00.01-2004 Part 1(IEC 61511-1 Mod), Functional safety: safety instrumented systems for the process industry Sector - Part 1: Framework, definitions, system, Hardware and Software requirements*, Instrumentation, Systems and Automation Society, 2004.

What risk should public accept from chemical process facilities?

Ernst Meyer, Gøran Andreassen, Dr. Cindy Wei, and Stephen Shaw
Det Norske Veritas (USA) Inc.
15340 Park Ten Place, suite 100
Houston, TX 77084
Phone: (281) 721-6600

Prepared for Presentation at
American Institute of Chemical Engineers
2006 Spring National Meeting
8th Process Plant Safety Symposium
Orlando, Florida
April 24-26, 2006

February 28, 2006

UNPUBLISHED

What risk should public accept from chemical process facilities?

Ernst Meyer, Gøran Andreassen, Dr. Cindy Wei, and Stephen Shaw
Det Norske Veritas (USA) Inc.
15340 Park Ten Place, suite 100
Houston, TX 77084
Phone: (281) 721-6600

Abstract

Major incidents in chemical process plants such as Bhopal have raised the questions of facility safety and security by stakeholders and interest groups. How will the facility be perceived as safe given the possible consequences? No one denies that there have been improvements in process safety since Bhopal and many safety regulations have come into effect. Public may still question whether it is safe to live or work near a chemical plant nowadays?

This paper discusses the risk that the public should accept under governmental leadership and guidance. Also discussed is how the chemical process industry should ensure risk acceptance criteria compliance and maintenance of compliance throughout the lifetime of the facility. Safety may be enforced by compliance with a pre-defined set of risk acceptance criteria. These criteria may be absolute and tangible, but in some cases are more abstract. Different practices are seen among different countries, states and regions as well as between different industry segments. This paper discusses the meaning of risk acceptance criteria and how exposed people and regulatory bodies should relate to the criteria.

1. Introduction

The Chemical Process industry has a relatively good safety record compared to other industries. However, there have been significant concerns on the potential catastrophic events. It is important to identify the risks of chemical process operations. By definition, risk is the complete set of scenarios, the likelihood and the consequences of each scenario. Unfortunately, the public and regulations tend to emphasize on consequences instead of risks.

In general, members of the public fear major incidents or disasters, no matter what the likelihood is, and are less concerned about real risks in everyday situations. Many agonize over the possibility that, in a thousand years or so, high level nuclear wastes in geological storage may leak, while paying almost no attention to the fact that every year about a thousand Americans are accidentally electrocuted. Commercial aircraft traffic continues to increase; however, the fear of flying lingers, despite statistics clearly demonstrating that plane travel is ten times safer than driving the same distance. The public's perception of risk posed from chemical process facilities

is often based on misleading information or worst case scenarios of negligible risks in-line with risks posed by cosmic or geological events.
In the United States many process safety regulations have come into place since the Bhopal incident. These requirements cover a wide range of aspects of chemical process facilities, including siting, design practices and hazard control systems. To implement the regulations and standards, consequence analysis is required, but comprehensive risk-based analysis is not necessary.

Therefore, a risk-based decision making approach is needed to take into account both the event likelihood and the effect of mitigation measures. Without proper risk criteria, decision making will, in many cases, be performed on the basis of calculating expected consequences. These worst case results with low probability and very high uncertainty can inappropriately drive operational decisions and system designs, or cause costly project delays and even non-approval of projects. This paper addresses the risks exposed by the public from chemical process facilities, and risk acceptance criteria for the public are proposed.

2. Importance of Risk Communication with The Public

The public demands zero risk, which is not attainable. The plant owners and responsible government agencies therefore must be open about risks (otherwise, they will be thought to be hiding something) and must accurately portray the risks and the benefits of the proposed activities.

The perception of risk is quite complex. A critical early study [1] compares the risk perception of four groups of people: League of Women Voters, college students, active club members, and risk experts. The perceptions of risk by these groups were dramatically diverse. Risk experts ranked risks according to their actual harm history or their estimated risk. Other groups, without access to such data, ranked risks more by two important factors: knowledge of the risk and dread factor. The dread factor clearly appears to be an issue for chemical process facilities, despite attempts by the government and plant owners to communicate the actual risk levels to the public.

This difference in thinking is fundamental to understanding how to communicate actual qualitative risk information to the public. A common way of judging the acceptability of calculated risk values is to use a comparison between calculated value and an average risk of which most individuals are familiar. Table 2-1 shows how such comparisons can be presented.

Table 2-1: Activities Associated with One in a Million Incremental Risk of Death in a Year[2]

Activity/Exposure	Type of Risk
Smoking 1.4 cigarettes	Cancer, heart disease
Spending 1 hour in a coal mine	Black lung disease
Living 2 days in New York or Boston	Air pollution
Traveling 300 miles by car	Accident
Traveling 10 miles by bicycle	Accident
Flying 1000 miles by jet	Accident
Living 2 months in Denver on vacation from New York	Cancer caused by cosmic radiation

3. Risk-Based Decision Making

A risk-based approach is an alternative to a pure consequence-based approach. Several studies [3, 4, 5, 6,7] show that a risk-based approach, taking into account both the event likelihood and the effect of mitigation measures, is a superior decision-making basis, compared to a consequence-based approach. An example of risk-based regulation is the European Norm Standard EN 1473 that requires risk analysis in design and modification decisions. This standard provides requirements for all types of LNG-related equipment. Events are evaluated based on their frequency and consequence, and risk mitigation measures are required for events that do not meet established acceptance criteria.

In the absence of risk criteria, decision making will, in many cases, be performed on the basis of calculating expected consequences. These worst case results with low probability and very high uncertainty can inappropriately drive operational decisions and system designs, or cause costly project delays and even non-approval of projects. This approach may also inhibit opportunities to shape and implement a true risk based emergency response system by the states. Expensive and demanding means may be implemented for high consequence, but low risk scenarios while high risk events, may be ignored. The consequences from Hurricane Katrina in New Orleans in 2005 are a good example of the wrong focus in emergency preparedness planning and execution.

4. Risk Acceptance Criteria

The risk acceptance criteria are necessary to make risk-based decisions. Decision-makers are not always familiar with numerical risk estimates; therefore a risk analysis must be reduced to qualitative terms before decision-makers can balance risks against the other relevant factors. Operational, economic, social, political and environmental factors are usually important as well. However, if the risks are very high, they can be the overriding factor. Establishment of risk criteria would reduce this obstacle. These qualitative terms are, in fact, value judgments about the significance of the risks.

Risk acceptance criteria are the standards used to translate numerical risk estimates (e.g. 10^{-7} per year) into value judgments (e.g. "negligible risk") which can be set against other value judgments (e.g. "desirable employment prospects") in a decision-making process. Risk acceptance criteria are also useful when a decision is to be made between two or more options purely on the basis of risk. Such decisions are sometimes complicated by the multi-dimensional nature of risk, e.g. rare high-fatality accidents may be exchanged for more likely low-fatality ones. Risk acceptance criteria can help the ranking of such options.

A requirement without a measurable limit is not as effective as it could be. It is a noble statement to require that the risk from any hazardous activity, like operating an LNG facility, must be low. But without a stated measurable value, it is like requiring vehicle operators to drive safely without providing a speed limit. The lack of risk criteria can result in misperception or even misrepresentation of the risk, and is the root of many significant uncertainties. Hence, governments should establish and regulate risk criteria on behalf of the public.

5. Principles of Setting Risk Acceptance Criteria

Given that society accepts hazardous activities in principle and does not have limitless resources to devote to their safety, the following set of principles might be appropriate when making decisions about their acceptability in specific cases:

1. The activity should not impose any risks which can reasonably be avoided.
2. The risks should not be disproportionate to the benefits (in terms of jobs, tax revenues and finished products) that the activity produces.
3. The risks should be equitably distributed throughout the society in proportion to the benefits received.
4. The risk should be driven by higher-frequency lower-consequence events (which can be adequately responded to by existing emergency services) rather than by lower-frequency high-consequence events. .

In reality, principles such as these are impossible to achieve. In fact, when resources are limited, such principles may be in conflict with each other. For example, reducing catastrophic risks may require an expenditure which could have saved more lives had it been used to prevent low fatality accidents.

The following approach is proposed by DNV for assessing the risks from chemical process facilities, or any hazardous activity, as the most ideal practical approach:

- Individual risk criteria should be used to limit risks to individual workers and members of the public. These criteria address the equity requirement (3) above insofar as it applies to individuals.
- Societal risk criteria should be used to limit risks to the affected population as a whole. These attempt to address requirement (2) above, although in a necessarily rudimentary fashion, since the benefits of hazardous activities are even more difficult to quantify than their risks. They also address the equity requirement (3) as it applies to communities. By expressing societal risk criteria on a frequency-fatality (FN) curve, the criteria can also address catastrophic risk in requirement (4) above.
- Cost-benefit analysis should be used to ensure that, once the above criteria are satisfied, an optimum level of safety measures is chosen for the activity, taking costs as well as risks into account. This addresses requirement (1) above.

An activity is said to have tolerable risks if it satisfies all of the above DNV approach, and intolerable risks if it fails to meet any of them.

5.1. Individual Risk Acceptance Criteria

Individual risk is defined by the Institution of Chemical Engineering as the frequency at which an individual may be expected to sustain a given level of harm from the realization of specified hazards [8]. The individual risk acceptance criteria measure the risk to a single person, real or hypothetical, from the hazard under consideration. This person may be representative of a group (such as the workers on a plant with a typical range of activities and locations) or it may be a theoretical marker such as the risk to a person spending 100% of their time at a fixed location (such as the boundary fence of a potentially hazardous facility).

A summary of individual risk acceptance criteria for members of the public used by various authorities around the world is given in Table 5-1[9].

Table 5-1: Summary of Existing Individual Risk Acceptance Criteria for Public[9]

Authority and Application	Maximum Tolerable Risk (per year)	Negligible Risk (per year)
VROM, The Netherlands (New plants)	10^{-6}	10^{-8}
VROM, The Netherlands (Existing plants or combined new plants)	10^{-5}	10^{-8}
Health & Safety Executive, UK (Existing hazardous industry)	10^{-4}	10^{-6}
Health & Safety Executive, UK (New nuclear power stations)	10^{-5}	10^{-6}
Health & Safety Commission, UK (Existing dangerous substances transport)	10^{-4}	10^{-6}
Health & Safety Executive, UK (New housing near existing plants)	3×10^{-6}	3×10^{-7}
Hong Kong Government (New plants)	10^{-5}	Not used
Department of Planning, New South Wales (New plants and housing)	10^{-6}	Not used
Environmental Protection Authority, Western Australia (New plants)	10^{-6}	Not used
Santa Barbara County, California, USA (New plants)	10^{-5}	10^{-7}

Based on the existing risk acceptance criteria and DNV's risk analysis experience, the maximum tolerable risk of existing activities for members of the public is recommended as 10^{-4} per year (one of 10,000 is killed per year in accidents). For new process activities, maximum tolerable risk for the public is recommended as 10^{-5} per year. The accumulated risk due to new activities would be still in the range of 10^{-4} per year.

5.2. Societal Risk Acceptance Criteria

Societal risk is defined as the relationship between the frequency and the number of people suffering a given level of harm from the realization of specified hazards[8]. Societal risk acceptance criteria measures the risk of multiple fatalities and takes into account the presence of a large number of people potentially affected and the public response to disasters. The societal risk from an activity or type of hazard is normally expressed as a cumulative risk that is plotted as the frequency of accidents causing N or more fatalities, against the number of fatalities, N.

A summary of the few existing official societal risk acceptance criteria is given in Table 5-2.

Table 5-2: Summary of Existing Societal Risk Acceptance Criteria

Authority	FN Curve Slope	Max. Tolerable Intercept with N=1	Negligible Intercept with N=1	LIMIT ON N
VROM, The Netherlands	-2	10^{-3}	NA	N starts at 10
Hong Kong Government (New plants)	-1	10^{-3}	10^{-5}	1000
Health & Safety Commission, UK (Existing ports)	-1	10^{-1}	10^{-4}	-

Based on the existing risk acceptance criteria and DNV's risk analysis experience, the maximum tolerable risk of multiple fatalities of the public is recommended as 10^{-6} per year.

5.3. As Low as Reasonably Practical (ALARP)

Figure 1 shows the framework of UK Health & Safety Executive (HSE) risk acceptance criteria. In the ALARP region, mitigating measures are desired, but only necessary if they are practical and cost beneficial to implement. While the risk acceptance criteria determine whether mitigation is required, the cost benefit analysis (CBA) determines how much is financially feasible. The standard approach to CBA in risk assessment involves placing a financial value on risks to human life. This is often described as placing a "value of life". It is ethically difficult to place a monetary number on a human life, and most companies are avoiding this discussion.

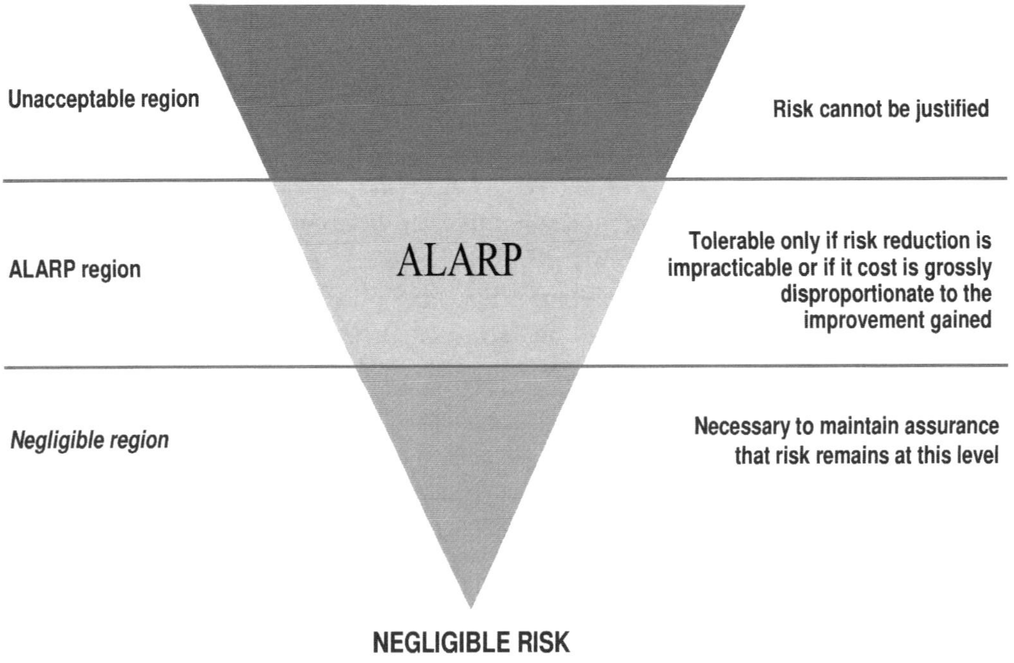

Figure 1: HSE Levels of Risk and ALARP Approach [10].

6. Discussions

Almost every activity involves risk, and we face the acceptable risk problem every day. It is believed that risk is never acceptable unconditionally, risk is only acceptable if some benefit or award can compensate for the risk [11]. In the situations where risk becomes unacceptable, risk reducing measures must be carried out. For instance, the risk of loss of life in a car accident can be reduced by wearing a seatbelt. However, risk can not be eliminated to zero, but only to the level of acceptance.

For the chemical process industry, total risk exposure to the public can not be eliminated to zero. Due to the benefits of chemical process facilities in the long term for the public, risk has to be accepted by the public at a reasonable level. When considering a hazardous industry, a common demand is for it to be "completely safe". This is an understandable desire, but it is virtually impossible to attain. The risks associated with accidents and terrorism can always be reduced by implementing additional safeguards, but unless the public is prepared to pay for them (in the form of subsidies to the industry or higher prices for its products), the industry will eventually become unprofitable and cease operation. Zero risk can only be achieved by not having the industry at all.

Society must decide whether the benefits from having the industry (in the form of jobs, tax revenues, standard of living, etc.) outweigh the risks involved. Therefore, the popular question "Is it safe?" must be interpreted in the risk analysis as meaning, "Are the risks low enough for the public to tolerate?" If so, the risks are said to be "tolerable." And sufficient safety is provided. This is what risk criteria attempts to establish.

The question arises as to who should establish the risk acceptance criteria for the public. Stakeholders, including facility owners, public and regulatory agencies usually have different points of views due to their own benefits and interests. When conflicts exist with the chemical process facilities, authorities including local government and federal agencies should balance the needs of the interested parties and make a decision on chemical process risk.

Risk criteria used by different countries and industries have been reviewed in this paper.
The judgments implicit in risk criteria should reflect a broad consensus of people in society, assuming they understand the risk estimates and take an objective view of the interests of people exposed to the risks relative to those of the society as a whole. The criteria, therefore, attempt to predict the judgments which an educated, objective, representative member of the public would make about the hazardous activity. Alternatively, they may be viewed as decisions that can be justified to the public, given an understanding of the issues involved.

Therefore, it is impossible for risk criteria to precisely represent what is or is not acceptable to the public. Value judgments vary between societies, and alter with time, accident experience and changing expectations of life. Some societies place great emphasis on individual rights, and it may not be accepted that anyone other than the person exposed can decide on the acceptability of an imposed risk. In such societies, the regulatory authority makes both a social and a political judgment when it determines whether or not a risk is tolerable. Such decisions can be guided by, but not solely based on, technical advice.

Based on this review and DNV's experience, DNV has proposed a set of risk acceptance criteria for the public. When risk acceptance criteria are established, the focus should be on the principle that incremental risk from new activities should be at least one order of magnitude lower than the existing risk of the affected groups. Following this principle, the new activity will not pose tangible risk to the public.

7. Conclusions

Defining risk acceptance criteria is a good practice for risk management. Quantitative safety goals may be achieved and safety may be enforced by compliance with a pre-defined set of risk acceptance criteria. Good acceptance criteria can help the public and chemical process industry to reach a consensus on the safety and security issues of chemical process facilities. Because the process plant owners and the public share different interests and goals, risk acceptance criteria must be established and enforced by regulations after balancing the needs of stakeholders. Good risk acceptance criteria will protect the public from unacceptable industrial risk. It should be the process plant owners' responsibility to document the total risk of their facilities, and the risk acceptance criteria should be applied to every phase of development.

8. References

1. B. Fischhoff et al., Acceptable Risk, Cambridge University Press, New York, 1981.
2. R Wilson and Edmund Crouch, Harvard University Press, Cambridge Massachusetts, 1982.
3. M. Hightower, L. Gritzo, A. Luketa-Hanlin, J. Covan, et. Al., Guidance on Risk Analysis and Safety Implications of a Large Liquefied Natural Gas (LNG) Spill over Water, SAND2004-6258, Sandia National Laboratories, Albuquerque, NM, December 2004.
4 R. Pitblado, J. Baik, G. Hughes, C. Ferro, and S. Shaw, Consequences of LNG Marine Incidents, Process Safety Progress, Volume 24, No 2, pp 108, June 2005
5 R. Pitblado, J. Baik, and V. Raghunathan, LNG Decision Making Approaches Compared, Proceedings, Mary Kay O'Connor Process Safety Center, Texas A&M University, College Station, TX, 2004.
6 M.L. Casada and D.C. Nordin, The current status of LNG facility standards and regulations, Process Safety Progress, 2005, 24, pp 152.
7 American Bureau of Shipping/Federal Energy Regulatory Commission (ABS/FERC), Consequence Assessment Methods for Incidents involving Releases from Liquefied Natural Gas Carriers, contract FERC04C40196, 2004.
8 Institution of Chemical Engineers, Nomenclature for hazard and risk assessment in the process industries, IChem.E., Rugby, UK, 1992.
9 Health & Safety Executive (HSE), Risk acceptance criteria for land-use planning in the vicinity of major industrial hazards, HMSO, 1989.
10 Health and Safety Executive, The tolerability of risk from nuclear power stations, HMSO, London, 1992.
11 B. Fischhoff et al., Acceptable risk, Cambridge University Press, New York, 1981.

Frequency and Consequence Modeling of Rare Events using Accident Databases

Meel A., O'Neill L. M., and Seider* W. D.
Department of Chemical and Biomolecular Engineering
University of Pennsylvania
Philadelphia, PA 19104-6393

Oktem U.
Risk Management and Decision Center, Wharton School
University of Pennsylvania
Philadelphia, PA 19104-6340

Keren N.
Department of Agricultural and Biosystems Engineering
Iowa State University
Ames, IA 50011-3080

Prepared for Presentation at
American Institute of Chemical Engineers
2006 Spring National Meeting
8th Process Plant Safety Symposium
Orlando, Florida
April 24-26, 2006

UNPUBLISHED

* Corresponding author: Email: seider@seas.upenn.edu,, Ph: 215-898-7953

Frequency and Consequence Modeling of Rare Events using Accident Databases

Meel A., O'Neill L. M., and Seider* W. D.
Department of Chemical and Biomolecular Engineering
University of Pennsylvania
Philadelphia, PA 19104-6393

Oktem U.
Risk Management and Decision Center, Wharton School
University of Pennsylvania
Philadelphia, PA 19104-6340

Keren N.
Department of Agricultural and Biosystems Engineering
Iowa State University
Ames, IA 50011-3080

Abstract:

Accident databases (NRC, ATSDR's, RMP, and others) contain records of incidents (e.g., releases and spills) that have occurred in United States chemical plants during recent years. For various chemical industries, Kleindorfer and coworkers [1] summarize the accident frequencies and severities in the RMP*Info database. Also, Anand and coworkers [2] use data mining to analyze the NRC database for Harris County, Texas.

Classical statistical approaches are ineffective for low frequency, high consequence events because of their rarity. Given this information limitation, this paper uses Bayesian theory to forecast incident frequencies, their relevant causes, equipment involved, and their consequences, in specific chemical plants. Systematic analyses of the databases also help to avoid future accidents, thereby reducing the risk.

More specifically, this paper presents dynamic analyses of incidents in the NRC database. Probability density distributions are formulated for their causes (e.g., equipment failures, operator errors, etc.), and equipment items are utilized within a particular industry. Bayesian techniques provide posterior estimates of the cause and equipment-failure probabilities. Cross-validation techniques are used for checking the modeling, validation, and prediction accuracies. Differences in the plant- and chemical-specific predictions with the overall predictions are demonstrated. Furthermore, the NRC database is exploited to model the rate of occurrence of incidents in various chemical and petrochemical companies using Bayesian theory. Extreme value theory is used for consequence modeling of rare events by formulating distributions for events over a threshold value. Finally, the fast-Fourier transform is used to estimate the risk within an industry utilizing the frequency and severity distributions.

---

* Corresponding author: Email: seider@seas.upenn.edu,, Ph: 215-898-7953

1. Introduction

Since the accidents at Flixborough, Seveso, and Bhopal, the reporting of abnormal events in the chemical industries has been encouraged to collect accident precursors. Efforts to increase the reporting of near-misses, with near-miss management audits, have been initiated by the Wharton Risk Management Center [3]. In addition, the AIChE Center for Chemical Process Safety (CCPS) has facilitated the development of a Process Safety Incident Database (PSID) to collect and share incident information, permitting industrial participants access to the database, while sharing their collective experiences [4]. Finally, the Mary Kay Safety Center at Texas A&M University (TAMU) [2, 5] has been gathering incident data in the chemical industries.

An incident/accident database, involving oil, chemical, and biological discharges into the environment in the U.S. and its territories, is maintained by the National Response Center (NRC) [6]. To record accidents, European industries submit their data to a common server [7], while chemical companies in the United States are required to report accident data every five years under the RMP Rule [1, 8]. The latter applies only when one or more of the chemicals on the EPA list is stored above the threshold amount indicated for each chemical.

Several researchers have been analyzing and investigating incident databases to identify common trends and to estimate risks. For example, Chung and Jefferson [9] have developed an approach to integrate accident databases with computer tools used by chemical plant designers, operators, and maintenance engineers, permitting accident reports to be easily accessed and analyzed. In addition, Sonnemans et al. [10] have investigated 17 accidents that have occurred in a petrochemical industry in the Netherlands and have demonstrated qualitatively that had accident precursor information been recorded, with proper measures to control future occurrences, these accidents could have been foreseen and even prevented. Furthermore, Sonnemans and Korvers [11] observe that even after observing accident precursors and disruptions, the operating systems inside companies often fail to prevent accidents. The results of yet another analysis feature the lessons learned from the major accident and near-miss events in Germany from 1993-96 [12, 13]. Finally, Elliott et al. [14] analyze the frequency and severity of accidents in the RMP database with respect to socioeconomic factors and found that larger chemically-intensive companies are located in counties with larger African-American populations and with both higher median incomes and higher levels of income inequality. Note that accident precursors have been studied also in railways, nuclear plants, health science centers, aviation, finance companies, and banking systems.

On the risk estimation frontier, Kirchsteiger [15] discusses the strengths and weaknesses of probabilistic and deterministic methods in risk analysis using illustrations associated with nuclear and chemical plants. It is argued that probabilistic methods are more cost-effective, giving results that are easier to communicate to decision and policy makers. In addition, Goossens and Cooke [16] describe the application of two risk assessment techniques involving: (i) formal expert judgment to establish quantitative subjective assessments of design and model parameters, and (ii) system failure analysis, with accident precursors, using operational evidence of system failures to derive the failure probability of the system. Furthermore, a HORAAM (human and organizational reliability analysis in accident management) method is introduced to quantify human and organizational factors in accident management using decision trees [17].

In this work, statistical methods are introduced to estimate the operational risk of individual chemical companies using the accident precursors reported by the NRC, with the risk estimated as the product of the frequency and the number of consequences. First, the frequency of abnormal events of the individual companies, on a yearly basis, is formulated using Bayesian theory. Later, a loss-severity distribution of the abnormal events is modeled using extreme value theory (EVT). Subsequently, the operational risk of the individual chemical industries is computed by performing fast-Fourier transforms (FFT) of the frequency and loss-severity distributions to obtain the aggregate loss distribution – although these calculations are being completed. The Bayesian theory upgrades any prior information using data to increase the confidence level in modeling the frequency of abnormal events; decreasing the uncertainty in decision-making with annual information upgrades [18]. Through EVT, both extreme and unusually rare events, which characterize incidents reported in the chemical industries, are modeled effectively. Note that EVT has been applied in structural, aerospace, ocean, and hydraulic engineering [19]. Herein, EVT is introduced to measure the operational risk in the chemical industries.

This approach to measuring risks in specific companies provides a quantitative framework for decision-making at higher levels. Using the platform provided, chemical industries should be encouraged to collect accident precursor data more regularly. Through implementation of this dynamic risk assessment methodology, improved risk management strategies should result. Also, the handling of third party investigations should be simplified after accidents.

1.1 Overview

More specifically, herein, the NRC database is explored for seven companies, including petrochemical and specialty chemical manufacturers, in Harris County, Texas. Figure 1 shows the algorithm to be described for calculating the operational risk of a chemical company. Initially, Bayesian models for the frequency of abnormal events are developed. Note that significant differences in the prediction of abnormal events are observed for the individual companies, as compared with predictions obtained when the incidents from all of the industries are lumped together. In addition, Bayesian models are developed to provide the frequency distribution of the day of the week on which the incidents occur, the equipment types involved, the causes behind the incidents, the chemicals involved, the equipment reliability, and the human reliability. Furthermore, the failure probabilities of the process units, as well as the causes of the incidents, are predicted. To obtain the loss-severity distribution, EVT is applied to the NRC database by formulating a quantitative index for the loss as a weighted sum of the different types of consequences. Then, FFTs are applied to obtain the aggregate loss distribution, also known as the aggregate severity distribution. Relevant information, like the Value at

Risk (VaR), is obtained from the loss distribution. Again, these calculations are being completed.

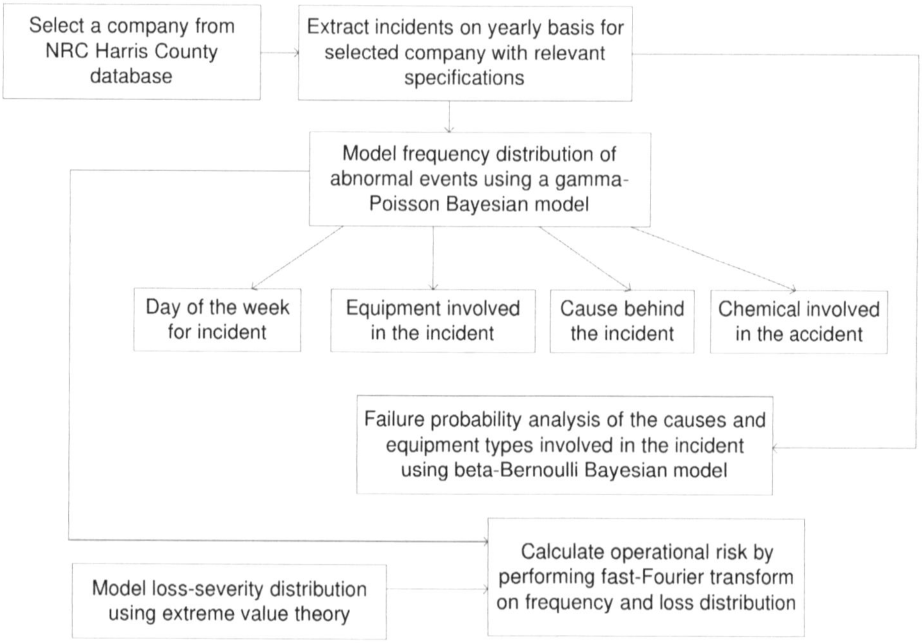

Figure 1. Algorithm to calculate the operational risk of a chemical company

2. Modeling the frequency of abnormal events

This Bayesian model formulates the frequency of occurrence of an abnormal event in a time interval for a company. The possible number of occurrences of an abnormal event in each time interval is a non-negative, integer-valued outcome that can be formulated using the *Poisson* distribution for y:

$$y \sim p(y = y_l) = \left\{ \frac{\lambda^{y_l} e^{-\lambda}}{y_l!} \right\}, \qquad y_l \in \{I^1\}, y_l \geq 0, \ \lambda > 0 \qquad \text{(1a)}$$

where y_l is the number of abnormal events in time interval l, and λ is the average number of abnormal events in the time intervals, with the expected value, $E(y)$, and variance, $V(y)$, equal to λ. Due to uncertainty, the prior distribution for λ is assumed to follow a *Gamma* distribution, $\lambda \sim Gamma(\alpha, \beta)$:

$$p(\lambda) \propto \lambda^{\alpha-1} e^{-\beta\lambda}, \qquad \alpha > 0, \ \beta > 0 \qquad \text{(1b)}$$

From Baye's theorem, the posterior distribution, $p(\lambda \mid Data)$, is:

$$p(\lambda \mid Data) \propto l(\lambda \mid Data) p(\lambda) \propto (\lambda^s e^{-N_t\lambda})(\lambda^{\alpha-1} e^{-\beta\lambda}) \propto \lambda^{(\alpha+s)-1} e^{-(\beta+N_t)\lambda} \qquad \text{(1c)}$$

where $Data = (y_0, y_1, \ldots, y_{N_t})$, $s = \sum_{l=0}^{N_t} y_l$, N_t is the number of times intervals, and $l(\lambda \mid Data)$ is the *Poisson* likelihood distribution. Note that $p(\lambda \mid Data)$ is also a *Gamma* distribution, $Gamma(\alpha+s, \beta+N_t)$, because λ is distributed according to $Gamma(\alpha, \beta)$, which is a conjugate prior to the *Poisson* distribution. The mean of the posterior distribution is the weighted average of the means of the prior and likelihood distributions:

$$\frac{\alpha + s}{\beta + N_t} = \frac{\beta}{\beta + N_t}\left(\frac{\alpha}{\beta}\right) + \frac{N_t}{\beta + N_t}\frac{s}{N_t} \qquad (1d)$$

and the variance of the posterior distribution is $(\alpha + s)/(\beta + N_t)^2$.

The predictive distribution to estimate the number of abnormal events in the next time interval, y_{N_t+1}, conditional on the observed *Data*, is discussed by Meel and Seider [20]. This gives a predictive mean, $(\alpha + s)/(\beta + N_t)$, and predictive variance, $(\alpha + s)/(\beta + N_t)[1+1/(\beta + N_t)]$, and consequently, the posterior and predictive means are the same, while the predictive variance exceeds the posterior variance.

Model-checking using predictive distributions. To check the accuracy of the model, the number of abnormal events in interval l, y_l, is removed, leaving the data, $y_{-l} = (y_0,\ldots, y_{l-1}, y_{l+1},\ldots, y_{N_t})$, over N_t -1 time intervals. Then, the current Bayesian model applied to y_{-l} is used to predict y_l. Finally, y_l and $E[y_l \mid y_{-l}]$ are compared, and predictive z-scores are used to measure their proximity:

$$z_l = \frac{y_l - E[y_l \mid y_{-l}]}{\sqrt{V[y_l \mid y_{-l}]}} \qquad (2)$$

For a good model, the mean and standard deviation of $z = (z_0,\ldots, z_{N_t})$ should approach zero and one, respectively.

3. Analysis of NRC database

The NRC database contains reports on all of the oil, chemical, radiological, biological, and etiological discharges into the environment anywhere in the United States and its territories [6]. A typical incident report includes the chemical involved, the cause of the incident, the equipment involved, the date of the incident, the volume of the chemical release, and the extent of the consequences. Herein, the incidents reported for Harris County, Texas, for fixed facilities during 1990-2002 are analyzed to determine their frequencies and consequences (loss or severity). This dataset was obtained from the Mary Kay Safety Center at TAMU, which filtered the NRC database for Harris County, taking care to eliminate duplications of incidents when they occurred. More specifically, the filtered dataset by Anand et al. [2] is used herein for further processing.

The equipment is classified into 13 major categories: electrical equipment (E_1), pumps/compressors (E_2), flare stacks (E_3), heat-transfer equipment (E_4), hoses (flexible pipes) (E_5), process units (E_6), process vessels (E_7), separation equipment (E_8), storage vessels (E_9), pipes and fittings (E_{10}), unclassified equipment (E_{11}), relief equipment (E_{12}), and unknowns (E_{13}). The Harris County database includes several causes of the incidents, including equipment failures (EF), operator errors (OE), unknown causes (U), dumping, and others, with the EF and OE causes being the most significant.

3.1 *Statistical analysis of incidents at chemical companies*

Table 1 shows incidents extracted from the NRC database for the seven companies located in Harris County. The total number of incidents, N_{total}, and the number of incidents of equipment failures, N_{EF}, operator errors, N_{OE}, and due to unknown causes, N_U, are listed during the years 1990-2002. In addition, from the 13 equipment categories, the number of process units, N_{PU}, process vessels, N_{PV}, storage vessels, N_{SV}, compressors/pumps, $N_{C/P}$, heat-transfer equipment, N_{HT}, and transfer-line equipment, N_{TL}, are included.

Table 1. Number of incidents for seven companies in the NRC database

Companies	Type	N_{total}	N_{EF}	N_{OE}	N_U	N_{PU}	N_{PV}	N_{SV}	$N_{C/P}$	N_{HT}	N_{TL}
A	Petrochemical	688	443	56	101	59	50	101	86	58	121
B	Petrochemical	568	387	48	88	110	37	69	127	47	56
C	Specialty chemical	401	281	35	46	45	93	61	10	28	77
D	Petrochemical	220	122	24	16	25	47	16	36	27	15
E	Specialty chemical	119	77	21	8	13	12	22	11	12	23
F	Specialty chemical	83	57	14	7	6	6	21	8	10	18
G	Specialty chemical	18	9	2	5	1	1	1	1	3	2

For each of the seven companies, several predictions of abnormal events for future years are carried out utilizing data from previous years, including the prediction of the total number of incidents, N_{total}, incidents associated with each equipment type, and incidents associated with each cause.

Figures 2a and 2b show the predictions of the number of incidents for companies B and F which are chosen arbitrarily to illustrate the variations in the predictive power of the models. In these figures, the number of incidents for the year n are forecasted using the gamma-Poisson Bayesian techniques based on the number of incidents from 1990 to n-1, where n = 1991, 1992, …, 2002. These are compared to the number of incidents that occurred in year n for companies B and F, respectively.

In the absence of information to model the prior distribution for the year 1990, α and β are assumed to be 0.001, providing a relatively flat distribution in the region of interest; that is, a non-informative prior distribution. Note that information upon which to base the prior parameters would enhance the early predictions of the models. This has been illustrated for a beta-Bernoulli Bayesian model, using informative and non-informative prior distributions, showing the sensitivity of the predictions to the prior values [20]. For company B, using non-informative prior distributions, either the numbers of incidents are close to the predicted numbers or higher than those predicted. However, for company F, the numbers of incidents are close to or less than those predicted.

When examining the results for the seven companies, the sizable variations in the number of incidents observed in a particular year are attributed to several factors, including the management and planning efforts to control the incidents, it being assumed that no significant differences occurred to affect the reporting of the incidents from 1990-2002. Therefore, when the number of incidents is less than those predicted, it seems clear that good incident-control strategies were implemented within the company. Similarly, when the number of incidents is higher than those predicted, the precursor data yields a warning to consider enhancing the measures to reduce the number of incidents in the future.

A good agreement between the numbers of incidents predicted and observed indicates a *stable equilibrium* is achieved with respect to the predictive power of the model. Such a state is achieved when the numbers of incidents and their causes do not change significantly from year-to-year. Note, however, that even as stable equilibrium is approached, efforts to reduce the number of incidents should continue. This is because, even when successful measures are taken year after year (that reduce the number of incidents), the predictive values are usually conservative, lagging behind until the incidence rates converge over a few years.

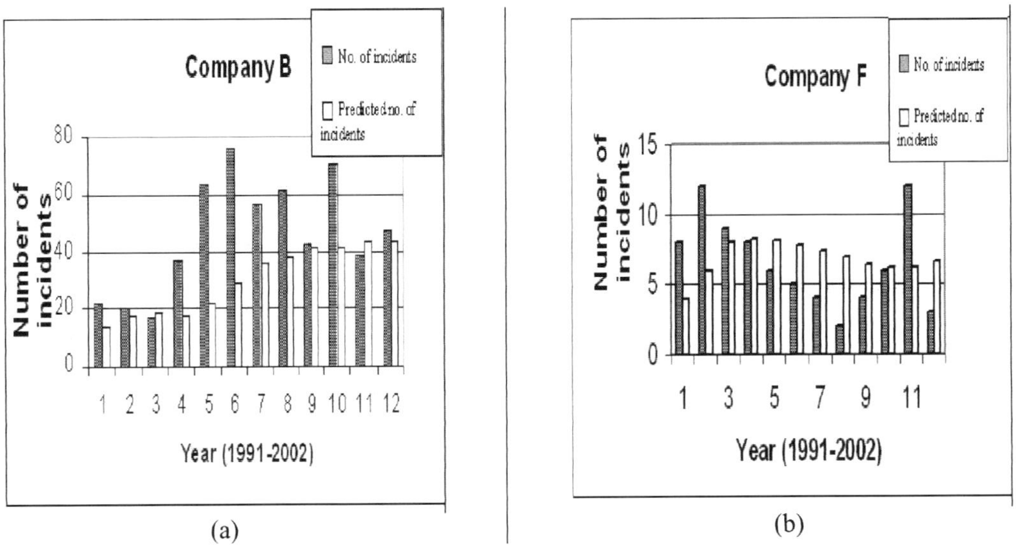

Figure 2. Total number of incidents: (a) Company B, (b) Company F

Next, the results of the Bayesian model checking using the *R* software package [21] to compute predictive distributions are presented in Q-Q plots. For company F, Figure 3a shows the density profile of incidents, while Figure 3b shows the normal Q-Q plot, which compares the distribution of z (Eq. (2)) to the normal distribution (represented by the straight line), where the elements of z are represented by circles. The sample quantiles of z (ordered values of z, where the elements, z_i, are called quantiles) are close to the theoretical quantiles (equally-spaced data from a normal distribution), confirming the accuracy of the model predictions. Most of the values are in good agreement, except for two outliers at the theoretical quantiles, 1.0 and 1.5.

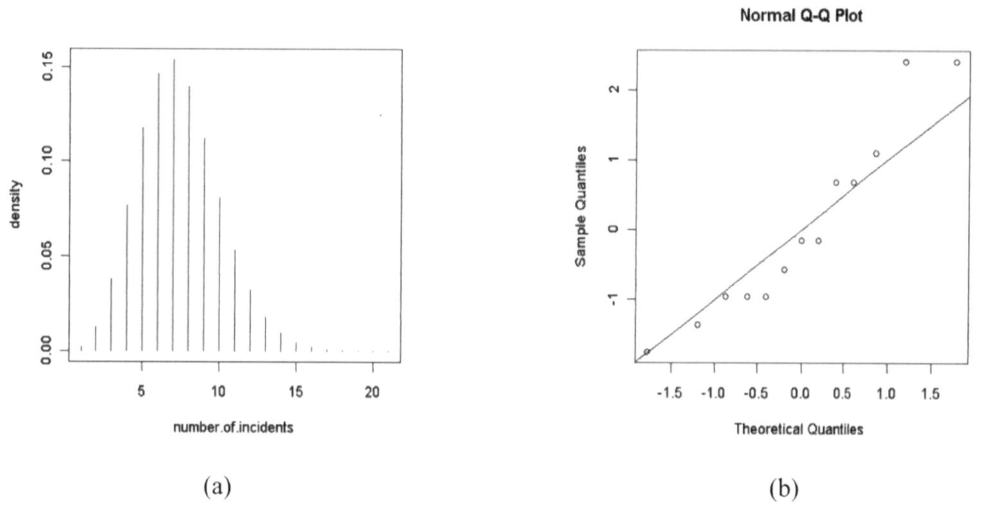

Figure 3. Company F: (a) Density of incidents, (b) Q-Q plot

Figures 4a and 4b show the density profile of incidents and the Q-Q plot for company B. Comparing Figures 4a and 3a, the number of incidents at company B are much higher than at company F. In addition, the variation in the number of incidents in different years is higher at company B (between ~25-65) than at company F (between ~0-15). Note that the circles on the Q-Q plot in Figure 4b depart more significantly from the straight line, possibly due to the larger year-to-year variation in the number of incidents as well as the appropriateness of the of gamma-Poisson distribution. The circles below the straight line correspond to the safe situation where the number of incidents is less than that predicted. However, the circles above the straight line, with the number of incidents higher than those predicted, provide a warning.

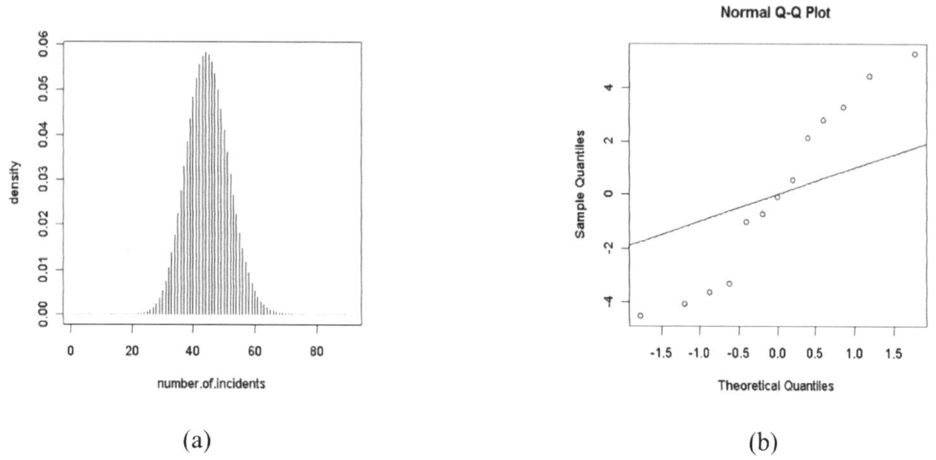

Figure 4. Company B: (a) Density of incidents, (b) Q-Q plot

To reduce the departures of the circles from the straight line, it is also possible to use a hierarchical Bayesian model. Alternatively, other distributions for the likelihood function, for instance, the negative binomial or the binomial distribution instead of the Poisson distribution can be explored. Note that the Poisson, negative binomial, and binomial distributions belong to the same family, but have a different relationship between the mean and variance of the distributions. The negative binomial distribution is useful when the data mean is less than the variance. The binomial distribution is suitable when the mean exceeds the variance, and the Poisson distribution is preferred when the mean and the variance are in close proximity. Furthermore, the higher number of incidents reported drifts the posterior mean towards the data mean. This is desirable, although poorer predictions may be due to the choice of the distribution. These effects will be examined more closely in future studies.

3.2 Statistical analysis of incident causes and equipment types

In this analysis, for each company, Bayesian models are formulated for each cause and equipment type. Because of the large variations in the number of abnormal events (incidents) observed over the years, the performance of the gamma-Poisson Bayesian models differ significantly. For company F, Figures 5a and 5b show the Q-Q plots for equipment failures and for operator errors, respectively. Figure 5a shows better agreement with the model because the variation in the number of incidents related to equipment failures is small, while the variation in the number of incidents related to operator errors is more significant. This is consistent with the expectation that equipment performance varies less significantly than operator performance over time.

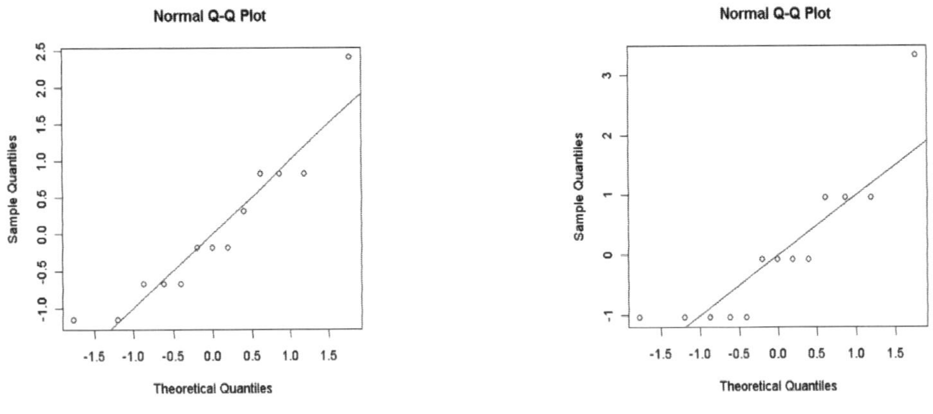

(a) (b)

Figure 5. Company F: (a) Equipment failures, (b) Operator errors

Figures 6a and 6b show the Q-Q plots for equipment failures and for operator errors, respectively, at company B. When comparing Figures 5a and 6a, the predictions of the numbers of equipment failures at company B are poorer than at company F. This is similar to the predictions for the total numbers of incidents at company B, as shown in Figure 4b, compared with those at company F, as shown in Figure 3b. Yet, the predictions for the operator errors are comparable at companies F and B, and consequently, the larger variation in reporting incidents at company B are attributed to the larger variation in the numbers of equipment failures.

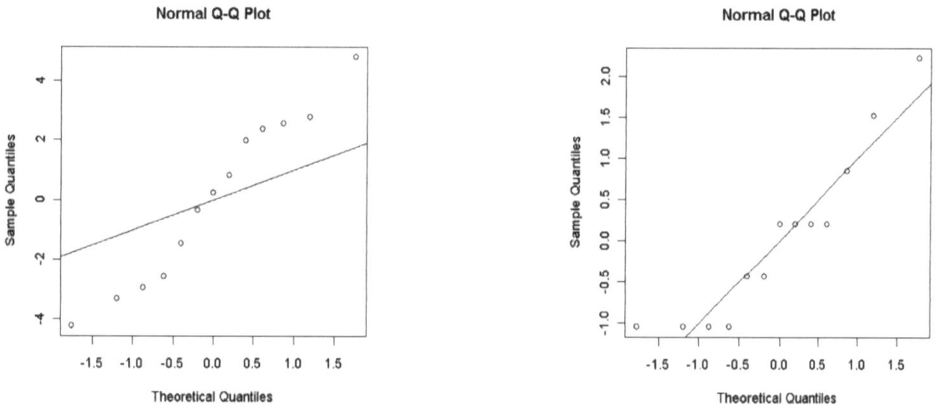

(a) (b)

Figure 6. Company B: (a) Equipment failures, (b) Operator errors

Figures 7a - 7d show the Q-Q plots for incidents associated with the process units, storage vessels, heat-transfer equipment, and compressors/pumps at company A. Figures 7a and 7c are consistent with the small variation in the numbers of incidents per year for process units and heat-transfer equipment, while the variations in the numbers of incidents per year associated with storage vessels and compressors/pumps are more significant in Figures 7b and 7d. Similar analyses show comparable trends for all of the equipment types and causes at each company.

3.3 *Statistical analysis of chemicals involved*

For each company, an attempt was made to identify trends for each of the top five chemicals associated with the largest number of incidents in the NRC database. However, no specific trends for a particular chemical associated with a higher number of incidents in all of the companies were observed. This could be because the different products are produced in varying amounts by the different companies. It might be preferable to carry out the analysis for a company that manufactures similar chemicals at different locations or for different companies that produce similar products.

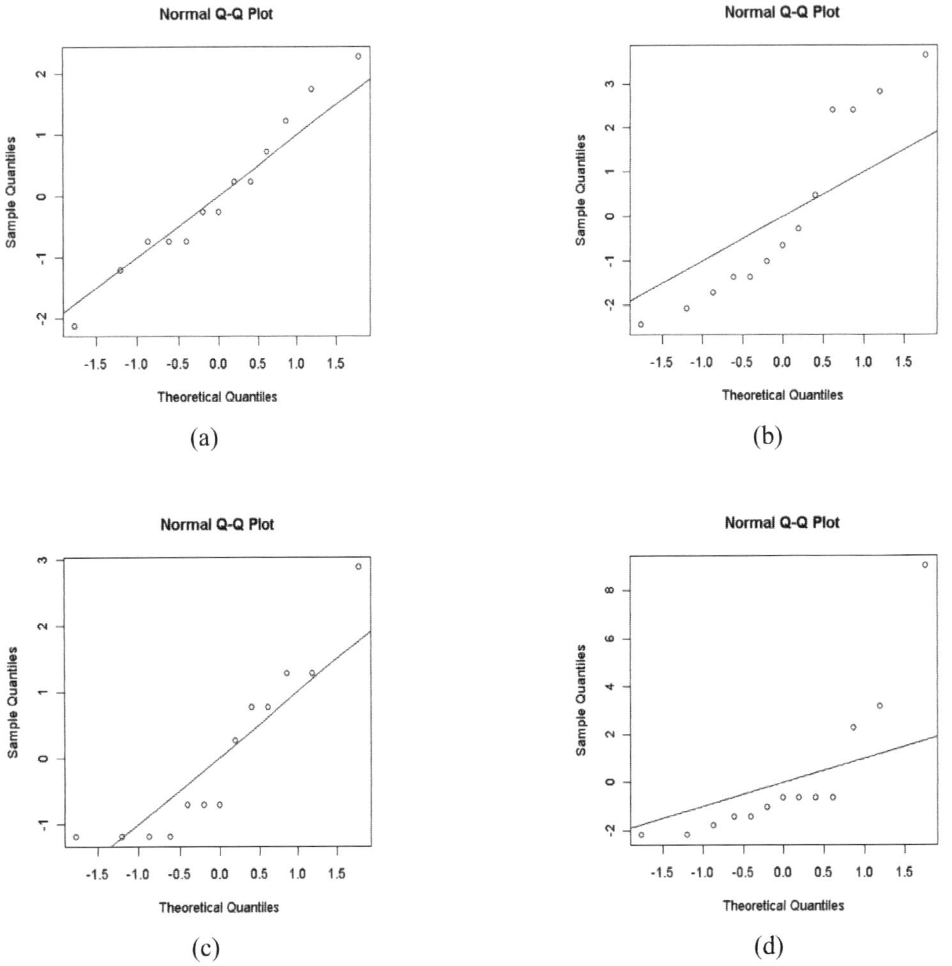

Figure 7. Company A: (a) Process units, (b) Storage vessels, (c) Heat-transfer equipment, and (d) Compressors/pumps

3.4 Statistical analysis of the day of the week

For each of the seven companies, Table 2 summarizes the model checking of the Bayesian predictive distributions, with the mean and variance of z displayed. Again, the predictions improve with the total number of incidents observed for a company. As seen, the mean and variance of z indicate that higher deviations are observed on Wednesdays and Thursdays for almost all of the seven companies. Lower deviations occur at the beginning of the week and over the weekends, which may be attributed to the alertness of the operators at the beginning of the week and fewer operations on the weekends.

To provide a better understanding of this important phenomenon, additional data through operator surveys should be obtained. Furthermore, the higher values of the means and variances for company G on Friday and Saturday suggest that the number of data points is inadequate to generate a reliable Bayesian model. Other analyses that relate the causes of the incidents to the days of the week can be carried out to identify more specific patterns

Table 2. Q-Q plot properties for day of the week analysis of incidents

	Mon	Tue	Wed	Thru	Fri	Sat	Sun
A	0.027, 1.5	0.015, 1.06	0.032, 1.55	0.046, 1.9	0.023, 1.31	0.022, 1.23	0.055, 1.93
B	0.032, 1.53	0.047, 1.8	0.06, 2.12	0.058, 2.05	0.035, 1.55	0.027, 1.25	0.033, 1.46
C	0.027, 1.28	0.024, 1.21	0.047, 1.67	0.048, 1.62	0.031, 1.33	0.019, 1.002	0.039, 1.48
D	0.15 2.3	0.165, 2.7	0.2, 2.96	0.2, 3.22	0.13, 2.44	0.126, 2.22	0.27, 3.4
E	0.038, 1.06	0.037, 1.19	0.086, 1.66	0.078, 1.64	0.11, 1.89	0.07, 1.46	0.036, 0.96
F	0.034, 1.06	0.06, 1.27	0.04, 1.08	0.87, 0.05	0.035, 0.98	0.043, 1.01	0.07, 1.22
G	0.06, 1.09	0.14, 1.29	0.14, 1.29	0.14, 1.29	7.84, 29.26	15.82, 58.48	0.23, 1.96

3.5 Rates of equipment failures and operator errors

In this section, for an incident, the probabilities of the involvement of each of the 13 equipment types and the probabilities of their causes (e.g., equipment failures, operator errors, and others) are modeled. The tree in Figure 8 shows, for each incident, the possible causes, and for each cause, the possible equipment types. Note that alternatively the tree could show, for each incident, the possible equipment types followed by the possible causes. x_1, x_2, x_3 are the probabilities of causes EF, OE, and O for an incident, and d_1, d_2, d_3 are the cumulative number of incidents at the end of each year. e_1, e_2, e_3, …, e_{13} are the probabilities of the involvement of equipment types, E_1, E_2, …, E_{13}, in an incident through different causes, where $M_1 + N_1 + O_1$, $M_2 + N_2 + O_2$, $M_3 + N_3 + O_3$, …, $M_{13} + N_{13} + O_{13}$ are the cumulative number of incidents associated with each equipment type.

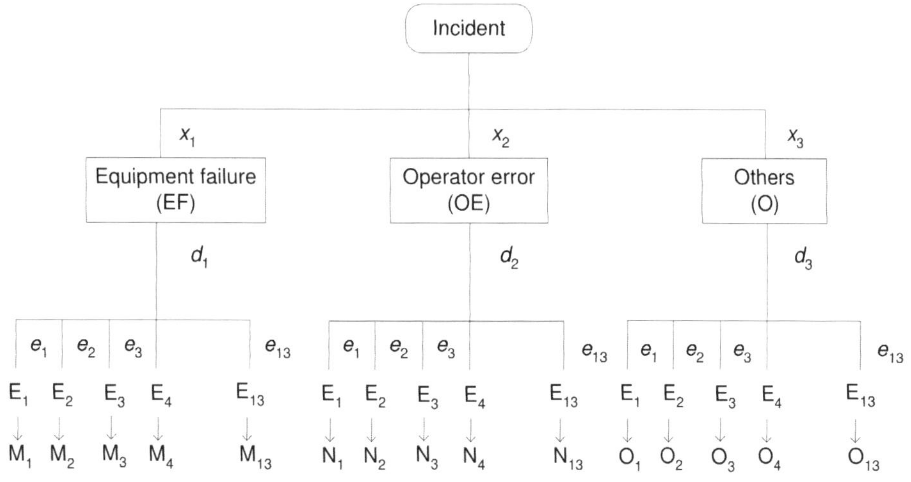

Figure 8. Equipment involved and cause analysis for an incident

The prior distributions of the probability of x_i are modeled using *Beta* distributions with parameters a_i, b_i:

$$f(x_i) \propto (x_i)^{a_i-1}(1-x_i)^{b_i-1}, \quad i = 1,\ldots,3 \qquad (3)$$

having means = $a_i/(a_i + b_i)$ and variances = $a_ib_i/(a_i + b_i)^2(a_i + b_i+1)$. These conjugate *Beta* prior distributions are updated using *Bernoulli's* likelihood distribution to obtain the posterior distribution of the probability of x_i:

$$f(x_i \mid Data) \propto (x_i)^{a_i-1+d_i}(1-x_i)^{b_i-1+\sum_{k=1,\neq i}^{3} d_k} f(x_i) \qquad (4)$$

The posterior distributions, which are also *Beta* distributions having parameters, $a_i + d_i$, and $b_i + \sum_{k=1,\neq i}^{3} d_k$, change at the end of each year as d_i change. a_1 and b_1 are assumed to be 1.0 and 1.0 to give a flat, non-informative, prior distribution; a_2 and b_2 are assumed to be 0.998 and 1.002 to give an non-informative, prior distribution; and a_3 and b_3 are 0.001 and 0.999. Consequently, the mean prior probabilities of EF, OE, and O are 0.5, 0.499, and 0.001, respectively. The posterior means and variances are obtained over the years 1990-2002 for each of the seven companies.

Figures 9a-c show the probabilities of the causes EF, OE, and O for an incident at company F. Using the data at the end of each year, the probabilities increase from 0.5 for equipment failures, decrease from 0.499 for operator errors, and increase from 0.001 for the others, with operator errors approaching slightly higher values than those for the others. Similarly, analyses for equipment types are carried out using *Beta* distributions, $f(e_i)$ and $f(e_i|data)$, with the *data*, $M_1 + N_1 + O_1$, $M_2 + N_2 + O_2$, $M_3 + N_3 + O_3$, …, $M_{13} + N_{13} + O_{13}$. Figure 9d shows, for an incident, that the probability of the involvement of the process vessels (PV) decreases over time. Similarly, the probabilities for the other equipment types approach stable values after a few years with occasional departures from their mean values.

3.5.1 Equipment and human reliabilities

By comparing the causes of incidents between the equipment failures and operator errors, insights regarding equipment and human reliabilities are obtained. In general, for all of the companies, incidents involving equipment failures exceed incidents involving operator errors, even though the OE/EF ratio changes at the end of each year. For petrochemical companies, the ratio is much lower than for specialty chemical companies, as seen in Table 3.

Table 3. OE/EF ratio for the petrochemical (P) and specialty chemical (S) companies

Company	A (P)	B (P)	C (S)	D (P)	E (S)	F (S)	G (S)
OE/EF ratio	0-0.3	0-0.22	0-0.75	0-0.5	0-0.667	0-0.667	0-0.5

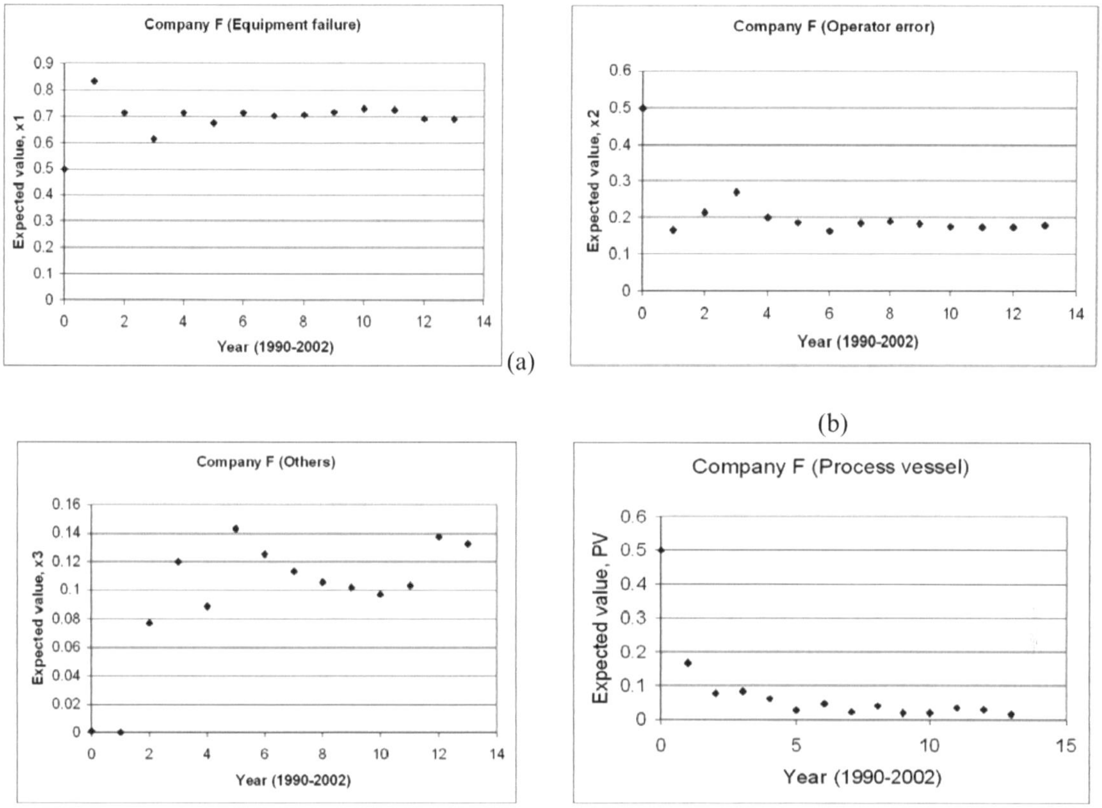

(c) (d)

Figure 9. Probabilities for company F: (a) EF, (b) OE, (c) others, (d) PV

3.6 Specialty chemicals and petrochemicals

To identify trends in the manufacture of specialty chemicals and petrochemicals, data for companies C, E, F, and G are combined and compared with the combined data for companies A, B, and D. Note that this is advantageous when the data for a single company are insufficient to identify trends, and when it is assumed that the lumped data for each group of companies are identically and independently distributed. For these reasons, all of the analyses in Sections 3.1 - 3.5 were repeated with the data for specialty chemical and petrochemical manufacturers lumped together. Because the number of data entries in each lumped data set are increased, the circles on the Q-Q plot lie closer to the straight line. However, the cumulative predictions for the specialty chemical and petrochemical manufacturers differ significantly from those for the individual companies. Hence, it is important to carry out company specific analyses. Nevertheless, when insufficient data are available for each company, the cumulative predictions for specialty chemical and petrochemical manufacturers are preferable. Furthermore, when insufficient lumped data are available for the specialty chemicals and petrochemical manufacturers, trends may be identified by combining the data for all of the companies.

3.7 *Loss-severity modeling using extreme value theory*

For rare events with extreme losses, it is important to identify those that exceed a high threshold. Extreme value theory (EVT) is a powerful and fairly robust framework to study the tail behavior of a distribution. Embrechts et al. [19] provide an overview of extreme value theory as a risk management tool, discussing its potential and limitations. In another study, McNeil [22] examines the estimation of the tails of the loss-severity distributions and the estimation of quantile risk measures for financial time-series using extreme value theory. Herein, EVT is employed to develop a loss-severity distribution for the seven chemical companies.

The distribution of excess values of losses, x, over a high threshold, u, is defined as:

$$F_u(y) = \mathbf{Pr}\{X - u \le y \mid X > u\} = \frac{F(y+u) - F(u)}{1 - F(u)}, \qquad x \in X \qquad (5)$$

which represents the probability that the value of x exceeds the threshold, u, by at most an amount, y, given that x exceeds the threshold u, where F is the cumulative probability distribution. For sufficiently high threshold u, the distribution function of the excess may be approximated by the generalized Pareto distribution (GPD), and consequently $F_u(y)$ converges to GPD as the threshold becomes large. The GPD is:

$$G(x) = \left\{ \begin{array}{ll} 1 - \left(1 + \xi \dfrac{x}{\beta}\right)^{-1/\xi} & \text{if } \xi \neq 0 \\ 1 - e^{-x/\beta} & \text{if } \xi = 0 \end{array} \right\} \qquad (6)$$

where ξ is the shape parameter and the tail index is $\alpha = \xi^{-1}$. Note that the GPD reduces into different distributions depending on ξ. The distribution of excesses may be approximated by the GPD by choosing ξ and β and setting a high threshold u. The parameters of the GPD can be estimated using various techniques; for example, the maximum likelihood method and the method of probability-weighted moments.

3.7.1 Loss-severity distribution of NRC database

A software package, Extreme Value Analysis in MATLAB (EVIM), is used to obtain the parameters of the GPD for the NRC database [23]. Because few incidents have high severity levels, the incidents analyzed for the seven companies are assumed to be identically and independently distributed (iid). The loss for an incident, L, is calculated as a weighed sum of the numbers of evacuations, injuries, hospitalizations, fatalities, and damages:

$$L = w_e N_e + w_i N_i + w_h N_h + w_f N_f + w_d N_d \qquad (7)$$

where w_e = \$100, w_i = \$10,000, w_h = \$50,000, w_f = \$2,000,000, and w_d = 1, with N_d reported in dollars. For the NRC database, the threshold value, u, is chosen to be \$1,000.

As expected, the NRC database has few incidents that have a significant loss. Only 92 incidents among those reported (~3,000) had monetary loss ($L > 0$) and even fewer exceeded the threshold. Note that to obtain a satisfactory prediction of the GPD parameters, usually 100 data points are needed. Therefore, the cumulative distribution of the losses for the NRC database in Figure 10 would be improved with additional data. To achieve this, data from additional companies in Harris County could be included. However, the performance in Figure 10 is considered to be satisfactory. The parameters, $\xi = 0.8688$ and $\beta = 1.7183\times10^4$, are computed using the maximum likelihood method.

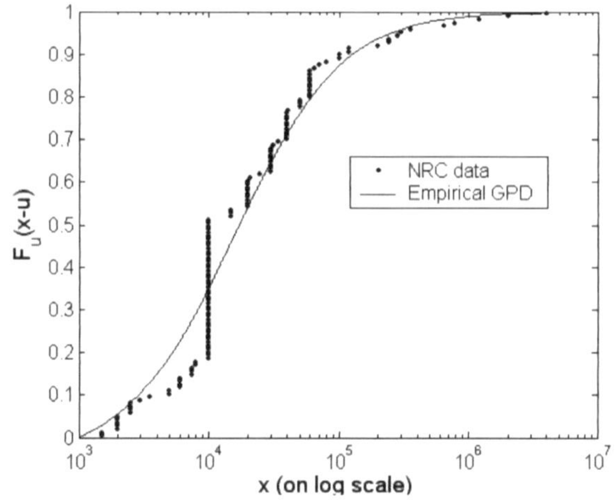

Figure 10. Loss distribution of NRC database

4. Operational risk

Calculations are being completed to estimate the Value-at-Risk (VaR) from the aggregate loss distribution by performing fast-Fourier transforms on the frequency and loss-severity distributions.

5. Conclusions

Statistical models to analyze accident precursors in the NRC database have been developed. They:

1. provide Bayesian models that facilitate improved company-specific estimates, as compared with lumped estimates involving all of the specialty chemical and petrochemical manufacturers.

2. identify Wednesday and Thursday as days of the week in which incidents are more likely to occur.

3. are effective for testing equipment and human reliabilities, indicating that the OE/EF ratio is lower for petrochemical than specialty chemical companies.

4. are expected to be beneficial for obtaining the Value-at-Risk (VaR) from the loss-severity distribution using EVT. Note that the VaR calculations are being completed.

Consistent reporting of incidents is crucial for the reliability of this analysis. In addition, the predictive errors are reduced when: (i) sufficient incidents are available for a specific company to provide reliable means, and (ii) less variation occurs in the number of incidents from year-to-year. Furthermore, to obtain better predictions, it helps to select distributions that better represent the data, properly modeling the functionality between the mean and variance of the data.

References

1. Kleindorfer PR, Belke JC, Elliott MR, Lee K, Lowe RA, Feldman HI. Accident epidemiology and the US chemical industry: Accident history and worst-case data from RMP*Info. Risk Anal. 2003; 23:865-881.
2. Anand S, Keren N, Tretter MJ, Wang Y, O'Connor TM, Mannan MS. *Harnessing data mining to explore incident databases. 7th Annual Symposium, Mary Kay O'Connor Process Safety Center.* 2004. College Station, TX.
3. Phimister JR, Oktem U, Kleindorfer PR, Kunreuther H. Near-miss incident management in the chemical process industry. Risk Anal. 2003; 23:445-459.
4. CCPS. Process Safety Incident Database (PSID). http://www.aiche.org/CCPS/ActiveProjects/PSID/index.aspx
5. Mannan MS, O'Connor TM, West HH. Accident history database: An opportunity. Environ. Prog. 1999; 18:1-6.
6. NRC. National Response Center http://www.nrc.uscg.mil/nrchp.html.
7. Rasmussen K. The experience with Major Accident Reporting System from 1984 to 1993. European Commission, Joint Research Center, EUR 16341 EN. 1996;
8. RMP. 40 CFR Chapter IV, Accidental Release Prevention Requirements; Risk Management Programs Under the Clean Air Act Section 112(r)(7); Distribution of Off-Site Consequence Analysis Information. Final Rule, 65 FR 48108 2000;
9. Chung PWH, Jefferson M. The integration of accident databases with computer tools in the chemical industry. Comp. Chem. Eng. 1998; 22:S729-S732.
10. Sonnemans PJM, Korvers PMW, Brombacher AC, van Beek PC, Reinders JEA. Accidents, often the result of an 'uncontrolled business process' - a study in the (Dutch) chemical industry. Qual. Reliab. Eng. Intern'l. 2003; 19:183-196.
11. Sonnemans PJM, Korvers PMW. Accidents in the chemical industry: Are they foreseeable? J. Loss Preven. Proc. Ind. 2006; 19:1-12.
12. Uth HJ. Trends in major industrial accidents in Germany. J. Loss Preven. Proc. Ind. 1999; 12:69-73.
13. Uth HJ, Wiese N. Central collecting and evaluating of major accidents and near-miss-events in the Federal Republic of Germany - results, experiences, perspectives. J. Hazard. Mat. 2004; 111:139-145.

14. Elliott MR, Wang Y, Lowe RA, Kleindorfer PR. Environmental justice: frequency and severity of US chemical industry accidents and the socioeconomic status of surrounding communities. J. Epidem. Commun. Health 2004; 58:24-30.
15. Kirchsteiger C. Impact of accident precursors on risk estimates from accident databases. J. Loss Preven. Proc. Ind. 1997; 10:159-167.
16. Goossens LHJ, Cooke RM. Applications of some risk assessment techniques: Formal expert judgement and accident sequence precursors. Safety Sci. 1997; 26:35-47.
17. Baumont G, Menage F, Schneiter JR, Spurgin A, Vogel A. Quantifying human and organizational factors in accident management using decision trees: The HORAAM method. Reliab. Eng. Sys. Safety 2000; 70:113-124.
18. Robert CP. The Bayesian Choice. Springer-VerlagNew York: 2001.
19. Embrechts P., Kluppelberg C., Mikosch T. Modelling Extremal Events. SpringerBerlin: 1997.
20. Meel A, Seider WD. Plant-specific dynamic failure assessment using Bayesian theory. Submitted to Chem. Eng. Sci. 2005;
21. Gentleman R, Ihaka R, Bates D, Chambers J, Dalgaard J, Hornik K. The R project for Statistical Computing. http://www.r-project.org/ 2005;
22. McNeil AJ. Estimating the tails of loss severity distributions using extreme value theory. ASTIN Bulletin 1997; 27:117-137.
23. Gencay R, Selcuk F, Ulugulyagci A. EVIM: A software package for extreme value analysis in MATLAB. Stud. Nonlin. Dynam. Economet. 2001; 5:213-239.

Utility Failures – Are You Prepared?

Katherine Pearson
Rohm and Haas Texas, Inc.
6519 State Hwy 225
Deer Park, TX 77536 U.S.A.
katherinepearson@rohmhaas.com

Prepared for Presentation at
American Institute of Chemical Engineers
2006 Spring National Meeting
8th Process Plant Safety Symposium
Orlando, Florida
April 24-26, 2006

UNPUBLISHED

Utility Failures – Are You Prepared?

Katherine Pearson
Rohm and Haas Texas, Inc.
6519 State Hwy 225
Deer Park, TX 77536 U.S.A.
katherinepearson@rohmhaas.com

Abstract

Many incidents have been triggered by unanticipated interactions with utility systems. Utility systems have many connections to processes, both physical and electrical. This interconnectivity results in common mode failures and the potential for cross-contamination. This paper will use "real-plant" incidents to illustrate what can happen when the hazards of interactions are not identified or understood. Common failures which occur in utility systems will also be discussed.

1. Introduction

Utility systems are not very glamorous and are often overlooked in Process Hazard Analyses (PHA's), operations, and maintenance; however, many incidents have been triggered by unanticipated interactions between processes and utility systems. Utility systems have many connections to processes, both physical and electrical. This interconnectivity results in the potential for cross-contamination as well as common mode failures.

Project design reviews, PHA's, and Management of Change reviews (MOC's) give us an opportunity to anticipate and plan for these interactions. The intent of this paper is to stimulate your thinking about utility systems in a different way, so you will ask the right questions in these reviews.

My first experience with a contaminated utility system came as a young engineer working in a unit with pneumatic instruments. One day, the instrument air system became contaminated with a very stinky and flammable material. The material made it back to the control room. The material was difficult to clear from all the instrument piping, and the control room smelled for weeks!

Thankfully, most of us do not have a lot of pneumatic instruments these days. But we still have many connections between utility systems and processes. Common mode utility failures such as sudden power outages, loss of instrument air, or loss of refrigeration can also result in incidents and near-misses.

Three incidents of increasing complexity will be reviewed in detail in the paper. The first incident involves unexpected materials in a sewer. The second incident involves contamination of a cooling tower water (CTW) system with an unexpected result. The third incident is a complex incident which resulted in damage to a scrubber during a power outage. Reactive chemistry played a part in each incident.

Evaluating utility systems will be covered in the last part of the paper. The reactivity matrix tool and issues to look for in individual utility systems will be covered.

2. Incident #1 - Flashback to "Waste Water" Tank:

This first incident is a simple example of unexpected contamination in a waste water tank and a chemical sewer. Steam jet condensate from a barometric condenser was collected in a waste water tank. The water was analyzed for several contaminants. If the waste water did not contain the specified contaminants, it was drained to a chemical sewer trench.

One day, hot work was going on in the unit to install new safety equipment. The job had started after the usual checks for flammability around the work piece. At some point, the waste water tank was drained to the chemical sewer. Nothing happened until the tank was almost empty. The investigation determined that an organic layer had built up inside the waste water tank. When the organic layer ran through the chemical sewer, vapors found the ignition source at the hot work, and flashed back to the waste water tank.

The ignition lifted the tank off of its foundation, propelling it to an elevation above the control room roof. The coils landed on the control room roof. In Figure 1, part of the coils can be seen on the roof (just to the left of the tank). The tank turned upside down and landed back on its foundation.

Figure 1: Tank Damage from Flashback

Before the incident, the NFPA diamond for the tank originally showed "0" in the flammability category. After the incident, the NFPA diamond for the sister tank was corrected to show "4" in the flammability category, based on the potential presence of the organic layer

This example highlights that barometric condensers constitute a direct connection between the process and the sewer system. Steam jet condensate usually contains some process material. If the process system is upset, the jet condensate can contain a lot more process material.

There are several key learnings from this incident:

- Condensate from barometric condensers can contain process chemicals.
- Sewer systems should always be treated as if they contain hazardous chemicals.
- Conditions can change during the job.

For hot work to be performed safely near a sewer, sewer openings have to be covered and conditions monitored during the job.

3. Incident #2 - Flashback to Cooler:

A catastrophic leak in an acid heat exchanger resulted in acid contamination of the carbon steel cooling tower system. On-line pH analyzer alarms in the cooling tower went off-scale. Operators did not realize the severity of the leak at first. The part of the system containing the leaking exchanger was identified and blocked off. The rest of the system was neutralized.

The part of the system containing the leaky exchanger stayed contaminated with the acid containing material and sat stagnant. Several exchangers were individually isolated, drained, and flushed; then they were opened for inspection and repair. Five days after being opened, a flashback occurred on one exchanger when a welder's torch was put near the tube sheet. Luckily, the welder was not severely injured.

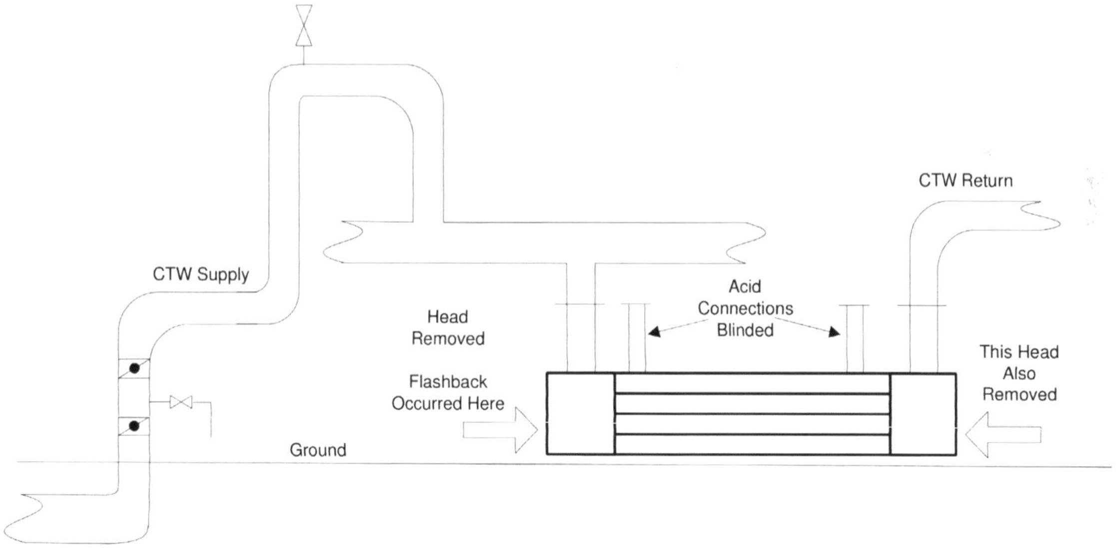

Figure 2: Schematic of Cooler System

The investigation team concluded the most likely cause of the incident was hydrogen gas generated by corrosion in the carbon steel CTW supply piping. There was liquid above the blocked valve used to isolate the cooler; the piping went up and over into the cooler, so there was about 8 ft of piping holding liquid. The liquid contained acid, and the pH was 1.25. Calculations showed that up to 10 m3 of hydrogen could have been generated from corrosion of the pipe in contact with the liquid during the 5 day period.

In addition, corrosion of the underground CTW piping could have also contributed some hydrogen gas. This piping was concrete-lined carbon steel, encased in concrete. The acid probably would have eaten though the internal concrete lining in at least some locations. Once the carbon steel was exposed, corrosion would result in hydrogen gas. The CTW supply piping was isolated from the exchanger using two block valves, but the vent between the valves was not open. The valves were not designed to be vapor tight. Gas had built up between the two valves, because when the vent valve was opened, witnesses heard it depressurize. The upstream block valve was also not liquid tight; liquid followed the depressurization.

This system had originally been locked out for opening and inspection, and the plant procedures did allow for a single block. However, for hot work, the procedures specified blinding for process and CTW connections. The process connections had been blinded, but the CTW connections were not blinded. In this case, the operating personnel did not recognize that the new scope of work required a different lockout. They also knew that the cooling water system was contaminated, but still treated it as "water."

This incident highlights several issues:

- Heat exchangers leak, and the leaks can result in cross-contamination.
- Utility systems should be treated as if they contain the materials in adjacent processing equipment.
- Hydrogen gas generation is a concern any time corrosion is an issue.
- Appropriate permitting procedures are critical to safe maintenance.

4. Incident #3 – Packing Damage in Off-gas Scrubber:

In an acid plant, an absorbing tower was used to absorb SO3 gas and the off-gas was sent to an off-gas scrubber (see Figure 3).

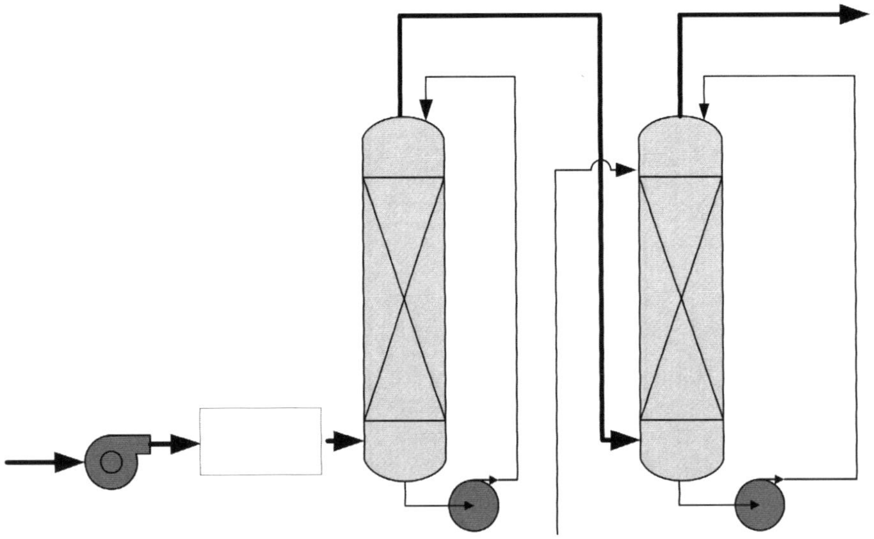

Figure 3: Schematic of Acid Tower and Off-Gas Scrubber

After a sudden power outage, the polyethylene packing in an SO3 off-gas scrubber was damaged when gas containing a high concentration of SO3 went through the dry scrubber. The main unit blower was a steam turbine and remained running after the power outage, because steam was not affected. Unit personnel were not prepared for the main unit turbine blower to stay running, because in previous power failures, the blower had always tripped.

Investigation of this incident revealed an interesting scenario. All the instrumentation had been converted to a DCS (Distributed Control System) with a UPS (Uninterruptible Power Supply) backup about ten years earlier, so why had the blower always tripped? It turned out that the system had been running on the "backup" electric lube oil pumps for years. When the system lost power, the blower lost its oil flow and a low pressure interlock (safety instrumented function) shut the blower down. When the turbine driven lube oil pumps were repaired, the blower maintained its oil supply and kept running.

Repairing the turbine driven lube oil pumps was part of an effort to improve reliability of the main unit blower. For most other unit shut downs, the plant wanted the blower to keep running. With the blower running, off-gas was scrubbed in the off-gas scrubber and pollution was minimized. If the blower tripped, vacuum would be lost in several loosely sealed parts of the process such as the furnace and boiler and process gas could be released at grade.

In several scenarios, however, keeping the blower running was not the appropriate action. If circulation was lost in the acid absorber, the high SO3 concentration would overwhelm the off-gas scrubber and result in a large environmental release. If circulation was lost in the off-gas scrubber, the scrubber would no longer be effective and could also result in an environmental release.

The unit blower HAZOP (Hazard and Operability Analysis) identified that the blower could either shutdown or keep running during a power outage depending on which lube oil pumps were used. Unfortunately, the off-gas scrubber HAZOP assumed the blower would always shut down.

This incident was very interesting because it also had reactive chemical elements. While the reaction between SO3 and polyethylene was well known in other parts of the company, it was not known in this business. While the unit had a reactivity matrix (see discussion in Section 5.1), materials of construction were not listed.

A longer account of this incident appeared in a paper given by my colleague, Al Ness [1]; the original version of Figure 3 appeared in that paper.

Key learnings from this incident:

- PHA's should consider utility failures across the unit as a whole as well as in individual sections.
- If systems have both turbine and electric pumps, there should be a clear understanding of why two power supplies are used, which is the "primary," and under what conditions and how long is it acceptable to run on the "secondary."
- If a system has been broken for years, fixing it should be considered a change and require an MOC review.
- Reactivity matrixes should include materials of construction.

5. Evaluating Utility Failures

These incidents illustrate why we need to evaluate interactions between processes and utility systems. In the incidents described, there are several common themes:

- Interactions between the process and utility were overlooked.
- Change was not identified.
- Reactive chemistry was not anticipated.

In this section, I will discuss some tools I use to evaluate interactions. The reactivity matrix is one tool. Then, I will take each utility system and discuss some issues I have found through experience (as a unit engineer, in PHAs, and in incident investigations).

5.1 Reactivity Matrix

One of our most powerful hazard recognition tools is the reactivity matrix. An example containing some of the chemicals discussed in the example incidents is shown below.

Table 1: Example Reactivity Matrix

Chemicals	H2SO4	SO3	NaOH (aq)
Sufuric Acid (H2SO4)	104, 107		
Sulfur Trioxide (SO3)	No Rxn	104, 107, 108	
Caustic Solution (NaOH aq)	B1,C1	B1,C1	--
Materials of Construction			
Carbon Steel	Highly Corrosive*, Rate depends on acid concentration	May react with water and corrode	Slightly Corrosive*
Polypropylene	B1,C1,D4	B1,C1,D4	No Reaction

Key to Hazard Codes

B1 - May cause fire
C1 - Heat generation by chemical reaction, may cause pressurization
D4 - Innocuous and nonflammable gas generation; causes pressurization

104 - Strong Oxidizing Agent
107 - Water-Reactive
108 - Air-Reactive

* Corrosion may generate hydrogen gas.

Most of the example matrix was created using the NOAA *Chemical Reactivity Worksheet* [2]. The worksheet, however, does not list most materials of construction. You will have to enlist chemists and corrosion specialists to understand your systems.

Good descriptions of constructing and using a reactivity matrix can be found in references [3] and [4].

All of the following should be included in your matrix:

- all the chemicals (or families of chemicals) in a process area,
- all the chemicals (or families of chemicals) in nearby process areas,
- other chemicals which could conceivably make their way into your processes (similar raw materials from a manufacturer, other material in drums used in the plant, etc.),
- common contaminants such as air, oil, water, rust,
- water treatment chemicals (note which systems they are used in),
- key temperatures in the process such as steam temperatures and tracing temperatures, and
- people.

The reactivity matrix is an excellent tool for use in PHAs and MOCs to evaluate possible effects of deviations such as: reverse flow where piping joins, unloading the wrong material to a tank, and potential reactions in the sewer.

I also like to review the adverse reactions identified in the matrix with a team, then brainstorm scenarios of how we could get the chemicals together. The temperatures at which materials degrade should be compared with key temperatures in the unit.

5.2 Utility Failures by System

Author's Disclaimer: I am most comfortable using HAZOP for PHAs, so this section will have a HAZOP bias. Our company uses HazardReview Leader® software for PHA documentation. I like to keep all intentions for a process in the same file to allow easy checking of interactions with other intentions. Many of our incidents seem to come from areas between intention boundaries. I try to make sure that the boundaries of each intention match up, so that the entire process area is covered.

In evaluating utility failures, I start with individual equipment failures such as loss of heating or cooling in individual exchangers, leaks, etc. Most HAZOPs do an excellent job with the individual failures.

I look at common mode utility failures at the end of each intention, but I have developed a preference of recording these failures in a separate HAZOP intention. After each process intention, I go back to the common mode failure intention. We add or review the most recent consequences and look for interactions between the process areas. In this way, the common mode failures are reviewed by every HAZOP team and are reviewed more frequently than once every 5 years.

To review common mode failures, I usually start with loss of steam, cooling tower water, refrigerated water, and any other heating or cooling media. Then I move on to loss of nitrogen or other inerting gas. Then I consider loss of instrument gas. Then I consider total power failures and, finally, partial power failures where a single substation or motor control center is lost.

Let's take a quick look at these systems one at a time. I've listed some common issues I look for in each system. The HazardReview Leader® software includes a Utility Failures template which lists common causes, consequences, and safeguards for each system.

Remember that any failure that results in low pressure in a utility system can also result in the opportunity for reverse flow through direct process connections or leaking indirect connections.

5.2.1 Water Systems

Water used for heating or cooling (CTW, RW, and tempered water) is usually connected to the process indirectly through a heat exchanger or jacket. Identifying exchangers which would normally leak from the process to the water system will help you figure out what kind of monitoring you should have in the water system for leak detection. If you have a leak, can you identify the location of leak fast enough to save your equipment? Will you have emissions from your cooling tower?

If water leaks into your process, do you know the consequence? How much water is bad? How will you find a leak? What effect will the water treatment chemicals have on the process or product? If you have water reactive chemicals, using water as the heating or cooling media can have disastrous results. The Napp Technologies fire and explosion was initiated by a water leak into the process [5].

You should make sure you know the consequences of exchangers leaking both directions as well as the "normal" direction of a leak. Shutdown of a key pump or system may result in leaks in the "wrong" direction (from normally high-pressure and to the normally low-pressure side).

Many heat exchangers also have connections to allow them to use two different water supplies, for instance, refrigerated water and CTW. One frequent contamination issue is one water system flowing into the other. A wrong line-up is all it takes. An exchanger that is lined to CTW on the supply side and RW on the return side can quickly overload a closed-loop cooling system. If multiple processing areas are using the same RW system, operators need to be able to identify which exchanger is incorrectly lined up quickly.

We also need to look at the unit as a whole and make sure we consider what happens when these entire systems are lost. Are interlocks set up to trip the system safely? Especially consider the warm cooling water scenarios. Your flare may be able to handle the increased emissions from one column, but what if multiple columns began sending increased organics to the flare?

Often the cooling water supply to refrigeration machines is from the cooling tower system. If you lose cooling tower fans or water flow, you will quickly lose cold refrigerated water. Refrigerated water systems often require their own HAZOPs or FMEAs to do a good job in understanding what can shut them down.

5.2.2 Steam & Condensate

Barometric condensers, as frequently used in steam jet systems, are a special concern. Steam jet condensate can contain process materials under normal conditions and higher concentrations under upset conditions.

Another potential hazard of vacuum system is degradation of a chemical in contact with the steam temperature in the jet or high temperatures in a vacuum pump. Vacuum is often used to reduce the process temperature because higher temperatures favor undesired reactions or degrade one of the system materials. First, make sure you understand what chemicals can reach the overhead system in normal and upset conditions. Upset conditions should include a lot of the HAZOP guidewords – more/less reactant, more/less catalyst, too long, etc. Then, evaluate how the chemicals will react to the steam jet or vacuum pump temperature. Especially look for the possibility that toxic gasses like chlorine or H2S might be formed.

Especially in a large plant, it's important to consider whether a shutdown of the unit will affect the plant steam balance. If the steam header pressure starts swinging, multiple units can be quickly affected. If another steam producing unit shuts down, the problem can get worse. In our largest plant, we completed a HAZOP on the entire steam system because of all the interactions.

5.2.3 Sewer Systems

Simultaneous spills can result in mixing many chemicals. I have heard several "old timers" talk about blowing the manhole covers off their sewer systems from acid/base reactions or deflagration of flammable chemicals. Every time I have heard one of these stories, someone else present has their own example to add.

What materials can end up in your sewer system? Consider reactor overhead systems, condensate, spills, etc. Under what conditions (e.g., temperature, pH, etc.) could these materials degrade and form toxic or flammable materials? If any sulfur compounds are used, formation of H2S may be possible. In 2002, two workers were killed when H2S fumes were formed in a chemical sewer at the Georgia-Pacific Naheloa mill [6].

A lot of newer sewer systems are closed systems that drain to tanks. If materials react in the system, is there enough relief? Could solid materials form which could plug lines or vents? If piping gets plugged, where will materials backup?

5.2.4 Inerting Gas

Gases received by pipeline may be very reliable. Inerting gasses brought in by rail or truck and stored on-site are worth special consideration. How reliable is the supply? If several units are purging at the same time, will the inerting gas system be able to keep up with the demand?

Cryogenic liquids with vaporizing systems have special hazards. Are these systems owned and maintained by the supplier or by the plant? How often are they maintained?

5.2.5 Instrument Gas

Plants may have instruments driven by either air or nitrogen. You first need to look at the unit and understand whether all the instruments are using the same instrument gas. If not, make sure you understand what systems are using each supply. Then, you can move on to looking at the valve failure positions and how the instruments will react with loss of gas. If you have some instruments on air and some on nitrogen, go through the exercise with each system.

Make sure the valve utility-failure (shelf) positions make sense. (The "shelf" position is the position the valve goes to when the instrument gas or instrument power is removed or the way it comes off the manufacturer's shelf. The utility-failure position does not imply the valve cannot fail while in service in another position. During operation, valves can fail in any position, open, closed, or in-between.). If the process system will produce vapors, is it vented? Sometimes water feeds fail open; can equipment flood?

Sometimes plants have systems where nitrogen can be used to supplement instrument air pressure. These systems are of special concern because nitrogen is an asphyxiation hazard. Is instrument gas used in enclosed areas such as shops or lab areas? Are there gas-powered diaphragm pumps (one example is SANDPIPER® pumps) within the area? The discharges of such pumps could be hazardous if nitrogen is used.

5.2.6 Power Failure

Power is the hardest utility to evaluate, because losing power is a common mode failure for the other utility systems. Knowing which substations supply power to key utility areas such as instrument air, refrigerated water, cooling tower water, and steam is critical to a thorough evaluation. Individual substation outages may present more challenges than a total plant power outage because of unit interactions.

Before the introduction of UPS's, evaluating what would happen to your control and shutoff valves on loss of power was relatively straightforward; all valves would go to their utility-failure positions just as they would on loss of instrument gas.

A UPS present many opportunities in making systems safer, but evaluating electrical failures becomes trickier. With their power supplied by a UPS, control valve and safety shutoff valves will not go to their utility-failure position during an outage until the backup power runs out or the valve receives a signal to close. Raw material, intermediates, and utility feeds not affected by the power outage can continue unless shut-off. Natural gas and steam frequently have power supplies separate from process units. This can be advantageous if you want to keep natural gas flowing to your flare, but not shutting off steam feeds to your reactors or columns can be a problem.

Let's consider a distillation column with UPS power to the controllers. If the pumps feeding the column are on the affected power system, they will shutdown and stop the feed. If the cooling water is affected, water pumped to the condensers will be lost. If the steam to the column is from a controller with a UPS backup, the steam flow may continue. This is a recipe for a release from a pressure relief valve, or worse, a potential column overpressure. An interlock with a separate steam shutoff valve may be required.

What if the column feed pump is on a different (unaffected) power supply? Then the feed can continue and fill up the column. Many columns are not designed to be liquid filled. For many situations, you should program distributed control and safety systems to shut key valves on loss of power. But shutting the valves can result in other problems. For example, if the feed valve is programmed to close, you may end up with a pump running deadheaded against a blocked valve. Power or amp monitors can be used to trip a deadheaded pump (power monitors are preferred).

If steam is not affected by a power outage, steam turbine pumps and/or compressors can continue to run during a power failure. Each turbine should be considered individually to determine whether it should be programmed to trip or not. A turbine pump that supplies oil to a compressor may be a good one to leave running. A compressor that supplies process air feed to equipment that could become flammable would be a good one to shutdown.

With a UPS, you do have the opportunity to keep key valves from going to their failure positions. If the instrument air supply is on the same substation as the process, critical valves may also need backup air bottles to prevent them from going to their utility-failure position unnecessarily. Examples of situations which could benefit from backup air bottles include: a safety valve which dumps material from equipment, the natural gas valve to a flare, or a reactor quench valve which will flood the equipment.

Control systems present special challenges too. What do the operators have to do to get back to the right part of a sequence? After a power outage, will the control system recipe need to be reloaded? A filter overpressure incident that occurred due to a batch recipe overwriting a cleaning sequence is reviewed in Reference [1].

Particularly difficult to analyze are systems that can have partial failures, such as: part, but not all, of an electrical substation looses power, or 1 or 2 phases of the 3 phase power are lost. The results can be sporadic and very confusing. A question to ask is, "What if this device continues to operate while most other devices are tripped or shutdown?" If the consequences are serious, then we may elect to trip that device for specific detected utility or process conditions.

Normally, sudden power outages are short-lived, but what if the outage extends for 3 hours or even days (e.g., post-hurricane or post-flood). Most plants have emergency generators, but have you done a thorough analysis of what you need on emergency power? Does the operator know which devices are on emergency power? As systems change, emergency power needs change also. What systems will need to be pumped out? What equipment is critical to prevent environmental releases?

Summary

In summary, to prevent incidents it is critical to take a special look at utility systems and their potential interactions with process systems. The interconnectivity between processes and utilities can result in incidents such as the ones presented here.

Reactivity matrices, coupled with project design reviews, PHA's, and MOC's give us an opportunity to anticipate and plan for these interactions. I hope this paper has you thinking about utility systems in a different way.

References

[1] Ness, Albert, "Unplanned Shutdown Plus Lack of Knowledge Equals Incidents", *CCPS International Conference Proceedings: 2004 International Conference and Wrokshop Emergency Planning Preparedness, Prevention & Response, June 29, 30, and July1, Orlando, Florida,* Center for Chemical Process Safety of the American Institute of Chemical Engineers, 2004

[2] *Chemical Reactivity Worksheet*, U.S. National Oceanic and Atmospheric Administration, http:/response.restoration.noaa.gov/chemaids/react.html

[3] Mosley, David W.; Ness, Albert I.; Hendershot, Dennis C.; "Screen Reactive Chemical Hazards Early in Process Development", Chemical Engineering Progress 96, 11 (November), pp. 51-63, 2000.

[4] Johnson, Robert W.; Rudy, Steven W.; Unwin, Stephen D. Essential Practices for Managing Chemical Reactivity Hazards, Center for Chemical Process Safety of the American Institute of Chemical Engineers, 2003

[5] Improving Reactive Hazard Management, U.S. Chemical Safety and Hazard Investigation Board Report No. 2001-01-H, Dec 2002, Napp Technologies Inc. Case Study

[6] "Georgia-Pacific Hydrogen Sulfide Release," January 16, 2002, U.S. Chemical Safety and Hazard Investigation Board Investigation Digest

A Three Level Approach
for the Assessment of Domino Effect

Ernesto Salzano
Istituto di Ricerche sulla Combustione
Consiglio Nazionale delle Ricerche
via Diocleziano 328, 80125 Napoli, Italy.
salzano@irc.na.cnr.it

Valerio Cozzani
Dipartimento di Ingegneria Chimica, Mineraria e delle Tecnologie Ambientali,
Università di Bologna, via Risorgimento 2, 40136 Bologna, Italy.
valerio.cozzani@mail.ing.unibo.it

Prepared for Presentation at
American Institute of Chemical Engineers
2006 Spring National Meeting
8th Process Plant Safety Symposium
Orlando, Florida
April 24-26, 2006

UNPUBLISHED

A Three Level Approach For The Assessment Of Domino Effect

Ernesto Salzano
Istituto di Ricerche sulla Combustione
Consiglio Nazionale delle Ricerche
via Diocleziano 328, 80125 Napoli, Italy.
salzano@irc.na.cnr.it

Valerio Cozzani
Dipartimento di Ingegneria Chimica, Mineraria e delle Tecnologie Ambientali,
Università di Bologna, via Risorgimento 2, 40136 Bologna, Italy.
valerio.cozzani@mail.ing.unibo.it

Abstract

Gas or Vapour Cloud Explosions in petrochemical plants may be highly destructive and may trigger escalation or domino effects. The assessment of damage to process equipment surrounding the vapour or gas cloud may be carried out at different levels of detail: i) threshold values based on simplified empirical models for the evaluation of both the explosion scenario and the loss of containment, which only require the estimation of the total energy of cloud; ii) damage probability models based on statistical functions for both mechanical damage and loss of containment, that require a more accurate modeling of the explosion, geometrical lay-out details, design characteristics of target equipment and fuel reactivity (flame speed); iii) distributed parameter models (e.g. Computational Fluid Dynamics codes based on Reynolds Average Navier-Stokes equation), which require a detailed description of the explosion environment, to be coupled with statistical functions for the assessment of escalation effects.
In this study, the evaluation of the time frame and a sensitivity analysis of the three approaches cited above for the analysis of domino events triggered by blast waves are presented. The comparison was made on the basis of the results of a case-study and more specifically for a Vapor Cloud Explosion occurring in a low pressure unit within a hydrocracker plant. The analysis has been carried out by using semi-empirical approaches as the Multi-Energy and Baker-Strehlow models, and by a Computational Fluid Dynamics (CFD) code.

1. Introduction

Gas or Vapour Cloud Explosion (VCE) in petrochemical plants may be highly destructive. Domino effects are likely to be triggered by the blast waves, which may propagate the damage at very high distance from the ignition point. In recent years, several methodologies were developed for the prediction of the maximum overpressure and impulse, for the evaluation of the possible structural damage to surrounding equipment loaded by heat and blast waves, and finally for the evaluation of probability and intensity of secondary scenarios triggered by the loss of containment derived from the mechanical failure of equipment [1-12].
The evaluation of these steps allows a complete assessment of risk due to VCEs in industrial units. However, the needs of the plant safety management and of the risk analysts may vary,

depending on the level of detail required for the analysis. Thus, several methodologies are available in the literature in order to carry out the different steps of the analysis at different level of detail.
The assessment procedure, summarized in Figure 1, may be divided in two main stages: the assessment of the explosion strength and of blast wave propagation (steps 1-3, Figure 1), and the assessment of possible escalation effect (steps 4-7, Figure 1). The first stage of the procedure requires the evaluation of explosion source strength (the pressure and the impulse of blast wave, step 2), and the modeling of the blast wave propagation model in air at near and far field distances (step 3). The second phase requires the identification of possible target equipment (step 4), the assessment of structural damage (step 5) and of the intensity of loss of containment following the damage (step 6), and the analysis of possible escalation effects (step 7).

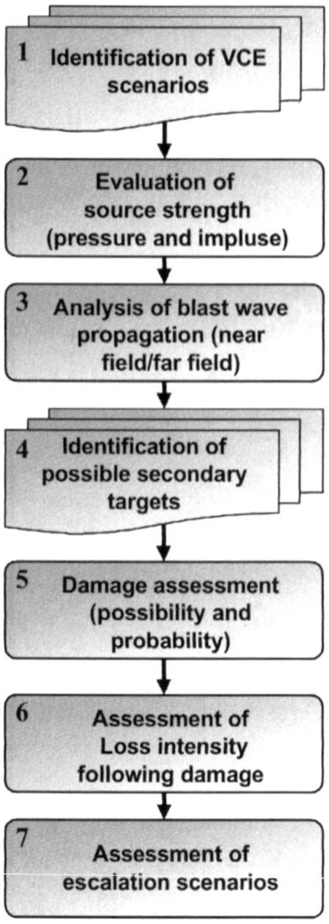

Figure 1. Steps required for VCE damage and escalation analysis in process plants.

The different level of detail used to approach the above described procedure is essentially related to the type of risk analysis that is required: a Worst Case Analysis (WCA) requires a simplified approach and is certainly conservative with respect to the Maximum Credible Accident (MCA) approach, based on probabilistic assumptions and on a thorough analysis of the explosion scenarios. On the other hand, Quantitative Risk Assessment (QRA) of an entire industrial plant or of an extended industrial area forces the analysts to reduce the complexity of analysis, often by the adoption of strong assumptions. As a consequence, also the costs (in terms of man-hours and computing time) of the analysis may vary strongly from MCA and WCA to QRA.
The above described procedure may be applied also to the early-design explosion analysis. Priorities for reinforcing the structure or for blast wall installation to protect residential units or

critical equipment may require the definition of several explosion scenarios (e.g. by varying the dispersion conditions and the position of the ignition point). Thus, the level of detail of the analysis may be addressed on the basis of the application: High details are usually required in congested structures as those typical of off-shore installations, whereas a less detailed analysis may be acceptable in less critical applications, as in the study of conventional on-shore plants.
The present study was addressed to sketch three possible approaches to carry out the procedure schematized in Figure 1, based on tools and methodologies at different levels of details. Each of the proposed approaches may be alternatively used for MCA, WCA, QRA and Early-Design Analysis, depending on the desired level of detail required for the assessment.

1.1 Low Level Methodology (Inherent Safety Distances)

The first level of detail is the preliminary analysis of possible escalation effects based on simplified explosion scenarios and threshold values for the identification of escalation targets. As discussed above, when the QRA analysis of a large industrial installation or of a large industrial area has to be undertaken, the number of possible accidental scenarios is so relevant that simplifying assumptions are needed to carry out the analysis. In this framework, an approach based on WCA may be used to limit the area of concern. To this regard, the assumption of a stoichiometric cloud having the maximum possible extension (in general filling the entire unit) is the usual, strong simplification for the gas or vapour dispersion from the process unit. Furthermore, the analysis of blast wave decay with respect to scaled distance is based on the maximum strength for the explosion, the latter evaluated easily by using the detonation curve in the energy-scaled or TNT mass-scaled plots.
Threshold values for structural damage and loss of containment may be coupled to this approach in order to identify all the possible targets of escalation. Table 1 reports the threshold values for damage, as obtained from an extended revision of literature data [7-10], and the corresponding scaled safety distances obtained by the Baker-Sthrelow approach [13] for the evaluation of blast wave decay. Specific thresholds for escalation were obtained by the analysis of equipment damage data, and assuming different levels of loss intensity triggered by equipment damage. Further details on the evaluation of these values are reported elsewhere [11-12].

Table 1: Overpressure (kPa) threshold values for structural damage and escalation caused by blast wave interaction with different categories of process equipment. In parenthesis the corresponding energy-scaled distance by Baker & Sthrelow [13].

Threshold type	Substance Hazard (target vessel)	Equipment category	
		atmospheric equipment	pressurized equipment
Structural Damage	All	5 (3.75)	35 (1.25)
Escalation	Flammable	22 (1.50)	16 (1.80)
	Toxic	22 (1.50)	16 (1.80)

Scaled inherent safety distances are plotted as a function of explosion energy in Figure 1. In WCA approach, this plot may be used to identify the inherent safety region with respect to escalation. As a matter of fact, the conservative assumptions used in the WCA coupled to the specific thresholds estimated for damage and escalation allow the conservative identification of the boundaries of the region where damage to equipment and/or following escalation is expected.

Thus, the identification of all possible equipment involved in escalation events triggered by the primary explosion of concern is possible. Clearly enough, more detailed approaches to explosion and escalation analysis are needed to estimate the credibility of the possible escalation events identified by this procedure.

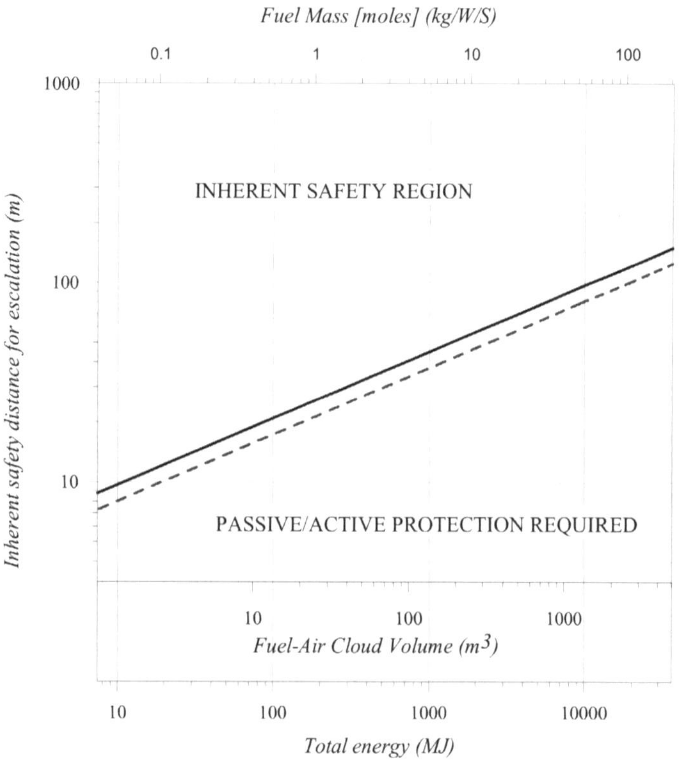

Figure 2: Inherent safety distances for escalation effects on equipment loaded by VCEs generated by common hydrocarbons. (W = molecular weight; S = stoichiometric concentration expressed in volume percent of fuel air mixture combustion). Distances should be calculated from the flammable cloud border. For more reactive fuels a factor 1.2 should be used to correct energy values based on cloud volume. Target vessel: i) Solid line: pressurised; ii) Dashed line: atmospheric.

1.2. Level 2: estimation of explosion strength and of damage probability

If higher levels of detail are required, as in the early design phase or for MCA or detailed QRA studies (e.g. for on-shore plants in densely populated or highly congested areas), statistical functions for mechanical damage of equipment may be used. This approach is useful if coupled to the use of explosion models based on higher level of accuracy, which require also geometrical details on plant lay-out and fuel reactivity (e.g. flame speed). In Europe, explosion analysis is typically carried out by the Multi-Energy Method (ME) [14] and useful GAME methodology [15]. This approach is derived from the MERGE and EMERGE experimental data set, and resulted in the following expression for open, industrial environment:

$$P^{\circ} = 0.84\left(\frac{VBR \cdot L_P}{D}\right)^{2.75} S_L^{2.7} D^{0.7} \qquad \textit{Equation 1}$$

where P° is the maximum overpressure produced by VCE, VBR is the volume blockage ratio, D is the average obstacle diameter, L_P denotes the distance traveled by the flame within the obstacle configuration to account for the influence of ignition location and S_L is the laminar burning velocity. The maximum overpressure allows recognizing a strength factor F, whose values identify the violence of explosion, which originally was derived by analyst – often arbitrary – choices on congestion and reactivity of fuels. Alternatively, Baker and Sthrelow methodology (B&S) [13] can be used, as it is common in US. In this approach, a flame Mach number is estimated, which is related to flame speed. Although based on simplified assumptions, the approach tries to reproduce the Schelkin effects by general rules based on congestion and fuel reactivity. ME and B&S approaches allow a more precise estimation of the explosion strength and of maximum peak overpressure.

In recent study, simplified damage probability models were developed for escalation assessment of different equipment categories [7,8]. These models allow the estimation of the equipment damage probability as a function of maximum peak overpressure. Moreover, the analysis of available literature data on the extension of equipment damage following a blast wave impact was carried using fuzzy set techniques [12].

It must be recalled that the characterization of the damage extension and the definition of loss intensity on the damaged equipment, are the starting point for the assessment of the secondary scenarios that may be triggered by the primary blast wave. Thus, following the approaches typically adopted for the damage caused by earthquakes, two damage states (DS) were defined and linked to loss intensity (LI) from the damaged equipment items:

i) Damage State 1 (DS1): light damage to the structure or to the auxiliary equipment, associated to a 10mm average release diameter, resulting in a moderate loss intensity (minor LI);

ii) Damage State 2 (DS1): intense, or catastrophic damage, or even total collapse of the structure, associated to an instantaneous release of the entire vessel content (major LI).

Figure 3 reports damage probability for atmospheric and pressurized equipment as a function of the scaled distance and explosion strength factor.

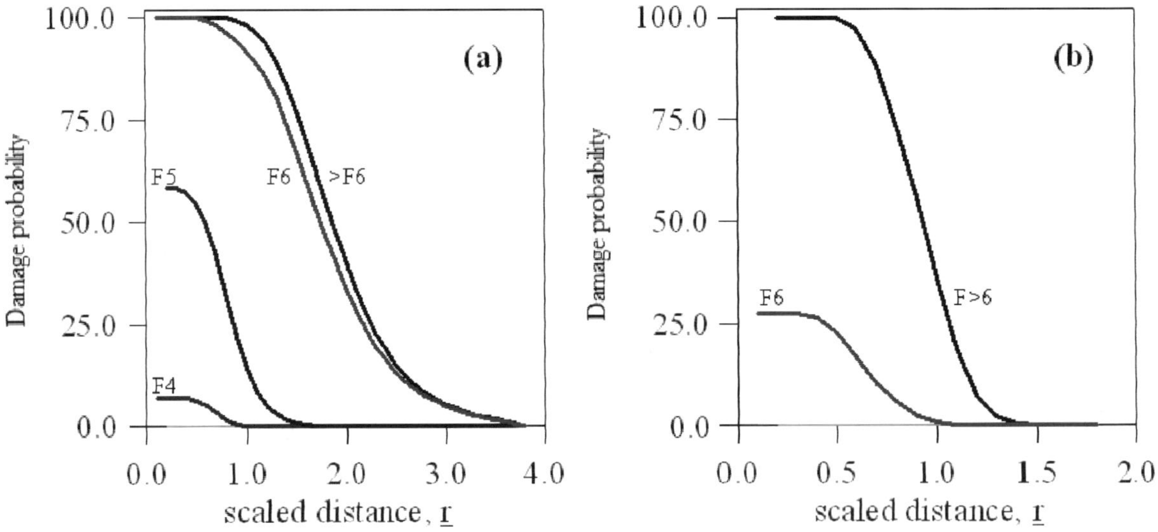

Figure 3. Damage probability as a function of scaled distance and explosion strength factor, as used by ME-GAME model. (a) atmospheric vessels; (b) pressurized vessels.

From the data reported in Figure 3, Table 2 reports the probability functions of yielding the minor and major loss intensity, given the pressure peak for the blast wave impact on target equipment (hence the corresponding damage state DS.

Table 2: Probabilities of minor LI and of major LI following equipment damage due to blast waves. (P°: maximum peak overpressure in kPa).

Atmospheric	minor LI	major LI
$P° < 15$	1	0
$15 \le P° \le 20$	$1 - 0.2(P° - 15)$	$0.2(P° - 15)$
$P° > 20$	0	1
Pressurized		
$P° < 40$	1	0
$40 \le P° \le 50$	$1 - 0.1(P° - 40)$	$0.1(P° - 40)$
$P° > 50$	0	1

Finally, Figure 4 shows the overall probabilities of minor and major loss of containment, as a function of the maximum peak overpressure impacting on the equipment item.

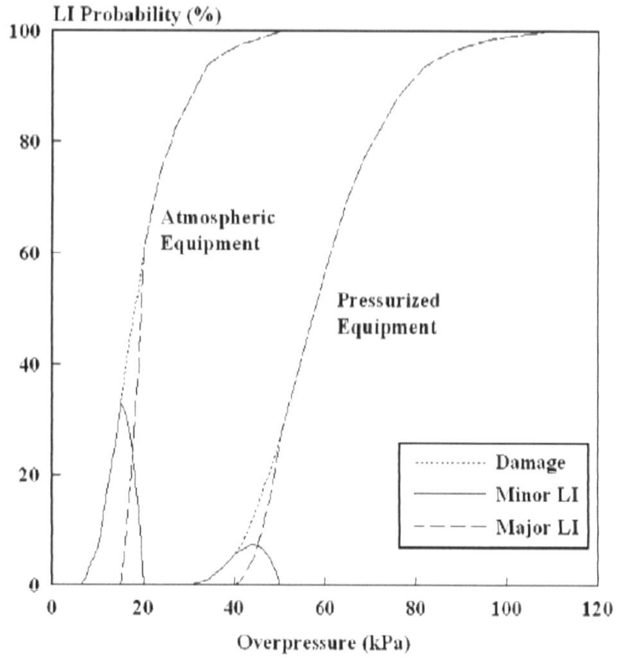

Figure 4: Probability of minor and major loss intensity (LI) as a function of peak static overpressure for atmospheric and pressurized equipment.

Damage data in Figures 3-4 have been obtained by using the strength factor in the ME model. However, the correspondence (Table 3) between the strength factor and the maximum peak overpressure at the ignition point may be used to estimate the strength factor from other approaches. Thus, similar plots to those in Figure 3 may be obtained for the Baker & Strehlow method [13].

Table 3: Correspondence between explosion strength factor in the ME model and maximum peak static overpressure at explosion centre obtained by GAME study [15].

F	P° (kPa)
1	1
2	2
3	5
4	10
5	20
6	50
7	100
8	200
9	500
10	1300

The approach described above, joined to the evaluation of the explosion strength, either by strength factor F or by flame Mach number M_f, allows: i) a reliable although simplified description of the primary explosion scenario; and ii) a detailed estimation of the expected frequencies and consequences of the secondary scenarios triggered by escalation following the blast wave. As a matter of fact, the use of Figures 3- 4 allows the direct estimation of the expected frequencies of escalation and of minor and major LI due to equipment damage. The use of the event tree technique and of consequence analysis models reported in literature may easily allow the analysis of the consequences of the secondary events. A specific methodology and dedicated software were developed to carry out this evaluation in a QRA framework and to allow the calculation of the risk due to the domino scenarios [23, 24].

1.3. Level 3: distributed parameters modeling

Distributed parameter models, which require a detailed description of equipment design and lay out involved in the explosion scenario, may be used when a high detail of the analysis is required, and in particular when explosion effects in the near field are addressed. These requirement are usual of offshore application, or, more generally, of critical installations whose structural integrity should be assessed with a high level of confidence, in particular in the early design phase in order to address the design procedure.

In recent years, few Computational Fluid Dynamics (CFD) codes based on Reynolds Average Navier-Stokes equation have been specifically developed for the consequence assessment of gas explosions in medium and large-scale scenarios. In the present study, a Computational Fluid Dynamics (CFD) code named AutoReagas [16, 17], was used. The code, developed by TNO (NL) and Century Dynamics Ltd. (UK), has been specifically developed to obtain predictions of the main effects of gas explosions on structures through the prediction of the pressure field. It solves the conservation equations of mass, energy and momentum in a finite volume formulation [18]. The turbulent flow field is described by the k-ε model [19] and the turbulent burning velocity, S_t, as calculated by Bray equation (1990) is used to characterize the mixture reactivity [20]. A "sub-grid" formulation has been considered to keep into account of the turbulence induced by obstacles when the volume of the same obstacle is smaller than the volume of the single computational cell. Hence, a characteristic turbulent length scale and a drag coefficient for these "sub-grid" objects are fixed and new turbulent source terms are introduced into the conservation equation [21].

The detailed pressure-time profile obtained by the CFD codes may be used as an input to analyze the structural resistance of process equipment to blast wave impact. Also in this case, distributed parameter models may be applied. In particular, the use of finite element codes may allow the analysis of the local stresses induced by the dynamic impulse caused by the blast wave. Figure 5 shows an example of the level of detail that may be obtained by this approach in the simulation of equipment items. Although finite element simulations by standard codes are usually limited to the elastic region, in the framework of safety assessment the possible plastic deformation of an equipment item may be assumed as a condition sufficient to cause a loss of containment. Thus, the use of finite element codes may be used to obtain a detailed simulation of the effect of blast wave loads on equipment items.

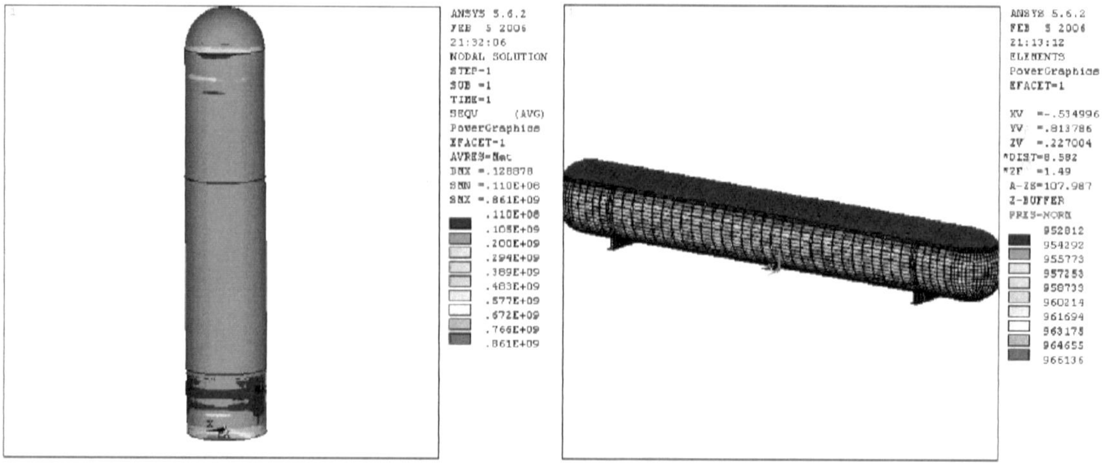

Figure 5: Finite element simulations of pressurized storage vessels used for the analysis of blast wave loads on process equipment.

Despite the refinement, it's quite clear that this approach has important limitation due to the computer time and man hours required to build the numerical domain and to carry out the simulations. In particular, it is not practical to use CFD in order to reproduce the behavior of blast wave propagation in the far field, due to the huge amount of computer power and time usually required for increasing the numerical domain. On the other hand, it is not possible to perform a specific finite element simulation for all the equipment items present on a plant. Moreover, the finite element simulation of an equipment item requires high amount of design details, which is usually not available in early design phases. Thus, while near field blast wave effects may be addressed by these tools, in particular in the case of early design, simplified methods seem more practical to apply when domino assessment is of concern.

As a matter of fact, in far-field blast wave and escalation analysis, coupling the above approach with the level 2 - lumped parameters methodology - is mandatory. Indeed, in the assessment of domino accidents, the main issue is the identification of the possible escalation events that may involve nearby units. Furthermore, as far-field escalation is usually of concern, the use of the available, distributed parameters modeling tools is actually extremely time consuming and scarcely practicable.

2. Results and discussion

In order to compare the results obtained by the application of the above defined methodologies, a case study was defined. All the three level methodologies were applied to the consequence analysis of possible gas explosions in a low pressure unit within a hydrocracker plant. Figure 6 shows the plant unit where the explosion consequence analysis was carried out. The plant section is relative to a depressurizing unit where the reaction products are cooled and fed to an atmospheric distillation. The plant lay-out, in Figure 6, evidences that the analyzed section is constituted by a beam structure supporting piping and pumps. Further details on the plant section analyzed are reported elsewhere [22].

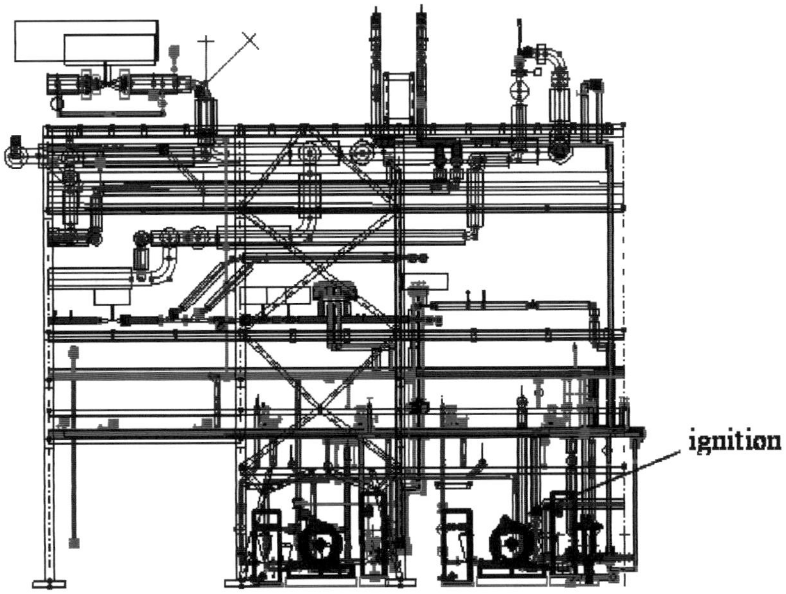

Figure 6. Piping, structure and equipment of a low pressure unit of a hydrocracker plant, front view [22].

A pure stoichiometric butane-air cloud completely filling the unit was considered. The total energy of cloud was estimated to be of 4800MJ. The blast wave was modeled using: i) a pure empirical approach and the TNT-equivalence model; ii) semi-empirical models as Multi-Energy (ME) and Baker-Strehlow (B&S) models; and iii) a Computational Fluid Dynamics (CFD) tool.
Due to the small turbulent length scale, the effects of objects with equivalent diameter < 0.1 m on the explosion severity were considered as negligible. Thus, these objects were not considered in the numerical simulation by semi-empirical models (ME, B&S) and CFD.
In the analyzed installation, the volume blockage ratio VBR was estimated to be equal to 0.04, the mean diameter D was estimated of 0.3m and the flame path LP was estimated assuming that the ignition was at the edge of the vapour cloud.
The application of the GAME approach to the ME model gives a peak overpressure P° of about 137 kPa, based on an explosion strength factor of F equal to 8. A similar value was obtained by the B&S method, by which maximum peak overpressure P° of 103kPa is calculated, the value corresponding to flame Mach number of 0.50 (low reactivity fuel, high congestion, 2.5D flame expansion).
The section of the hydrocracker unit was reproduced by means of CFD AutoReagas by using a

computational domain of 35 x 45 x 41 planes respectively along the x, y and z axes (a cell volume of 0.125 m3 was used). Pipelines, equipment and objects having a volume lower than the computational cell size were treated by a sub-grid formulation. Ignition was considered to occur within the pumping area of the plant unit (see Figure 6).

Figure 7 reports the main results of the simulations. The figure shows the overpressure profile at the ignition point, at the center of the vapour cloud and at a height of 1m above the ground level and along the central axis of the vapour cloud. The high degree of confinement, the strong congestion and the presence of several objects within the unit result in a very intense peak overpressure, and in a destructive blast wave that propagates to the surrounding environment. The maximum calculated overpressure P° at the vapour cloud center (which clearly is not the ignition point) is 380 kPa, which is higher than the maximum overpressure calculated by both the ME and the B&S approaches.

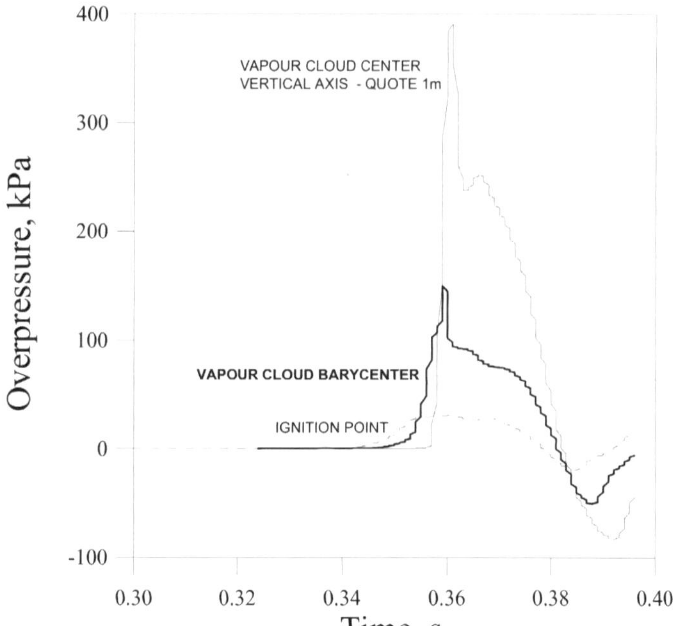

Figure 7. CFD analysis for the explosion on the hydrocracker unit shown in Figure 6: overpressure as a function of time at different locations within the unit.

Due to the limitation of numerical domain, the peak pressure reported in Figure 4 has to be coupled with other methods for the evaluation of blast wave propagation in the far field, where escalation effects on equipment are needed. Table 4 reports escalation threshold distances as calculated by coupling the peak overpressure obtained by CFD with ME (hence by using blast wave propagation only), by ME (GAME) and B&S (and M_f), and by over-simplified TNT methodology (level 1, inherent safety). Loss of containment has been obtained by using threshold values as reported in Table 1.

The table evidences that all the methods yield similar results, in the case of atmospheric equipment, regardless the method for the evaluation of peak pressure. Indeed, as a matter of fact, the pressure differences among blast wave of different initial intensities, at least for industrial explosion, tend to cancel out or at least fall within the strong uncertainties which characterize typically pre-accident explosion analyses. Moreover, in this case, inherent safety distance is similar as even low intensity loss of containment are considered in the case of atmospheric tanks.

This observation does not fit for pressurised equipment, as even for low structural damage it is likely an escalation, thus producing higher inherent safety distances for the detonation assumption on the explosion modeling. These results confirm further the validity of the level 1 approach (WCA + threshold values) for a first screening aimed to the identification of possible domino targets.

Table 4: Safety distance [m] from ignition point, for escalation effects (i.e. loss of containment) of atmospheric and pressurised equipment as calculated by the three methodology.

Equipment	Level 1: Inherent safety distances	Level 2: ME	Level 2: B&S	Level 3: CFD+MEM
Atmospheric, damage	67.0	65.0	66.5	65.0
Pressurised, damage	76.5	40.0	39.0	40.0

Finally, Table 5 shows the results obtained by level 2 and level 3 approaches in the estimation of escalation effects at positions that fall respectively in the near field and in the far field. Results confirm that level 3 approach is coincident to any other method in the far field, due to the need of integrating CFD predictions with semi-empirical models as ME. On the contrary, CFD predictions allow, in the near field, exact identification of peak overpressure (which in our case vary between 300 KPa and 10 KPa in the cloud hemisphere for coordinates x,y,z), i.e. structural damage and loss of containment probabilities.

Table 5: Probabilities of minor and major loss intensities following blast wave damages in the near field and in the far field. CFD + ME initial peak pressure = 380 kPa (F = 8).

Cloud border (Lp = 13 m from ignition point)	Level 2 - ME	Level 2 - B&S	Level 3 – CFD
Atmospheric, minor LI	0.00	0.00	(x,y,z) specific
Pressurised, minor LI	0.00	0.00	(x,y,z) specific
Atmospheric, major LI	1.00	1.00	(x,y,z) specific
Pressurised, major LI	1.00	1.00	(x,y,z) specific
10 m from cloud border	**Level 2 - ME**	**Level 2 - B&S**	**Level 3 - CFD+ME**
Atmospheric, minor LI	0.07	0.08	0.07
Pressurised, minor LI	0.68	0.07	0.68
Atmospheric, major LI	0.93	0.92	0.93
Pressurised, major LI	0.32	0.93	0.32

4. Conclusions

Three approaches at different level of detail were presented for the assessment of escalation triggered by blast waves. An application to a case-study was performed, in order to allow the comparison the methods and the analysis of the results.

Level 1 approach, based on Worst Case Analysis for explosion phenomena and by threshold values for escalation, proved to be effective for the preliminary identification of any possible domino target. With respect to the more detailed methods, the results were strongly dependent on the position of the possible escalation targets. It must be recalled here that ME and B&S methods were set on the basis of medium-scale experiments and further investigations are needed to verify the effect of scale on the calculated peak overpressure. However, high uncertainties are also present in the CFD calculations, and the detailed method still needs energy-scaled plots for blast wave analysis in the far-field.

The results has evidenced that - in the near field - CFD analysis is worth only for safety-oriented structural design. In this framework, also the use of finite element models for the analysis of the possible damages caused by escalation may be attractive. On the other hand, less detailed methods seem to be sufficient to address MCA, WCA and more generally QRA purposes, in particular if domino assessment is carried out in the far field. Indeed, the time frame is still a fundamental parameter in the choice of assessment level: the three methods are separated by more than an order of magnitude either in costs or in time: Level 1 approach – starting from dispersion analysis – requires only few minutes, whereas level 2 approach, which needs the evaluation of blockage ratio and statistical function for structural damage and loss intensity, thus requiring several man-hours. In level 3 approach, the construction of the computational domain and of the mesh of the equipment items may require several man-months, even for the simplified case-study described in the present study.

4. References

[1] Lees F.P., "Loss Prevention in the process industries", II Ed., Butterworth-Heinemann, Oxford (UK), 1996.

[2] CCPS, Center for the Chemical Process Safety, "Guidelines for chemical process quantitative risk analysis", II E., AIChE, New York, 2000.

[3] CCPS, Center for the Chemical Process Safety, "Guidelines for evaluating the characteristics of Vapour Cloud Explosions, Flash Fires, and BLEVEs", AIChE, New York, 1994.

[4] Baker W.E., Cox P.A., Westine P.S., Kulesz J.J., Strehlow R.A., "Explosion Hazards and Evaluation", Elsevier, Amsterdam, 1983.

[5] Uijt de Haag P.A.M., Ale B.J.M., "Guidelines for Quantitative Risk Assessment (Purple Book)", Committee for the Prevention of Disasters, The Hague (NL), 1999.

[6] Van Den Bosh C.J.H., Weterings R.A.P.M., "Methods for the calculation of physical effects (Yellow Book)", Committee for the Prevention of Disasters, The Hague (NL), 1997.

[7] Cozzani V., Salzano E., The Quantitative Assessment Of Domino Effects Caused By Overpressure. Part I: Probit Models, *Journal of Hazardous Materials*, 107, 67-80, 2004.

[8] Cozzani V., Salzano E., The Quantitative Assessment Of Domino Effects Caused By Overpressure. Part II: Case Studies, *Journal of Hazardous Materials*, 107, 81-94, 2004.

[9] Cozzani V., Salzano E., Threshold values for domino effects caused by blast wave interaction with process equipment, *Journal of Loss Prevention in the Process Industries*, 17, 6, 437-447, 2004.

[10] Salzano E., Cozzani V., The Analysis of Domino Accidents Triggered by Vapour Cloud Explosions, *Reliability Engineering and System Safety*, 90, 2-3, 271-284, 2005.

[11] Cozzani V., Gubinelli G., Salzano E., Escalation thresholds in the assessment of domino accidental events, *Journal of Hazardous Materials*, 129, 1-21, 2006.

[12] Salzano E., Cozzani V., A fuzzy set analysis to estimate loss of containment relevance following blast wave interaction with Process Equipment, *Journal of Loss Prevention in the Process Industries*, 19, 343-352, 2006.

[13] Tang M.J., Baker Q.A., A new set of blast curves from Vapor Cloud Explosion, *Process Safety Progress*, 18, 3, 235-240, 1999.

[14] Van den Berg A.C, Lannoy A. Methods for vapour cloud explosion blast modelling, *Journal of Hazardous Materials*, 34, 151-171, 1993.

[15] Eggen J.B.M.M., "GAME: Development of Guidance for the Application of the Multi-Energy method", TNO-PML, 1995.

[16] Tufano V., Maremonti M., Salzano E., Russo G., Simulation of VCEs by CFD modelling: an Analysis of sensitivity, *Journal of Loss Prevention in the Process Industries*, 11, 169-175, 1998.

[17] AutoReagas, Century Dynamics, http://www.century-dynamics.co.uk. Vers.2.015.

[18] Patankar S.V., Numerical heat transfer and fluid flow, Taylor and Francis, Minnesota, 1980.

[19] Launder B.E., Spalding D.B., Mathematical models of turbulence, Academic Press, London, 1972.

[20] Bray K.N.C., Proc. Royal Society of London, A431, 1990, 315-325.

[21] Popat N.R., Catlin C.A., Arntzen B.J., Lindstedt R.P., Hjertager B.H., Solberg T., Saeter O., Van den Berg A.C., *Journal of hazardous Materials*, 45, 1-25, 1996.

[22] Russo G., Bucciero A., Ditali S., Salzano E., Evaluating the gas explosion consequences in hydrocracker unit by CFD simulation, Combustion Meeting 1999, 22st Event of the Italian Section of the Combustion Institute, Firenze, Italy, 1999, 45-146.

[23] Cozzani V., Gubinelli G., Antonioni G., Spadoni G., Zanelli S., The assessment of risk caused by domino effect in quantitative area risk analysis, *Journal of Hazardous Materials*, 127, 14-30, 2005.

[24] Cozzani, V., Antonioni, G., Spadoni, G., Quantitative assessment of domino scenarios by a GIS-based software tool, *Journal of Loss Prevention in the Process Industry*, in press, 2006.

106b Dust Explosion Consequence Categories for Semi-Quantitative Risk Assessmen

David F. Tarverdi and Colin S. Howat. University of Kansas, 4132 Learned Hall, 1530 W. 15th Street, Lawrence, KS 66045

Semi-Quantitative Risk Assessment (SQRA) is an effective tool for the chemical process industries. SQRA relies on acceptable risk criteria; release, capital loss and downtime consequence levels; and failure frequencies in the assessment of risk. Since this set of tools was derived from process industry experience, the release consequence level is focused on liquid and vapor processing. This limits its application for industries that process solids.

This paper proposes an addition to the consequence levels. The focus is on dust release. The framework for the table is inspired by the liquid/vapor release table as published by the Center for Chemical Process Safety. The release characteristic and release size distinctions for dust release were derived by comparing dust explosion mechanisms, existing literature categorizations, and energy released to those for vapor explosions. In this way, the proposed table is consistent with the CCPS vapor/liquid release consequence table.

The proposed table has been used in the food process industry to assess risk. The LOPA application led to the identification of recommended action items for operating dust handling processes. These action items are consistent with case-specific recommendations found in the literature. This table can be added to the CCPS LOPA release, downtime and capital loss tables to extend the breadth of LOPA applications.

Facility Siting and the Texas City Incident, What It Means for Our Industry

Pat Berwanger
Berwanger, Inc.
4615 Southwest Freeway, Suite 900
Houston, TX 77027 U.S.A.
Phone: 713-570-2900
pc.Berwanger@berwanger.com

Prepared for Presentation at
American Institute of Chemical Engineers
2006 Spring National Meeting
8th Process Plant Safety Symposium
Orlando, Florida
April 24-26, 2006

UNPUBLISHED

Facility Siting and the Texas City Incident, What It Means for Our Industry

Pat Berwanger
Berwanger Inc.
4615 Southwest Freeway, Suite 900
Houston, TX 77027 U.S.A.
Phone: 317-4333-0950
pc.berwanger@berwanger.com

Abstract

Fifteen people lost their lives in the 23 March 2005 Texas City accident. Most of those killed were in a trailer close to the epicenter of the explosion. None of those who died were operating personnel.

Government, industry, and academic groups have each responded to the incident in their own way. The ultimate impact of how this incident plays out for industry will largely depend on the interaction, or lack thereof, between these various groups. The author offers his viewpoint, based on his years of working with several of the groups involved, how this interaction is now playing out and is likely to play out in the future.

The incident also laid out in plain view a number of factors that should be taken into account in facility siting for which, at least in retrospect, there was little in the way of definitive guidance provided by the industry's leading facility siting standard, API Recommended Practice 752, "Management of Hazards Associated with Location of Process Plant Buildings".

The author will also present a framework for evaluating several of these factors.

1. Introduction

During start-up of an Isomerization Unit at BP's Texas City Refinery, explosions and fires occurred killing fifteen and harming over 170 persons. The deaths resulted from the overfilling of a distillation column with hot gasoline-like fluids that discharged to atmosphere from an antiquated blowdown drum. Most of those killed were working in a trailer located within 150 feet of the drum. None of those killed were operating personnel.

This unfortunate accident was the most significant accident to occur at a US process facility since the accidents of the late 1980s and early 1990s which lead to promulgation of the process safety management regulation, 29 CFR 1910.119. As a result, this incident has garnered a tremendous amount of attention from government, industry, and academia.

The incident's impact on industry will ultimately be determined by the activities of various interest groups. In this paper, the author will discuss the activities and, where applicable, the interactions of several groups representing government, industry, and academia with which the author has had personal experience. In particular, the author will discuss,

- Government – Occupational Safety and Hazard Administration (OSHA)
- Government – US Chemical Safety and Hazard Investigation Board (also known as the "Chemical Safety Board" or "CSB")
- Industry – American Petroleum Institute (API)
- Academia – Mary Kay O'Connor Process Safety Center (MKOPSC)

In addition, several factors will be addressed that should be taken into account in facility siting for which definitive guidance had not been provided by the industry's leading facility siting practice, API 752, "Management of Hazards Associated with Location of Process Plant Buildings".
All opinions contained herein are strictly those of the author and are not intended to speak for representatives of the various organizations involved.

2. Government, Industry, and Academic Organizations Involved in the Aftermath of Texas City

2. 1 Government – OSHA

To understand OSHA's role, one must understand that, first and foremost, OSHA is a law enforcement agency. As such, OSHA personnel were tasked with determining which, if any, laws were violated by the company or companies (i.e. BP AND its contractors) involved in the accident.

OSHA's primary means of expressing its findings is through the issuance of citations (OSHA can also hold press conferences, issue reports, and refer matters to the Department of Justice for criminal prosecution). OSHA issues citations for what OSHA personnel believe are specific violations of the law.

In the author's experience, OSHA has tremendous latitude when it comes to issuing citations. Just like the local building inspector, if OSHA wants to "throw the book" at a company by nit-picking each and every violation they find, they are quite capable of doing so. Thus, given the amount of latitude OSHA has, one might expect OSHA to attempt to craft a "message" that signals to industry what OSHA will give priority to in the future.

OSHA did precisely that by choosing to classify as "egregiously willful" an enormous number of individual citations related to the location of the trailers involved in the Texas City incident. The vast majority of these citations were for electrical code violations. In the author's opinion, these violations were chosen as the basis for egregiously willful citations since electrical code violations are easy to understand and hard to argue with. Furthermore, there was already a legal precedent for issuing a citation for each instance of an electrical code violation – thereby enabling OSHA to issue lots of citations for purposes of arriving at a large fine. But, make no mistake about it; in the opinion of the author, the message to industry was that OSHA investigators were extremely unhappy about where the buildings, and more importantly, the personnel were located.

Based on the above, one would expect OSHA to give particular attention to facility siting issues in the future – particularly as it regards the location of personnel who have nothing to do with operating or starting up a unit.

2.2 Government – Chemical Safety Board (CSB)

Unlike OSHA, the CSB is not a law enforcement agency. Rather, its mission is to determine causes of accidents and to make recommendations to stakeholders (primarily government, industry, and labor groups) on how such accidents might be avoided in the future. From a legal standpoint, the stakeholders are not strictly obligated to embrace the CSB's recommendations. The degree to which they do embrace CSB recommendations is dependent on several factors such as: the merits of the recommendation, the "political" and "public relations" climate, as well as the persistence and determination of the current leadership of the CSB itself (more on this later).

CSB's findings and recommendations cannot be used in a court of law. Thus, plaintiff's attorneys cannot use CSB findings and recommendations as a means to facilitate collection of damages in lawsuits. The intention of this restriction is to encourage parties to cooperate with the CSB without fear that the information they provide will later be used against them in court.

Like OSHA, the CSB mobilized a team to investigate the Texas City incident. Unlike OSHA, however, CSB's mission is industry specific (the process industries, which is the entire world to CSB, are but a small part of the entire portfolio of industries that OSHA has to cover). As a consequence of its narrowed scope, CSB has at its disposal a more focused cadre of process industry specific experts than OSHA does. The personnel making up CSB's investigative team reflected this fact.

The CSB has added a new arrow to their quiver – the "urgent safety recommendation". The "urgent safety recommendation" is intended to expedite industry compliance with the CSB's suggestions. The first ever such "urgent safety recommendation" the CSB made was in the aftermath of the Texas City incident.

To appreciate the potential importance and effectiveness of CSB's strategy one needs to look at how the CSB made prior recommendations and how the industry responded to them. A case in point was CSB's 2002 recommendation calling for expansion of the OSHA Process Safety Management (PSM) standard's "covered list" of chemicals to encompass a much wider array of potentially reactive chemicals. That recommendation was taken under advisement by an array of industry groups and continues to be addressed in a deliberative fashion. OSHA itself elected not to support revision of the PSM standard.

On the other hand, CSB has made three "urgent safety recommendations" (at the time of this writing) related to the Texas City incident. The first recommendation was for BP to form an "independent panel" to review safety practices at its US refineries. The second and third recommendations were directed at the API (American Petroleum Institute) and the NPRA

(National Petrochemical and Refiners Association) "to ensure the safe placement of occupied trailers and similar temporary structures away from hazardous areas of process plants". The CSB news release also highlighted the fact that BP had already agreed to locate their trailers at least 500 feet away from potential hazards and move nonessential workers into office space outside the refinery. The clear implication was that the CSB board would not object if everyone else in industry did the same thing.

Significantly, the CSB did not wait for completion of its Texas City investigation to issue any of these recommendations. It would seem that it was the intention of the CSB board to get an immediate response from BP specifically and the refining and petrochemical industry in general. If that was indeed their intention, they were soon not to be disappointed. Essentially simultaneous with the first CSB recommendation, BP agreed to form an independent panel. And, immediately following the CSB's second recommendation, the API agreed to reconvene their API RP 752 task force even though a revision of the recommended practice was not due for several more years. In addition, refiners and related industries have not waited for the API committee to provide guidance to respond to the CSB's recommendation themselves.

Thus, it would seem the CSB has found an effective tool for cutting through the industry's, and government's, normal processes for adjusting their practices and regulations.

This author's recommendation: Keep an eye on the CSB and its energetic Chairman and CEO, Carolyn Merritt. As a relatively new government agency, the CSB has the luxury of being able to shape the rules of the game as it goes. This is a fact Ms. Merritt is surely not unaware of.

2.3 Industry – American Petroleum Institute (API)

As regards facility siting issues, the primary role of the API is to facilitate the promulgation of a consensus practice - specifically API RP 752. As a rule, reaching consensus on a matter as complex as facility siting is particularly challenging for committee members representing individual API member companies because the practices vary so much from company to company. Not surprisingly, the latest edition of API RP 752 – issued in November 2003 - provides much in the way of qualitative guidance but relatively little in the way of specific instructions. For example, API RP 752 suggests that one take into account how many people may be in a building but does not give specific limits on the number or function of personnel. This contrasts with many other recommended practices where specific instructions, right down to the specific equation to be used, are provided.

Largely as a result of the CSB's second urgent recommendation, we can expect the next addition of API RP 752 to be much more prescriptive in nature. Later in this article, the author discusses specific facility siting parameters that are likely to become more explicit.

2.4 Academia – Mary Kay O'Connor Process Safety Center (MKOPSC)

The Mary Kay O'Connor Process Safety Center, or "Center" as it is known, has as its mission to improve safety in the process industries through education and research. The Center is affiliated with Texas A&M University and is lead by Professor Sam Mannan – who is well known and respected in process safety management circles.

The Center also provides a "safe" forum for representatives of industry, government, and academia to come together to discuss process safety management topics of interest. None of the other organizations we discussed above provide such a direct forum. Also, unlike the other organizations, the Center does not lobby as an advocate for particular policy positions. Thus, by design, the Center can be expected to improve communication and thereby reduce tension between the government and industry.

A case in point was the Center's response to the Texas City incident as regards facility siting. In fall 2005, the Center sponsored a facility siting workshop that was widely attended by representatives of industry, government, and academia. The workshop provided an excellent forum for industry to educate government representatives about the complex issues surrounding facility siting as well as industry's earnest attempts to tackle them. Likewise, government representatives had an opportunity to express their priorities.

The Center is also sponsoring a web portal (www.facilitysiting.com) where stakeholders can access news and research as well as express their views on this important topic. Proceedings from the facility-siting workshop may also be downloaded from the portal.

3. Facility Siting Factors to Consider for Which API RP 752, "Management of Hazards Associated with Process Plant Buildings" Lacks Definitive Guidance

One of the findings that came out of the Mary Kay O'Connor Process Safety Center Facility Siting Workshop was that API RP 752, "Management of Hazards Associated with Process Plant Buildings" did not provide explicit guidance on several important aspects of facility siting which must, or at least should, be taken into account by the practitioner.

Now, one might argue there are, or at least were, good reasons for the API recommended practice not to provide explicit guidance – not the least of which is how difficult it is to arrive at a consensus at committee meetings. However, what became clear to investigators of the Texas City incident was that it would have been possible to apply API RP 752 but still end up with an aggregation of trailers containing non-essential personnel adjacent to a unit being started up that had a history of start-up incidents.

Whatever people's opinions are in industry, this situation is not acceptable to the regulators. Thus, one might expect that if we in industry fail to tighten up the rules ourselves, the government will do it for us - and certainly with less regard for the industry's cost or inconvenience. The opening salvo in this regard was the CSB's urgent recommendation for operating companies to move temporary trailers away from process units without regard to specific manning levels, specific hazards of the processes (how about trailers near lube oil units?), the blast resistance of the trailers (how about blast resistant trailers?), or the alternative of simply clearing out "non-essential" personnel during hazardous operations such as start-up.

Much technical work has been focused on blast modeling and the requisite blast resistance of structures. There does not seem to be a lack of guidance in that regard. However, there are several other factors for which more specific consensus guidance is desirable. A few of these are described below.

3.1 Occupancy Loading

Common sense would dictate that a building could only present an additional hazard to personnel if it is occupied. Of course, this begs the question, when should a building be designated as "occupied"? There are two extremes. On the one extreme, it would make sense not to be concerned about an analyzer shelter that does not even have a chair to sit down on. On the other extreme, a building containing office space for ten people that is used regularly for this purpose may justify the added expense of being blast resistant. And, of course, there is a lot of middle ground. For example, how about a temporary building that will only be occupied when the plant is down?

The approach suggested by API RP 752 is to establish "occupancy criteria" and, if those criteria are exceeded, to consider further hazard evaluation. Conversely, the implication is that if the "occupancy criteria" are not exceeded, no further specific evaluation is required, and presumably, the building does not need to be made blast resistant.

API RP 752 does not, however, establish specific occupancy criteria figures or specific ways to calculate the occupancy loads.

3.2 Trailer Park Syndrome

A plant has trailer park syndrome when individual trailers come and go but the area where the trailers are parked always happens to be occupied by some group of trailers. This situation can develop when a company makes its trailer siting decisions on a trailer-by-trailer basis. For example, the risk of putting a two-person trailer near a process area may meet occupancy criteria. However, having 10 two person trailers may not. A trailer-by-trailer management of change process might miss this very important point. The latest edition of API RP 752 does not preclude such a situation from developing. And, indeed, a trailer park did exist at the Texas City site.

3.3 Location of Personnel During Non-normal Operations (e.g. Start-up)

The author knows of no study that provides a definitive answer on just how much more hazardous a start-up is than normal operations. But, there is little doubt that a high proportion of incidents occur during start-up. Most engineers would not disagree with an assertion that perhaps a quarter or more of serious incidents occur during start-ups.

Let's assume, for the sake of argument, that one out of four incidents occurs during start-up, and that a typical unit spends about 2% of its life in start-up mode. It would follow that an individual sitting in a building next to a unit for say, one hour, during a start-up, is 12-1/2 times more likely to be killed than someone sitting in the same building for an hour during normal operations.

API RP 752 does not provide specific guidance on how to address non-normal operations. And, as is well known, the Texas City incident did occur during start-up.

3.4 Essential vs. Non-Essential Personnel

That none of the personnel killed in the Texas City Incident were operators is a fact that was not lost on government investigators. Again, whatever the opinion of industry representatives, it would not be far off the mark to assert that it is the uniform opinion of government investigators that none of the people killed in the Texas City Incident should have been there.

There are myriad ways to define "essential" personnel. One way would be to classify as essential all personnel whose job function in a unit is 24/7. Another way to define essential personnel would be anyone whose work product was required in the operating unit.

Once again, API RP 752 does not provide concise guidance in this area.

3.5 Overpressure Radii and Spacing Guidelines

There is no doubt that overpressure radii are very dependent on the specific geometry, process conditions, and potential loss of containment scenarios. In fact, one can justifiably conclude that each and every unit is inherently different. And, that even if two units were precisely the same, then the overpressure radii would still depend on the loss of containment scenario the facility siting practitioner happened to select as a basis for his or her decision making. And, if we really want to be completely honest with ourselves, we would have to admit that even if two practitioners selected precisely the same loss of containment scenario (say, a rupture of a 4" line) for exactly the same unit, they are quite likely to calculate quite different overpressure radii.

Obviously, this situation leads to inconsistent results and is subject to practitioners succumbing to wishful thinking in the face of day-to-day pressures.

Although API RP 752 provides clear guidance on what level of overpressure is expected to result in a particular level of damage to a building, it does not provide objectively clear guidance on how to arrive at overpressure radii in a consistent and reliable manner.

4. Conclusion

This paper described the unique roles several organizations (OSHA, CSB, API, and the Mary Kay O'Connor Process Safety Center) are playing in the aftermath of the Texas City Incident. We can expect:

- OSHA to give increased scrutiny to facility siting matters – particularly the location of non-essential personnel during start-ups.
- The CSB to continue to wield its "urgent recommendation" weapon to elicit prompt reaction from industry.
- The API to revise API RP 752 ahead of schedule and to provide more definitive facility siting guidance (lest regulators do it for us).

- The Mary Kay O'Connor Process Safety Center to continue to educate stakeholders on this important topic as well as to provide a forum for industry, government, and academia to collaborate rather than litigate.

127b Lessons Learned through Osha's Enforcement of the Facility Siting Provisions in the Psm Standard

Mark S. Dreux, Arent Fox PLLC, 1050 Connecticut Ave., N.W., Washington, DC 20036

On March 23, 2005, there was a tragic incident at a Texas City refinery in which 15 employees died and approximately 170 employees were injured. OSHA issued a lengthy citation which, among other things, alleged numerous facility siting deficiencies as “egregious and willful,” focusing on the placement of temporary trailers, the location for relieving blowdown stacks, and the lack of a blast resistant shelter.

This presentation will focus on the OSHA citations concerning facility siting, particularly those issued following the Texas City incident, and discuss the patterns of recurring hazardous conditions and the relevant industry standards. A checklist of facility siting issues developed from OSHA citations and other OSHA guidance documents will be discussed to assist employers in conducting future facility siting studies. Finally, the presentation will review the concept of “acceptable risk” (including qualitative and quantitative risk analysis) that has arisen in OSHA enforcement actions and review its application to facility siting issues.

127c Facility Siting Following a Merger of Two Large Oil Companies

Gregg A. Sapp, Conoco Phillips, P. O. Box 1267, 650-15 South Tower, Ponca City, OK 74602-1267 and John A. Alderman, RRS Engineering, 2525 South Shore Boulevard, Suite 206, League City, TX 77573.

Conoco Inc. and Phillips Petroleum Company merged in late 2002. ConocoPhillips Company is a company that has brought many refining cultures together and with that merger, the facilities siting practices of several petroleum companies. Corporate management was very interested in the facilities siting efforts that were in place for the combined company at the time of the merger and created a task force to address this subject on a timely basis. An initial assessment indicated that facility siting had been accomplished many different ways with results that varied widely.

This paper discusses the efforts that ConocoPhillips undertook and have developed to address facility siting within the downstream businesses. One of the key elements in the ConocoPhillips program is consistency between sites. The paper will explain the development of key processes which have been standardized within the company. There will be a discussion of the methodologies employed, preliminary results and review processes. Also covered is a discussion of the methodology to prioritize remediation of buildings. The methodologies and processes for structural upgrades are also included. Management of change is also a component which is included in the paper that will speak to the revalidation of facilities siting studies along with the use of temporary buildings within the refineries.

Techniques for Siting New Buildings in Petrochemical Facilities

Martin L. Goodrich
John N. Dyer
Donald E. Ketchum
J. Kelly Thomas, Ph.D.
Baker Engineering and Risk Consultants, Inc. (BakerRisk)
San Antonio, Texas, 78218
(210) 824-5960
mgoodrich@bakerrisk.com

Prepared for Presentation at
American Institute of Chemical Engineers
2006 Spring National Meeting
8th Process Plant Safety Symposium
Orlando, Florida
April 24-26, 2006

UNPUBLISHED

Techniques for Siting New Buildings in Petrochemical Facilities

Martin L. Goodrich, John N. Dyer, Donald E. Ketchum and J. Kelly Thomas
Baker Engineering and Risk Consultants, Inc. (BakerRisk)
San Antonio, Texas, 78218
(210) 824-5960
mgoodrich@bakerrisk.com

Abstract

The choice of an optimum location for placing new structures in petroleum refining and chemical processing plants is subject to a host of considerations, of which the structure's blast resistance is one of the most significant. The decision centers on defining an affordable structure that maximizes occupant safety. This paper discusses the considerations and alternative methods of assessing blast and structural response siting issues by using an example building. The assessment involves systematically defining potential blast loads, the construction details necessary to withstand those loads, and identifying locations that make the most of an existing facility.

1. An Approach to Siting New Buildings

This paper presents a stepwise approach to siting new buildings at petrochemical facilities. An assessment of the relative merits of multiple locations can identify options that increase the effectiveness and decrease the cost of the building. Among the most significant factors influencing the selection of a building location (and the structural requirements of the building) is its exposure to potential blast loads. The AIChE's Guidelines for Facility Siting and Layout[1] offers a limited discussion on the influence of explosion hazards. This paper addresses how blast considerations can be integrated into the siting process.

The approach begins with first considering buildings of conventional construction at several locations. As the cost of hardened structures are significantly greater than conventional construction, it is prudent to determine whether a conventional building design might suffice, or if a conventional design can be practically upgraded to withstand the design basis blast loads. Knowing the construction requirements (on a conceptual level) at a number of locations will permit construction costs to be included with other considerations when determining an optimum location.

Siting a new building at a petrochemical facility includes the following steps:

1) Choose several potential locations
2) Estimate blast loads for a building at each of those locations
3) Determine the response of conventional buildings at those locations
4) Explore the possibility of upgraded conventional construction
5) Select the best location

2. Selecting Potential Locations

Choosing a building location begins with its function. Personnel not directly involved in the operation of process units should be located further away, where explosion hazards are less severe. Should the building serve as a safe haven, areas where personnel seeking shelter will come from should be considered.

A number of other factors should be included when selecting a building location:

- Proximity to Fire/Toxic Hazards: determine whether the location is subject to acute fire or toxic exposures.
- Proximity to the Process Units: distancing the building from the process units will generally lower its exposure to blast loads, but will increase the length of the connections and the cost of control systems. A substantial distance between the units and the buildings can also strain communications between field and board operators.
- Prevailing Wind: wherever possible it is preferable to place a building upwind of potential release sources. Being downwind of a source might allow flammable vapors to disperse towards the building with greater frequency, placing a potential vapor cloud explosion (VCE) nearer the building. Being downwind of release sources may also expose the building and its occupants to increased odors.
- 3rd Party Impact/Security: if the building is near the property line, the building should be placed with an awareness of potential explosion hazards from neighboring plants.
- Traffic/Parking/Accessibility: thought should be given to traffic, parking and personnel accessibility.
- Future Growth: if the building is sited near an open area available for future plant expansion, consider how the new construction might limit the placement of new units (whose potential blast loads may exceed those of existing units). That is, while a new building should be designed to withstand the potential blast loads from existing units, it might be designed to also withstand estimated loads from future units.

With these factors in mind, two or three potential sites should be identified. As an example, Figure 1 shows a typical chemical plant with three alternative locations for a new control building identified. Systems handling flammable materials from throughout the plant are also noted in the figure.

3. Estimate Blast Loads

3.1 Analysis Procedure

The analysis procedure for defining blast load at the potential building locations and the corresponding building structural response are shown in Figure 2, with the major steps described in the following paragraphs.

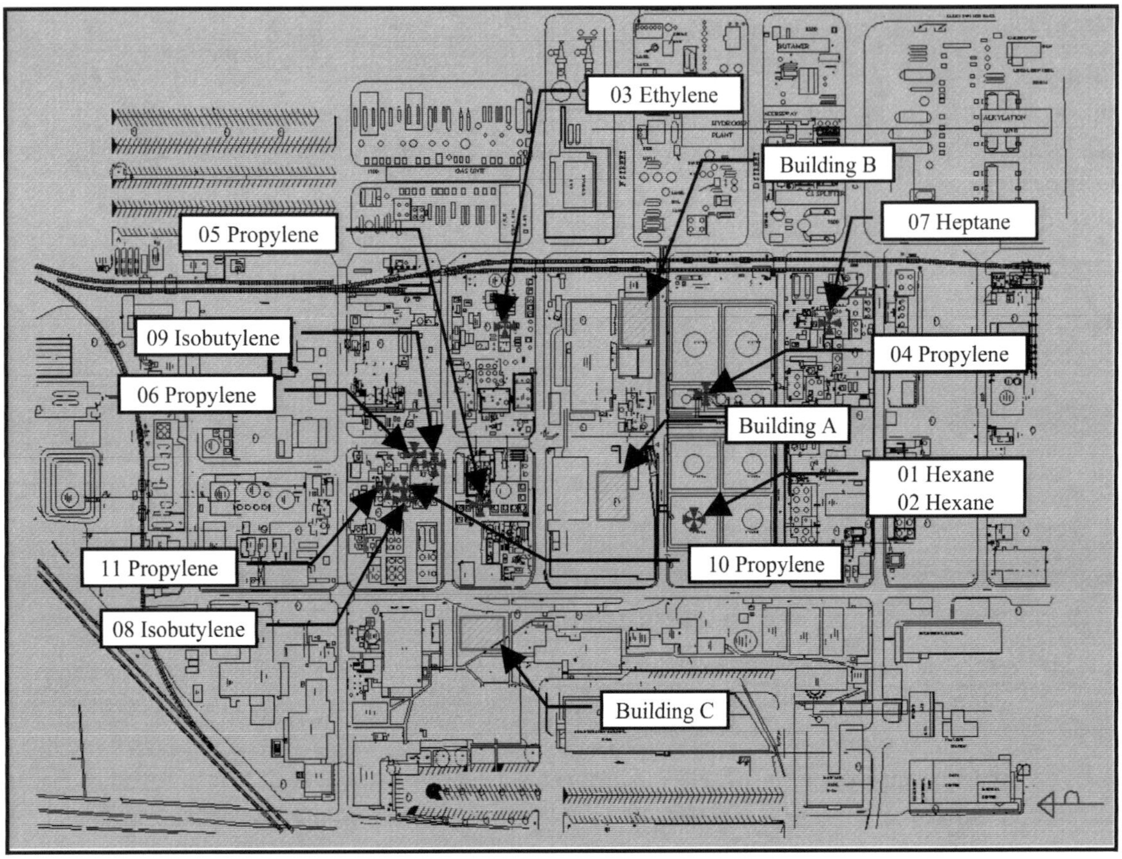

Figure 1. Building Siting Example With Three Alternate Locations

The assessment quantifies the estimated building response to a range of vapor cloud explosion (VCE) and bursting pressure vessel (BPV) scenarios. Scenarios are developed based on the process conditions of the units with consideration of factors that would increase the blast load severity, such as:

- Fast burning, or high reactivity, materials (e.g., hydrogen, ethylene, etc.),
- Process conditions tending to produce large flammable clouds such as high temperature liquid streams with a high flash fraction, and
- Flammables in close proximity to congested/confined areas near the candidate locations.

Example areas of equipment congestion and/or confinement are identified and mapped as shown in Figure 3 (note that actual zoning for most analyses would be more complex than shown in this figure). Flammable clouds released into uncongested areas will tend to produce fireballs with little overpressure. Obstacles in congested areas will create turbulence in a burning cloud, which will accelerate the flame speed, the energy release rate, and consequently, the blast loads. Surfaces restricting the expansion of burning clouds, such as concrete decks, also increase flame speeds and blast loads. The map shown characterizes the equipment into three levels of congestion (low, medium, and high) and two levels of confinement (3- and 2-dimensional).

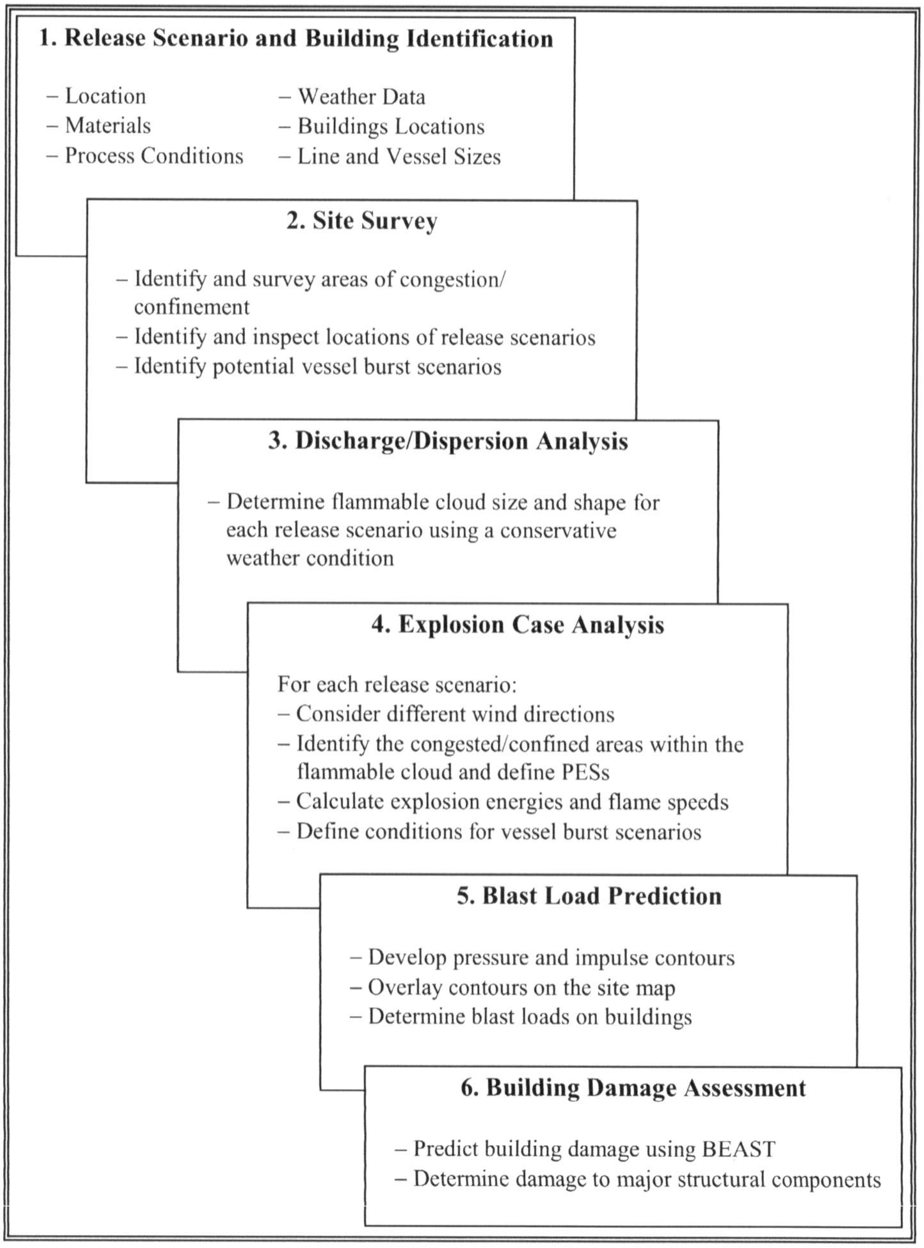

Figure 2. Analysis Procedure

VCE cases are developed by considering the congested/confined areas to either be filled with a flammable mixture, or by performing discharge and dispersion analyses for specific releases scenarios. Discharge and dispersion analyses may be performed with a number of commercially available codes (e.g., SafeSite$_{3G}^{TM}$, PHAST, etc.) using equivalent leak sizes consistent with industry practice and plant operating history. The results of the dispersion analysis are used to determine the flammable portion of vapor clouds. VCE cases are developed by overlaying the resulting flammable cloud footprints on the site map in order to identify congested/confined areas that could be reached by the flammable clouds. Explosion energies are calculated based on the portions of the flammable clouds within the congested/confined areas. An example is shown in Figure 4. The analysis should consider wind directions that could place flammable clouds in areas of high congestion and/or confinement near the potential building locations.

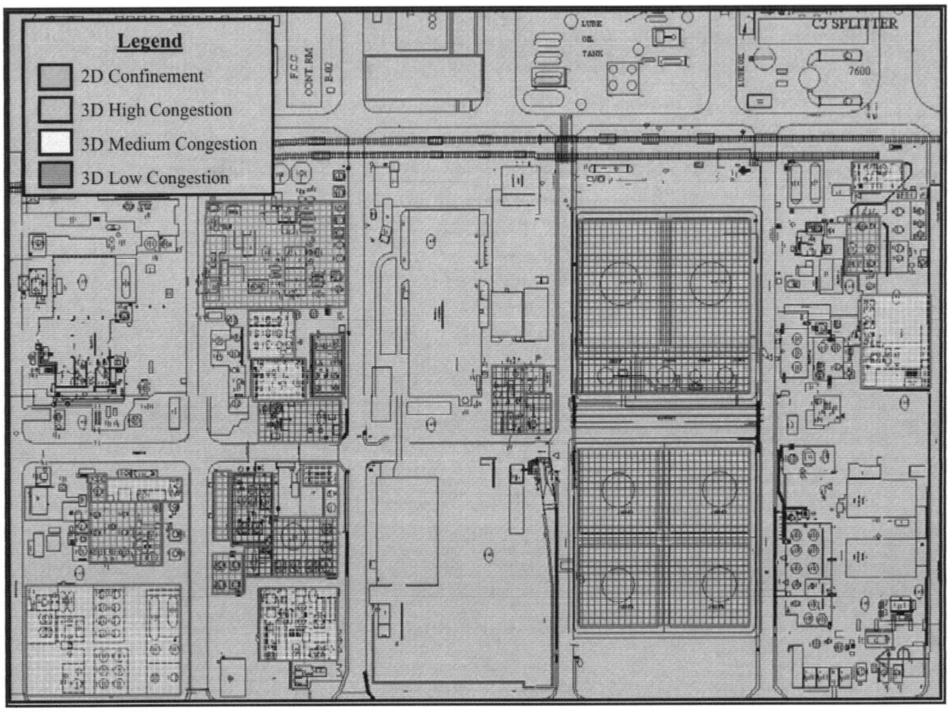

Figure 3. Congestion and Confinement Mapping

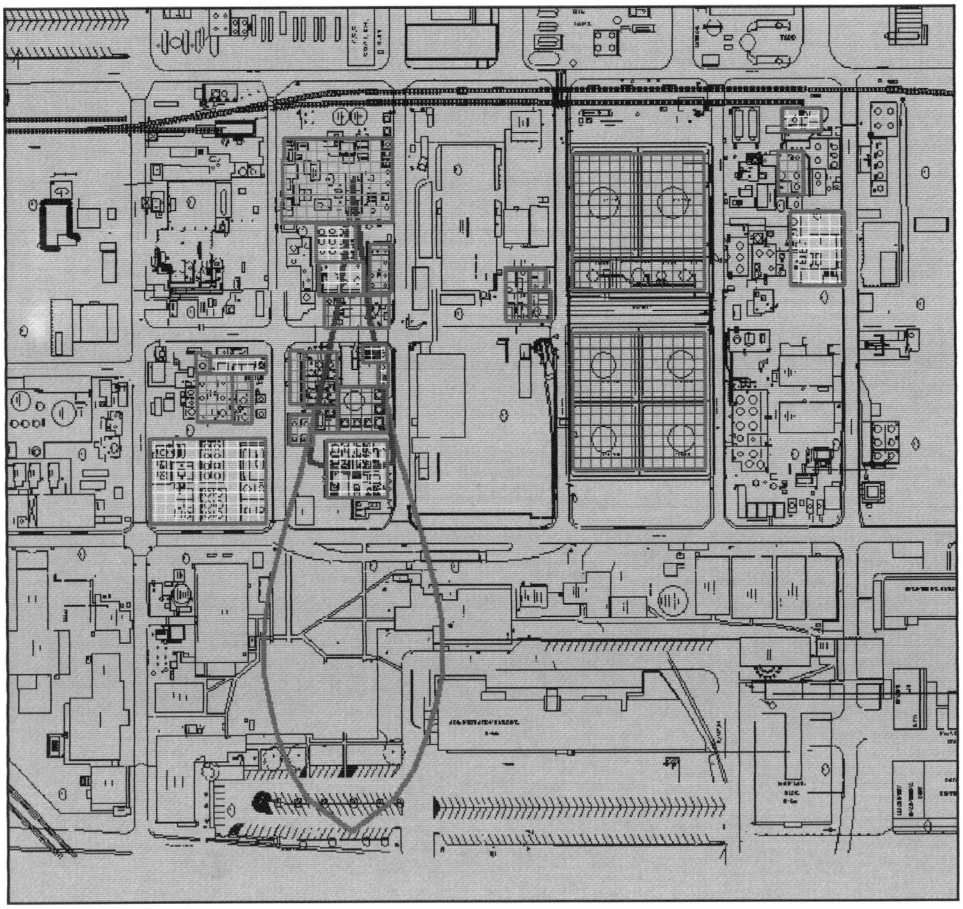

Figure 4. Flammable Vapor Cloud Region from Dispersion Analysis

Simplified VCE blast load prediction methodologies, such as the Baker-Strehlow-Tang (BST) method[2,3,4] or TNO multi-energy method, are normally utilized to determine VCE blast loads for siting purposes. The TNT-equivalent method is not sufficiently accurate for this purpose and should not be used. However, turbulent combustion computational fluid dynamics (CFD) VCE blast load prediction codes, such as GexCon's FLACS code, can be used to refine the blast load estimates developed using these simplified methods if the additional effort is warranted (e.g., if the potential building location is very near a congested/confined area). In addition, CFD codes capable of modeling the interaction of blast waves with structures, such as BakerRisk's Blast Wave Target Interaction (BWTI) code[5], can be used to examine the impact of blast wave shielding and focusing if the potential building location is in or near a group of buildings.

Bursting pressure vessel (BPV) explosions should also be considered in the assessment, as applicable. The blast load from a BPV is primarily a function of the vessel gas pressure and volume. Generalized BPV blast load charts have been developed based on one-dimensional numerical analysis[6].

For the purposes of the illustrative example, flammable clouds were developed from a discharge and dispersion analyses assuming leak areas equivalent to a 2-inch diameter hole. Multiple wind directions were considered so that the resulting flammable clouds reached as near as possible to congested/confined areas in the vicinity of all potential building locations. The BST method was used in this example to calculate the resulting VCE blast loads. The discharge, dispersion, VCE blast load, and BPV blast load analyses were all carried out using BakerRisk's SafeSite$_{3G}$™ computer code. SafeSite$_{3G}$™ provides an integrated suite of discharge, dispersion, blast load, structural response and occupant vulnerability models, and hence is an efficient tool for siting analysis.

3.2 Example Blast Results

Blast loads are typically expressed in terms of a peak pressure and impulse. The impulse is the area under the pressure-time curve, and hence is related to both peak pressure and duration. The results of the blast analysis are usually displayed in two forms. The first is contour plots. Figures 5 and 6 provide the blast pressure and impulse contour plots generated for the illustrative problem overlaid on a site plot plan. These are enveloping contours representing all the VCE and BPV explosion cases considered. The contour plots clearly show the low and high blast load areas within the facility, and can be utilized to identify additional potential building locations as needed. The blast pressure and impulse contour plots should be used in conjunction, since both pressure and impulse influence building response. As can be seen in these figures, the potential location B (see Figure 1) would be the most favorable with respect to blast loads.

The second blast load representation is tabular, as shown in Table 1. The table provides a matrix of blast pressure and impulse for each scenario at each candidate building location. The values presented in this table are free-field or side-on blast loads (e.g., the load applied to the building roof). The results provided in the table shown that the potential location B (see Figure 1) would be the most favorable with respect to blast loads, and that Scenario 3 (ethylene release) provides the controlling blast load case. The blast loads for this scenario are much higher than for any other scenario, which suggests that the conditions associated with this scenario should be closely scrutinized to ensure they are credible and that this scenario represents a reasonable design basis.

The applied blast loads on each building face are determined considering reflection factors; blast loads on building walls facing the source of the blast will be increased by a factor of 2 to 3 due to reflection of the blast load. The applied loads at Location B for Scenario 3 (i.e., the controlling case) are shown in Table 2.

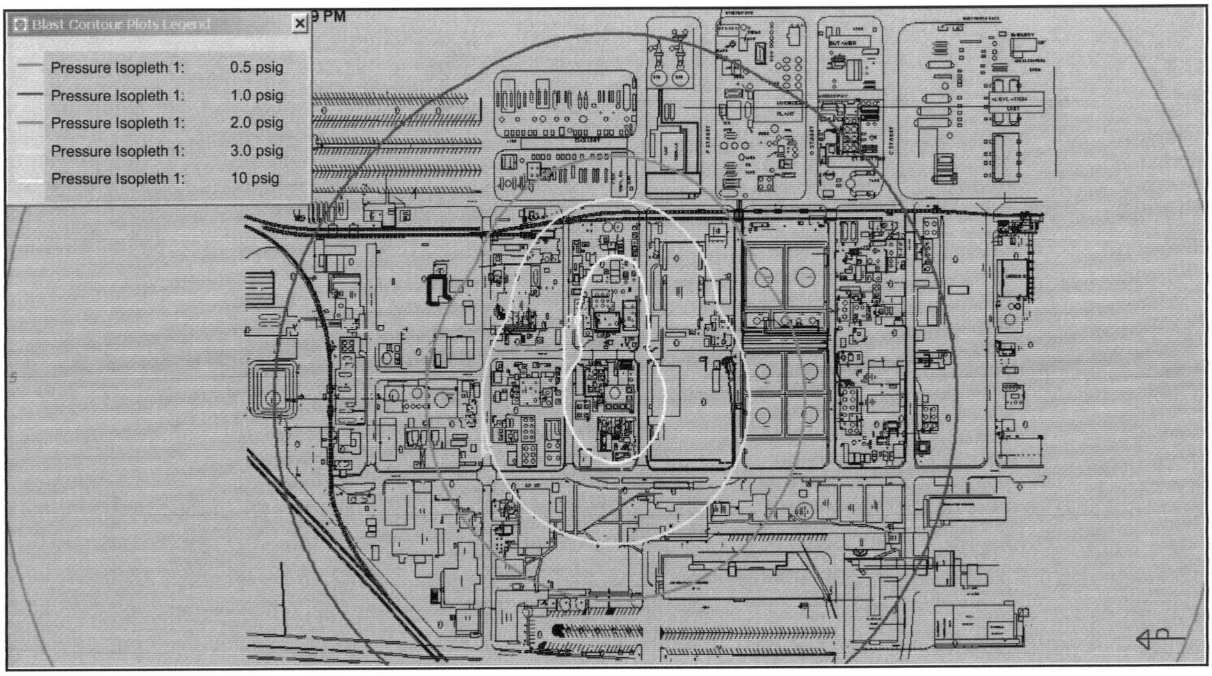

Figure 5. Example Pressure Contour Map

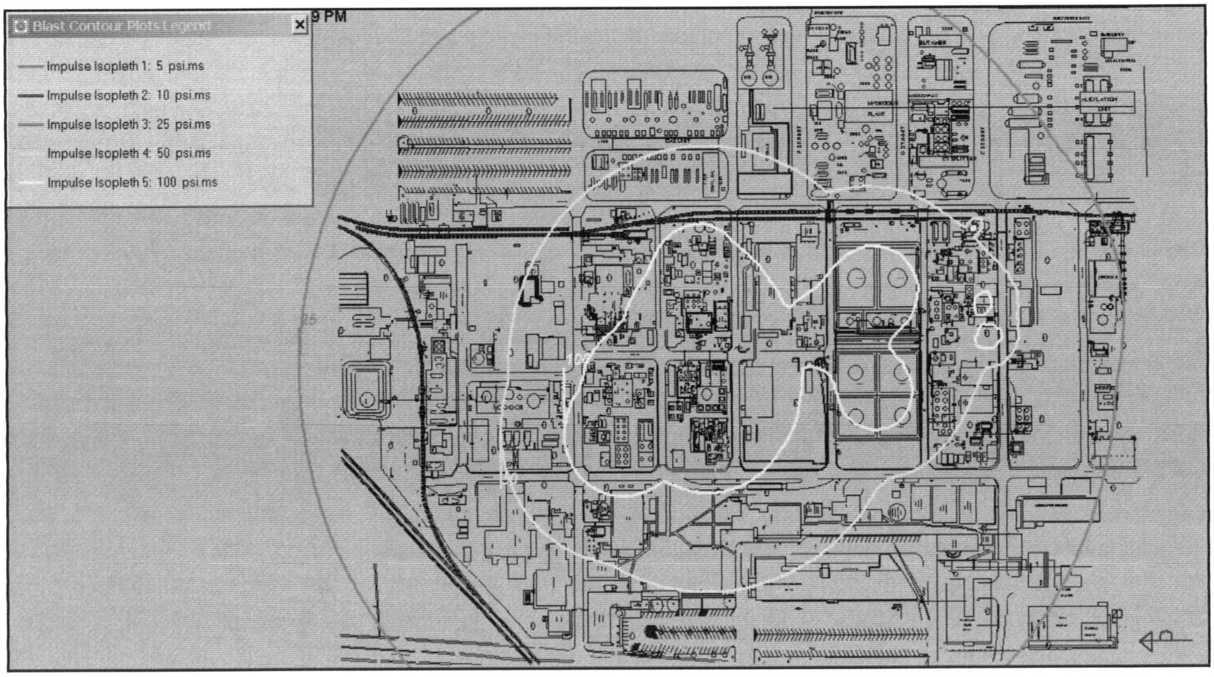

Figure 6. Example Impulse Contour Map

Table 1. Example Blast Loads Table

Scenario	Wind (from)	Blast Type	Building A		Building B		Building C	
			Pressure (psig)	Impulse (psi-ms)	Pressure (psig)	Impulse (psi-ms)	Pressure (psig)	Impulse (psi-ms)
01 Hexane 5psi	237	VCE	<0.1	1	<0.1	<1	<0.1	<1
02 Hexane 10psi	242	VCE	1.0	43	1.0	35	0.3	14
03 Ethylene	90	VCE	5.2	108	2.6	69	4.4	86
04 Propylene	174	VCE	0.6	11	0.7	14	0.2	4
04 Propylene	237	VCE	<0.1	3	<0.1	8	<0.1	2
05 Propylene	153	VCE	0.5	50	0.4	31	0.8	68
05 Propylene	276	VCE	0.7	50	0.8	45	0.5	42
06 Propylene	30	VCE	0.8	53	0.4	28	1.0	58
06 Propylene	319	VCE	0.8	50	0.8	52	1.0	41
07 Heptane	141	VCE	1.0	66	1.0	77	0.3	26
08 Isobutylene	0	VCE	0.8	43	0.5	23	1.1	40
08 Isobutylene	113	VCE	0.7	23	0.5	16	1.0	39
08 Isobutylene	22	VCE	0.9	45	0.5	24	1.1	58
08 Isobutylene	293	VCE	0.9	20	0.6	15	1.2	25
08 Isobutylene	315	VCE	0.8	27	0.6	21	1.1	28
08 Isobutylene	338	VCE	0.8	34	0.5	22	1.1	31
08 Isobutylene	45	VCE	0.7	19	0.5	13	1.0	30
08 Isobutylene	68	VCE	0.7	24	0.5	16	1.0	43
08 Isobutylene	90	VCE	0.7	23	0.5	16	1.0	41
10 Propylene	320	VCE	0.3	4	0.2	3	0.4	5
10 Propylene	59	VCE	0.3	6	0.2	4	0.5	8
11 Propylene	0	VCE	0.6	8	0.4	6	0.8	10
11 Propylene	70	VCE	0.5	11	0.3	7	0.7	19
14 Reactor	n/a	BPV	0.1	1	0.1	1	0.1	1
15 Catalyst Vessel 1 Bar	n/a	BPV	0.2	2	0.2	2	0.1	1
16 Catalyst Vessel Pmax	n/a	BPV	0.4	7	0.5	8	0.2	4
Maximum Blast Load			5.2	108	2.6	69	4.4	86

Note: North=0, East=90, South=180, West=270

Table 2. Controlling Blast Loads for Potential Building Location B

Building Surface	Applied Blast Load	
	Pressure (psi)	Impulse (psi-msec)
South	2.4	65
East	2.4	64
North	6.1	155
West	5.2	149
Roof	2.6	69

4. Structural Assessment

Damage assessments for conventionally constructed buildings at the potential building locations can be readily performed using a tool such as the "Building Evaluation and Screening Tool" (BEAST)[7], which was developed for the Explosion Research Cooperative (ERC). Alternatively, Single-Degree-of-Freedom (SDOF) analysis, a well-recognized methodology in the blast-design industry, or a more refined analysis method such as finite element analysis may be used to estimate building response.

The results of a damage assessment can be expressed in terms of building damage level (BDL). BDLs are qualitative assessments of the damage, ranging from cosmetic damage to complete collapse of the building. The BDL classification is given in Table 3.

Table 3. Description of Building Damage Levels (BDL)

Building Damage Level	Damage Description
1	Onset of visible damage to building walls facing explosive source. Repairs only needed for cosmetic reasons. Building is reusable following an explosion.
2a	Localized building damage, primarily to components on building walls facing the explosive source. Building performs function and can be used; however, major repairs are required to restore integrity of structural envelope. Total cost of repairs is moderate.
2b	Widespread building damage. Building cannot be used until major repairs are completed. Total cost of repairs is significant, approaching replacement cost of building.
3	Building has lost structural integrity and may collapse due to environmental conditions (i.e. snow or wind). Total repair cost exceeds replacement cost of building.
4	Building fails completely. Repair is not feasible.

To illustrate this approach, consider the example of a hypothetical control building subjected to the blast loads given in Table 1. The building is assumed to have a rectangular floor plan and a height of 14 feet. The structure is assumed to consist of a steel frame with 12-foot tall masonry infill-walls. The roof is taken to be a corrugated-metal deck with a 4-inch thick concrete slab supported by 18-inch deep open-web steel joists.

A table of BDL estimates for the matrix of explosion scenarios and building locations can be produced, similar to that for the blast loads. This table will show which scenarios pose the greatest threats to the new building, and which locations provide the most favorable response. The results for the example problem are provided in Table 4. The table shows the least severe building response (BDL 2b) for the most severe explosion scenario (Scenario 3, ethylene release) is achieved at Location B. Since most companies consider this level of response unacceptable (there is about a 5% chance of serious injury for building occupants in this case), a more detailed component level response should be considered.

A component level assessment may be used to determine how the design might be modified to provide an acceptable response to blast loads. A substantial increase of the blast resistance of a conventionally designed building can be achieved by modifying its original design within reasonable bounds, both financial and in constructability. The objective of the modified design is to reduce the building damage level (BDL) by assessing (and increasing the strength of, as needed) critical structural components.

Table 4. Building Damage Results for Example Problem

Scenario	Wind (from)	Blast Type	Building A	Building B	Building C	Maximum Damage
01 Hexane 5psi	237	VCE				
02 Hexane 10psi	242	VCE	2a	2a		2a
03 Ethylene	90	VCE	3	2b	3	3
04 Propylene	174	VCE		2a		2a
04 Propylene	237	VCE				
05 Propylene	153	VCE	2a		2a	2a
05 Propylene	276	VCE	2a	2a		2a
06 Propylene	30	VCE	2a		2a	2a
06 Propylene	319	VCE	2a	2a	2a	2a
07 Heptane	141	VCE	2a	2a		2a
08 Isobutylene	0	VCE	2a		2a	2a
08 Isobutylene	113	VCE	2a		2a	2a
08 Isobutylene	22	VCE	2a		2a	2a
08 Isobutylene	293	VCE	2a		2a	2a
08 Isobutylene	315	VCE	2a		2a	2a
08 Isobutylene	338	VCE	2a		2a	2a
08 Isobutylene	45	VCE	2a		2a	2a
08 Isobutylene	68	VCE	2a		2a	2a
08 Isobutylene	90	VCE	2a		2a	2a
10 Propylene	320	VCE				
10 Propylene	59	VCE				
11 Propylene	0	VCE				
11 Propylene	70	VCE			2a	2a
14 Reactor	n/a	BPV				
15 Catalyst Vessel 1 Bar	n/a	BPV				
16 Catalyst Vessel Pmax	n/a	BPV				
Maximum Damage			3	2b	3	

Note: North=0, East=90, South=180, West=270

Component Damage Level (CDL) assessments can be done qualitatively, similarly to the BDL assessments; however, a quantitative approach can be used by performing dynamic analyses of the components and comparing the calculated maximum displacement to response criteria such as those published in the American Society of Civil Engineers[8].

For the purpose of this example, it is assumed that the BDL is controlled by the masonry walls on the building north and west sides and by the roof joists. Therefore, the steel columns and girders, and the roof slab are assumed to be significantly stronger than the masonry walls and the joists. Typical construction for the walls and roof joists with the corresponding modifications are summarized in Table 5.

Table 5. Summary of Typical and Modified Construction Response to Controlling Blast Loads (Location B, Scenario 3)

Component	Construction	Typical Construction	Predicted Component Response
Roof joist	Typical	18K7s spaced 48 inches	Failure
	Modified	18K7s spaced 36 inches	Major Damage
Walls	Typical	8-inch CMU with #5s at 48 inches o.c.	Failure
	Modified	12-inch CMU with #6s at 32 inches o.c.	Moderate Damage

Pressure-impulse (P-i) diagrams are used to illustrate the capacity of a component. The diagrams are produced from a structural dynamic analysis, which predicts peak component displacement or damage as a function of the blast pressure and impulse. The roof joint and walls of the example building were assessed using dynamic structural analyses based on the Single-Degree-of-Freedom (SDOF) method. SDOF models of the components may be used to develop P-i diagrams that show contours of peak component displacement (or damage) in terms of both peak blast pressure and impulse. The P-i curves define the boundaries of the CDL regions and correspond to blast loads that result in Low Response (Level 1), Medium Response (Level 2), and High Response (Level 3) according to the ASCE definitions[8]. The response definitions and corresponding damage are described in Reference 8. Similarly, P-i diagrams can be constructed for the roof joists.

The P-i diagram for a conventional 8-inch masonry wall is shown in Figure 7 and an upgraded 12-inch masonry wall is shown in Figure 8. The figures show how modifications, evaluated using component level assessment, can upgrade a conventional building to an acceptable level of blast resistance. As seen in Figure 7, the blast loads on the south and north sides of the building correspond to points on the diagram that exceed any pressure and impulse combination even on the Level 3 curve. The walls on these sides would therefore fail under the applied blast load for the controlling scenario. The upgraded design (Figure 8) would only sustain minor damage on the south side and moderate damage on the north. These component modifications would then lower the overall building damage level to an acceptable level.

5. Selecting the "Best" Option

Construction cost estimates for each location can be developed based on the assessment of blast design requirements. These cost estimates should be considered along with other relevant considerations, such as those noted in Section 2. A quantitative or qualitative (i.e., high, medium, low) ranking of the relevant factors is a convenient way to present the advantages and disadvantages of each potential building location and allow management to make a more informed building siting decision. A detailed structural analysis of the final building design should be performed to verify the response estimates made during the siting study.

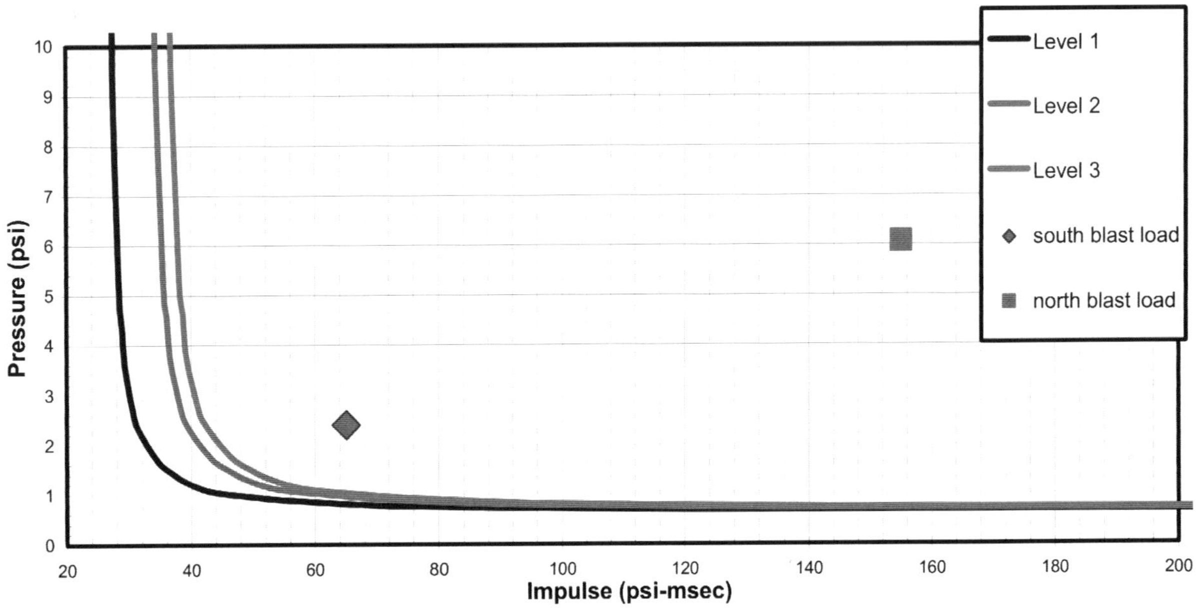

Figure 7. P-i Diagram for 8-inch CMU Wall

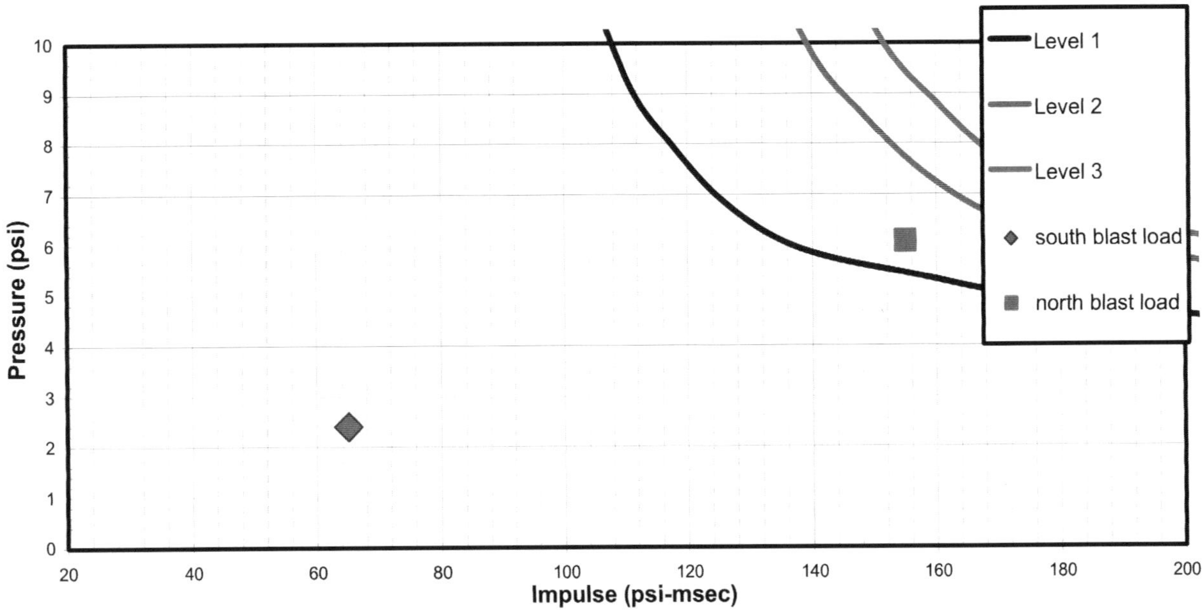

Figure 8. P-i Diagram for 12-inch CMU Wall

6. Summary

A summary of the recommended siting approach for new buildings is as follows:

- Assess the merits of several locations when siting a new building.
- Consider the factors influencing the building functionality.
- Estimate the potential blast loads at the candidate sites for explosion scenarios throughout the facility.
- Consider the response of a conventional building, which may be sufficient for locations distanced from potential explosion hazards.
- A component level analysis may be used to develop design modifications needed to provide adequate structural response.
- If the predicted blast loads overwhelm the structure, a specially designed hardened structure should be considered or alternative locations should be considered.
- Once the blast resistant requirements are defined, rough construction costs may be developed to evaluate the merits of alternate building locations.

Of course, the impact of other hazards (i.e., fire and toxic) at the proposed building location should be considered as a part of this evaluation process.

7. References

[1] Grounds, C. and J. Natale "Guidelines for Facility Siting and Layout", Center for Chemical Process Safety, AIChE, 2002.

[2] Baker, Q. A., M.J. Tang, E.A. Scheier and G.J. Silva, "Vapor Cloud Explosion Analysis," 28th Annual Loss Prevention Symposium, April 19, 1994.

[3] Baker, Q. A., C.M. Doolittle, G.A. Fitzgerald and M.J. Tang, "Recent Developments in the Baker-Strehlow VCE Analysis Methodology." 31st Annual Loss Prevention Symposium, March 9-13, 1997.

[4] Pierorazio, A.J., J.K. Thomas, Q.A. Baker and D.E. Ketchum, "An Update to the Baker-Strehlow-Tang Vapor Cloud Explosion Prediction Methodology Flame Speed Table," 38th Loss Prevention Symposium, American Institute of Chemical Engineers, New Orleans, LA, 26-29 April 2004.

[5] Thomas, J.K., J. Geng and A.D. Hallgarth, "Evaluation of Door Blast Shields," 31st Department of Defense Explosive Safety Seminar, San Antonio, TX, August 2004.

[6] Baker, Q.A. and M.J. Tang, "Effect of Vessel Temperature on Blast Loads," 2004 ASME/JSME Pressure Vessels and Piping Conference, July 25-29, 2004, San Diego, CA.

[7] Oswald, C., "Design Information to Increase Blast Resistance of Conventional Buildings," prepared by Baker Engineering and Risk Consultants (BakerRisk) for the 1999 Explosion Research Cooperative.

[8] American Society of Civil Engineers, "Design of Blast Resistant Buildings in Petrochemical Facilities," ASCE Task Committee on Blast Resistant Design, New York, NY, 1997.

Infrastructure Impacts During Hurricanes Katrina and Rita and Facility Siting Issues

David A. Moore, PE, CSP
President and CEO
Dorothy Allen Kellogg, MPA
AcuTech Consulting Group
2001 N. Beauregard Street, Suite 700
Alexandria, VA 22311
www.acutech-consulting.com

Prepared for Presentation at
American Institute of Chemical Engineers
2006 Spring National Meeting
8th Process Plant Safety Symposium
Orlando, Florida
April 24-26, 2006

UNPUBLISHED

Infrastructure Impacts During Hurricanes Katrina and Rita and Facility Siting Issues

David A. Moore, PE, CSP
President and CEO
Dorothy Allen Kellogg, MPA
AcuTech Consulting Group
2001 N. Beauregard Street, Suite 700
Alexandria, VA 22311
www.acutech-consulting.com

American Institute of Chemical Engineers
Process Plant Safety Symposium
April 23-27, 2006
Walt Disney World Dolphin Resort, Orlando, Florida

Abstract

Hurricanes Katrina and Rita affected the U.S. chemical and petroleum industries sectors as they did everyone and everything else along the Gulf, highlighting profoundly the strengths and weaknesses of plans to prepare for and recover from a devastating area-wide catastrophe. The direct impacts to the industries were more economic; no fatalities or serious injuries occurred from releases at chemical or petroleum facilities during the storms. Hurricane plans and release prevention efforts were largely successful. Nevertheless, the facilities, companies, and their employees were left with significant physical and economic recovery challenges resulting from the impacts on the facilities themselves and the supporting infrastructure.

The massive scale of the storms highlighted numerous weaknesses in facility as well as government and private sector preparations for such a truly catastrophic emergency – and opportunities for improvement. The lessons from these storms must be leveraged to improve the readiness of the U.S. industry for natural, accidental, and intentional events. This paper will examine some of the gaps in traditional facility siting considerations and risk assessments and suggest elements that could be added to those activities to better prepare for catastrophic events – whether natural disasters or intentional acts of terrorism.

1. Gulf Coast Chemical and Petroleum Industries, Hurricanes, and Katrina

Oil and natural gas production in the Gulf coast area as well as offshore in the Gulf itself coupled with the mouth-of-the Mississippi multi-modal transportation infrastructure fueled development of one of the world's largest concentrations of refining, chemical and petrochemical production along the Gulf Coast from Corpus Christi, Texas to Pascagoula, Mississippi. Proximity to raw materials and inexpensive transportation spurred not only the chemical and petroleum industries themselves, but also their suppliers and customers, as well as the suppliers and customers suppliers and customers. The industries are dependent on the region and the region is dependent on the industries.

Hurricanes, however, are a fact of life for anyone living or operating in Gulf of Mexico coastal areas. Since the U.S. Weather Service began keeping records in 1851, over half of the Atlantic storms struck at least one of the five Gulf coast states.[1] While the nation's chemical and petroleum industries are certainly familiar with hurricane preparedness and recovery, no one was fully prepared for the catastrophic one-two punch Katrina delivered to Louisiana, Mississippi and Alabama or for the follow-up blow from Hurricane Rita.

Hurricane Katrina was the first Category 5[2] hurricane of the 2005 Atlantic hurricane season and only the forth Category 5 storm on record.[3] The storm made landfall as a Category 1 hurricane near Miami, Florida on August 25, 2005, then again as a Category 4 storm, increasing to Category 5, on August 29 along the Central Gulf Coast just east of New Orleans. While initially striking communities from the mouth of the Mississippi to Mobile Bay, the levee system that protected New Orleans from Lake Pontchartrain to the north of the city succumbed in the 20-foot storm surge allowing the lake's waters to flood most of the city. At the request of the Governors of Louisiana, Mississippi and Alabama, President Bush issued disaster declarations for over 90,000 square miles of the Gulf coast in those states.

By all accounts, Hurricane Katrina was 'the most destructive and costliest natural disaster in the history of the United States.[4] The official death toll from Katrina exceeds 1300 and damage estimates surpassed $200 billion[5]. To put Katrina's devastation in perspective:

- Hurricane Andrew had been the most costly US hurricane before 2005 with damages estimated at $26.5 billion ($44 billion in 2004 $ based on U.S. DOC Implicit Price Deflator for Construction).[6]
- The only hurricanes comparable in loss of life occurred before 1930 -- 1900 Galveston – 8000 dead (could be as high as 12,000); 1928 southeast Florida/Lake Okeechobee – 2500 dead (could be as high as 3000); 1893 Louisiana – 1100 – 1400 dead (offshore losses increases death count to near 2000); 1893 Sea Islands, South Carolina & Georgia – 1000 – 2000 dead.[7]

---

[1] *U.S. Mainland Hurricane Strikes by State, 1851-2004* (http://www.nhc.noaa.gov/paststate.shtml?) derived from *The Deadliest, Costliest, and Most Intense United States Tropical Cyclones from 1851 to 2004;* NOAA Technical Memorandum NWS TPC-4; Blake, Jarrell, Rappaport, and Landsea; August 2005. http://www.nhc.noaa.gov/Deadliest_Costliest.shtml.

[2] Hurricanes are rated on the Saffir-Simpson Hurricane Scale, a 1-5 rating based on a hurricane's intensity. Category 5 hurricanes are characterized by winds greater than 155 miles per hour. The rating is used to give an estimate of the potential property damage and flooding expected along the coast from a hurricane's landfall. Wind speed is the determining factor as storm surge values are highly dependent on the slope of the continental shelf and the shape of the coastline. *See* http://www.nhc.noaa.gov/aboutsshs.shtml

[3] *U.S. Mainland Hurricane Strikes by State, 1851-2004* (http://www.nhc.noaa.gov/paststate.shtml?).

[4] *The Deadliest, Costliest, and Most Intense United States Tropical Cyclones from 1851 to 2004;* NOAA Technical Memorandum NWS TPC-4; Blake, Jarrell, Rappaport, and Landsea; August 2005; Table 2 and Table 3a. http://www.nhc.noaa.gov/Deadliest_Costliest.shtml.

[5] *The Federal Response to Hurricane Katrina: Lessons Learned*; Frances Fragos Townsend, Assistant to the President for Homeland Security & Counterterrorism; February 2006; Pg. 33.

[6] *Deadliest, Costliest, Most Intense* Table 3a.

[7] *Ibid.* Table 2

- Only three Category 5 hurricanes had struck the US – Andrew (1992), Camille (1969), and the 1935 hurricane striking the Florida Keys.[8]
- An estimated five million people were without power, with parts of New Orleans still without power five months later.[9]
- 95 % of daily Gulf of Mexico oil production and 88% of daily Gulf of Mexico natural gas production shut down[10]
- Almost three million customer phone lines were knocked out, telephone-switching centers were seriously damaged, and 1,477 cell towers were incapacitated.[11]

In addition to loss of oil and natural gas production, some chemical and petroleum industry facilities experienced:

- physical damage to facilities, such as from wind and flooding;
- extended plant closures in anticipation of, during, and following the storm;
- inability to contact employees during and following the storm;
- loss of supporting infrastructure – communications, transportation, power, police/fire services;
- disruption to supply chain from damage to suppliers and customers;
- reentry prohibitions, delaying damage assessment and repair;
- lack of housing for critical employees and contract workers;
- disruptions due to the need to attend to employee's individual needs, and
- unique security and safety demands caused by the changing environment.

Even facilities that were not directly struck by the storm – primarily those north and west of New Orleans – were affected by the damage to supporting infrastructure, especially transportation and communications. Fortunately, due to robust design and the emergency procedures and systems that were deployed, there were minimal releases of hazardous materials[12].

---

[8] *Ibid.*

[9] Office of Electricity Delivery & Energy Reliability; US Department of Energy; Final Gulf coast Hurricanes Situation Report #46; January 26, 2006 (3:00 p.m. EST).

[10] *Hurricane Katrina's Impact on the U.S. Oil and natural Gas Markets*; Special Report from the Energy Information Adminstration, US Department of Energy; Tuesday, August 30, 3:00 p.m. http://tonto.eia.doe.gov/oog/special/eia1_katrina_083005.html

[11] *Federal Response* pg. 34.

[12] Only one facility, Murphy Oil Meraux Refinery in St. Bernard Parish, LA, reported a release to the environment. See Hurricane Response 2005 Test Result, US Environmental Protection Agency; http://www.epa.gov/katrina/testresults/index.html

2. Lessons Learned

The events surfaced many expected and unexpected issues including the importance of considering chemical and petroleum facilities dependencies and interdependencies on other infrastructure – both private and public. While facilities are prepared for short-term loss of power or other services, few if any were prepared for the widespread and long-term loss of power, water, communications and other utilities; all modes of transportation; feedstocks and oil production; common suppliers and services; and the extensive flooding and loss of life.

Numerous lessons were learned including the need for:

- improved knowledge and coordination on industry facility and facility interdependency/dependency characterization with Government;
- upgraded national and regional contingency planning and emergency response coordination for major events;
- improved operational readiness and coordination between the Department of Homeland Security (DHS) and other Federal and local agencies during crises of this magnitude;
- the need for improved catastrophic event business continuity planning and crisis management; and
- facility and equipment siting to minimize impacts of events.

3. Facility Siting Pre-Katrina

In the context of chemical and petroleum facilities, the term "facility siting" can be applied in two ways. First, it may relate to locating a facility itself. Economic considerations such as proximity to raw materials and customers, availability of transportation infrastructure, other economic considerations as well as safety largely drive such decisions. More recently, environmental and local political considerations factor into siting decisions. For example, the Environmental Protection Agency (EPA) and states have promulgated siting regulations for solid waste landfills[13], and local citizens frequently organize to support or oppose many changes in their communities.

Facility siting may also refer to facility layout -- the locating of structures within a facility. Federal regulations – particularly EPA's Risk Management Program[14] and the Occupational Safety and Health Administration's (OSHA's) Process Safety Management Standard[15] – require consideration of the location of structures in close proximity to hazardous and flammable materials. These regulations also impose an obligation to consider utilities and other systems essential to preventing releases of materials. Such considerations, however, are primarily considered within the context of a relatively short-term (a few hours to a few days) loss of power or water – not for weeks or even months.

---

[13] 40 CFR 258 Subpart B – Location Restrictions applies to solid waste landfills); 40 CFR 264.18 applies to hazardous waste treatment, storage, or disposal facilities.

[14] Chemical Accident Prevention 40 CFR Chapter I, Subchapter C, Part 68.

[15] 29 CFR 1910.119.

The chemical and petroleum industries have also developed voluntary guidelines to address both facility siting and layout. The Center for Chemical Process Safety (CCPS) publishes 'Guidelines for Facility Siting and Layout'[16], and the American Petroleum Institute (API) and the Chemical Manufacturers Association (CMA), now the American Chemistry Council (ACC), developed Recommended Practice 752 (RP-752), 'Management of Hazards Associated with Location of Process Plant Buildings'.[17] A recent CCPS document entitled Guidelines for Evaluating the Characteristics of Vapor Cloud Explosions, Flash Fires, and Boiling Liquid Expanding Vapor Explosions (BLEVES) is also a useful reference for facility siting considerations.

Such facility siting and layout requirements and guidelines were developed to protect employees, facilities, and the surrounding environment and community primarily from the consequence of an event emanating from the facility. As a result, they have appropriately focused on the facility itself and what it must do alone or in partnership with local responders. The CCPS guidelines do recommend facilities consider whether they may be subject to hurricanes or other severe weather conditions, but do not provide significant additional guidance on the issues that were realized during the storms.

4. Emergency Response, Value Chain, and Business Continuity Planning

Chemical and petroleum facilities have developed plans to respond and recover from different emergency situations – from fires to releases to hurricanes. Some, such as the emergency planning and notification requirement under EPCRA,[18] are mandatory; state or local law, industry standards, or standard business practices may require others. Again, such plans tend to focus on the facility itself and its obligations and capabilities. In the case of hurricanes, they focused on preparing the facility itself and its personnel to ride out and recover from a storm, not on recovering when the entire surrounding region was devastated.

Chemical and petroleum facilities have invested significant effort and capital in understanding and working with value chain partners both to support just-in-time manufacturing and to improve their environmental, health and safety performance. Since 9/11, efforts have focused on the security of a facility's value chain and value chain partners.[19] Again, these assessments and subsequent value chain management were insufficient to respond to some of the new challenges brought on by Hurricane Katrina.

---

[16] Guidelines for Facility Siting and Layout, Center for Chemical Process Safety, AIChE, 2003.

[17] RP 752, Management of Hazards Associated with Location of Process Plant Buildings, American Petroleum Institute and Chemical Manufacturing Association, 1996.

[18] The Emergency Planning and Community Right to Know Act of 1986 42 USC 11011 et seq. Regulations at 40 CFR Chapter I, Subpart J—Superfund, Emergency Planning, and Community Right-to-Know Programs.

[19] For example, participants in the Department of Homeland Security's Customs Trade Partnership Against Terrorism (C-TPAT) program ensure the integrity of their security practices and communicate and verify the security guidelines of their business partners within the supply chain to US Customs and Border Protection. http://www.customs.gov/xp/cgov/import/commercial_enforcement/ctpat/ctpat_faq.xml. The American Chemistry Council's (ACC) Responsible Care® Security Code also requires ACC members to assess value chain as well as facility and cyber security. http://www.responsiblecaretoolkit.com/security_valuechain.asp

There are numerous references and guidelines for business continuity planning and emergency response that give general direction on the value of assessing risks and addressing risk mitigation as a precursor to emergency planning. They also refer to site selection as a critical component of contingency planning. In 2004 the National Fire Protection Association published the NFPA 1600 Standard on disaster/emergency management and business continuity.[20] The objective of this standard is to 'provide those with the responsibility for disaster and emergency management and business continuity programs the criteria to assess current programs or to develop, implement, and maintain a program to mitigate, prepare for, respond to, and recover from disasters and emergencies.' While the 9/11 Commission endorsed the NFPA standard[21], it is doubtful that implementation of the current standard would have sufficiently prepared a facility or company for Katrina.

5. Facility Siting Post-Katrina

Reports on lessons learned in the aftermath of Hurricane Katrina abound and will continue to be issues for years to come. To date, much of the study has focused on 'preparedness' and 'response' for disasters. It is generally understood that perfection in preparedness and response is impossible to achieve, but must continually be sought. In this paper we are suggesting enhancements to the front end of the disaster preparedness process -- the risk assessment and risk mitigation steps inherent in facility siting and layout – since they may pay dividends in responding to subsequent regional or national catastrophes.

By in large, we must assume that the chemical and petroleum industries will remain on the Gulf coast, largely in their existing locations and with many of their existing layouts. We must also acknowledge that hurricanes will continue to strike the region. Nevertheless, as facilities prepare for the 2006 hurricane season and beyond, certain risk reduction benefits can be gained by entertaining choices in facility siting and layout, equipment selection, value chain design, and other inherent features. The mere recognition of these issues in the risk assessment step of the process may lead to better preparedness. Such changes may enable a facility to avoid potential impacts of future catastrophic events, maintain a high level of performance, or recover faster.

While few if any manufacturing facilities can be expected to relocate, companies may consider impacts to corporate offices, sales offices, or other facilities less dependent on currently invested capital equipment or surrounding infrastructure. In addition to traditional location considerations, companies may want to consider the expected operating environment and geospatial situation in which a prospective office or facility may need to operate during a

---

[20] NFPA 1600' Standard on Disaster/Emergency Management and Business Continuity Programs; 2004 Edition

[21] "We endorse the American National Standards Institute's recommended standard for private preparedness. We were encouraged by Secretary Tom Ridge's praise of the standard, and urge the Department of Homeland Security to promote its adoption. We also encourage the insurance and credit-rating industries to look closely at a company's compliance with the ANSI standard in assessing its insurability and creditworthiness. We believe that compliance with the standard should define the standard of care owed by a company to its employees and the public for legal purposes. Private-sector preparedness is not a luxury; it is a cost of doing business in the post-9/11 world. It is ignored at a tremendous potential cost in lives, money, and national security." *The 9/11 Commission Report: Final Report of the National Commission on Terrorist Attacks upon the United States*, July 2004Pg. 398

regional disaster. Even facilities that will continue operating in their current location should consider the following and begin incorporating them into future planning:

- Availability of resources during emergencies:
    - the city and county law enforcement and other government services that may be distracted or unavailable;
    - the infrastructure that may be unavailable for extended periods, such as electrical, sewer, or water;
    - feedstock and supply disruptions on the supply side and customer disruptions on the demand side;
    - qualified personnel that may not be available during worst credible events due to personal needs or inability to reach the facility;
- Localized effects:
    - expected flooding of key roadways;
    - disablement or overdependence of evacuees or government services/military on critical infrastructure including roadways or bridges;
- The expected assignment of government resources, such as National Guard, by government authorities and the response time for providing security;
- The chain of command to get assistance when required – the importance of the chemical sector facilities was not entirely known to government so when resources to reestablish power and dewater were needed there wasn't a clear system for prioritization or communications;
- The potential for maintaining independence on most critical functions and to sustain during the duration of the emergency;
- Passive survivability design – the potential for inherently safer design features to allow for survivability during extreme events; this may include:
    - sufficient buffer zone to avoid impacts to neighbors,
    - higher elevations to avoid flooding,
    - avoidance of cutoff locations which are dependent on vulnerable bridges or roadways.

These are but a few of the lessons learned and that have to be incorporated into industry guidance for facility siting for future consideration and considered by industry during business continuity planning.

6. Risk Management Approach

The chemical and petroleum industries have learned that the systematic risk management approaches used to plan for accidents are adaptable to security assessments. A similar systematic, analytical process can be developed to better plan for regional or national disasters whether acts of nature or acts of man.

Classically, organizations plan for managing 'emergency situations' based on the residual risk following engineering and operations considerations. Organizations plan for some level of loss and disruption, so the emphasis of emergency planning is typically the following objectives to those identified events:

- Response
- Control
- Restoration
- Repair

Business continuity planning (BCP) is a process of considering means to minimize the impacts to an organization of anticipated events. Emergency planning is different from BCP in that BCP should involve a wider variety of risk management strategies than only to respond to emergencies and to restore operations. An emergency plan generally assumes that the objective is to minimize impacts of an emergency situation as efficiently as possible.

Some BCP plans do not properly consider the potential to change the current situation through risk mitigation including modifying the system to be more resilient or less consequential should an event occur. These considerations should be examined in a systematic process such as a vulnerability/risk assessment. A vulnerability assessment is a process that identifies weaknesses in physical structures, personnel protection systems, processes, or other areas and where options to eliminate or mitigate those weaknesses are evaluated.[22]

Organizations have four options for managing their risks:

- Treat - engineering or administrative controls to prevent and mitigate risks;
- Tolerate – accept the risk;
- Transfer - transferring risk to 3rd parties through insurance or partnerships;
- Terminate – no longer operate the system.

AcuTech promotes the concept of preventing incidents through a proactive and systematic approach aimed at identifying the hazards and vulnerabilities of processes and work activities, evaluating the associated risks, and controlling such risks using a hierarchy of controls (in order of preference based on degree of effectiveness). Risk management concepts applicable to facility siting include the following:

- Avoidance –
  - site selection to avoid as many critical issues as is feasible;
  - discontinuance of operations if risks are unmanageable or intolerable;
- Risk reduction –
  - Resilience - hardening facilities to withstand initial impacts;
  - Consequence minimization – active or passive controls to limit primary or secondary impacts;
- Contingency services – duplication or substitution of critical functions and support services, from electrical power backup to offsite duplication of the entire operation;

[22] *Homeland Security - Key Elements of a Risk Management Approach*; Statement of Raymond J. Decker, Director, Defense Capabilities and Management; Testimony Before the Subcommittee on National Security, Veterans Affairs, and International Relations; House Committee on Government Reform; United States General Accounting Office; GAO; October 12, 2001;

- Emergency restoration – repair and restoration of loss as expeditiously and efficiently as is possible.

All of the above concepts should be evaluated as issues and options for risk reduction during the vulnerability/risk assessment. To the extent that facility siting can address these issues it should be fully explored.

7. Industry Guidelines for Business Contingency Planning/Emergency Response Management

As described above, most facility siting and layout guidelines and regulations focus on avoiding impacts from releases from industrial facilities to neighboring personnel, equipment, the public, or the environment rather than focus on how large scale regional events can be anticipated and vulnerabilities reduced to the organization. Similarly, emergency response planning and value chain studies and assessments have not considered circumstances in which the facility partners in response or logistics are as devastated by an event as the facility itself.

NFPA 1600[23], referenced above, was written to 'provide those with the responsibility for disaster and emergency management and business continuity programs the criteria to assess current programs or to develop, implement, and maintain a program to mitigate, prepare for, respond to, and recover from disasters and emergencies.' While still insufficient to address all the issues raised by Hurricane Katrina, the standard begins to raise business continuity planning on par with emergency response planning. The standard recommends four phases of disaster/emergency management:

- Mitigation
- Preparedness
- Response
- Recovery

These phases are embodied in the recommended program elements of NFPA 1600 as shown in Figure 1.

[23] NFPA 1600' Standard on Disaster/Emergency Management and Business Continuity Programs; 2004 Edition

Figure 1 **NFPA 1600,** ***Standard on Disaster/Emergency Management and Business Continuity Programs*** **Program Elements**
5.2 Laws and Authorities 5.3 Hazard Identification, Risk Assessment, and Impact Analysis 5.4 Hazard Mitigation 5.5 Resource Management 5.6 Mutual Aid 5.7 Planning 5.8 Direction, Control, and Coordination 5.9 Communications and Warning 5.10 Operations and Procedures 5.11 Logistics and Facilities 5.12 Training 5.13 Exercises, Evaluations, and Corrective Actions 5.14 Crisis Communications and Public Information 5.15 Finance and Administration

In particular, the standard addresses the concept of '5.4 - Hazard Mitigation' following '5.3 Hazard Identification, Risk Assessment, and Impact Analysis'. NFPA 1600:5.4.1 "The entity shall develop and implement a strategy to eliminate hazards or mitigate the effects of hazards that cannot be eliminated." NFPA 1600:5.4.3 – "The mitigation strategy shall consider, but not be limited to, the following:

(1) The use of applicable building construction standards
(2) Hazard avoidance through appropriate land-use practices
(3) Relocation, retrofitting, or removal of structures at risk
(4) Removal or elimination of the hazard
(5) Reduction or limitation of the amount or size of the hazard
(6) Segregation of the hazard from that which is to be protected
(7) Modification of the basic characteristics of the hazard
(8) Control of the rate of release of the hazard
(9) Provision of protective systems or equipment for both cyber or physical risks
(10) Establishment of hazard warning and communication procedures
(11) Redundancy or duplication of essential personnel, critical systems, equipment, information, operations, or materials."

The NFPA Standard appears to be enlightened compared with other chemical process industry risk management concepts. Items 2, 3, and 6 are direct applications of typical facility siting concepts. Items 4, 5, 7, and 8 are similar to the four strategies of inherent safety as published by the American Institute of Chemical Engineers (AIChE), Center for Chemical Process Safety

(CCPS). [24] The CCPS Inherent Safety guidelines recommend that industry minimize risk by applying four key strategies:

1. Inherent
2. Passive
3. Active
4. Procedural

Inherent safety involves four key inherently safer strategies:

1. Minimizing,
2. Substituting,
3. Moderating, and/or
4. Simplifying

Each of these strategies can accomplish hazard reduction or elimination. Hazard is defined as a "physical or chemical characteristic that has the potential for causing harm to people, the environment, or property." The concept is based on the belief that if one can eliminate or moderate the hazard, not only is the risk reduced, it may be possible to remove the risk altogether from consideration. Alternatively, an inherently safer system would make the hazard less likely to be realized and less intense if there is an accident.

To the extent that facility siting can contribute to these concepts it should be applied. This is imbedded in the NFPA 1600 standard, although not necessarily expressed in the same way. As such industry is cautioned to realize the full benefits of inherent safety in overall risk reduction (perhaps during the Elements 5.3 and 5.4 above) rather than to only follow NFPA 1600:5.4.3 for business continuity/disaster and emergency planning guidance as the list is incomplete. This could be an area of improvement for the NFPA 1600 standard.

8. Capturing Lessons Learned

The chemical and petroleum industries will remain on the Gulf coast, largely in their existing locations and with many of their existing layouts and hurricanes will continue to strike the region. Plans to prevent catastrophic releases of hazardous or flammable chemicals worked – chemical and petroleum facilities maintained their integrity. However, site planning that focuses primarily on the facility itself is insufficient to address future regional calamities – whether from natural or man-made causes.

In light of Hurricanes Katrina and Rita, improvements could be made to all of the industry guidance. Primarily this is to broaden the guidance to include more catastrophic risks effecting more than the facility in question, and to clarify how to integrate a management systems model for facility siting considerations into one for contingency planning and emergency response. The

[24] Center for Chemical Process Safety, Inherently Safer Chemical Processes: A Life Cycle Approach, Center for Chemical Process Safety, AIChE, 1996.

NFPA 1600 Standard has a sequence of steps (program elements) that resemble a management systems approach but is incomplete on this issue.

A set of questions or risk modeling guidance would be constructive to upscale the magnitude of events to be managed. The very definition of what constitutes facility siting issues should be broadened to include such challenges as major disruptive events that effect whole regions or the nation, not only localized plant releases.

Facility Siting Database and Analysis

Alex M. See
James M. Hudson
Jon D. Chrostowski

ACTA Inc.
2790 Skypark Drive #310
Torrance, CA 90505
Phone: 310-530-1008

Prepared for Presentation at
American Institute of Chemical Engineers
2006 Spring National Meeting
8th Process Plant Safety Symposium
Orlando, Florida
April 24-26, 2006

UNPUBLISHED

Facility Siting Database and Analysis

Alex M. See, James M. Hudson and Jon D. Chrostowski
ACTA Inc.
2790 Skypark Drive #310
Torrance, CA 90505

Abstract

The OSHA Code of Federal Regulations requires a facility siting analysis as part of the Process Hazard Analysis requirements. Recognizing that the petrochemical community needed guidance on the preparation of a facility siting study, the American Petroleum Institute and the Chemical Manufacturer's Association, provide procedures and methods for screening and calculating the potential impacts of explosions, fires and toxic gases to personnel and buildings.

ACTA developed a Facility Siting Database and Analysis software (FSDAn) tool to automate the process of performing a facility siting study. The code provides consequence analysis and risk-based results. The tool uses a database approach to manage building data, potential blast sources, hazardous material inventories, and occupation information. The approach and complexities of the technical tasks performed by FSDAn are described in this paper.

The tool provides instant access to information for everyday decision-making and a cost effective means to update the study; the impact of process and other changes can be instantly assessed.

1. Structure of Facility Siting Analysis Tool

The Facility Siting Database and Analysis software (FSDAn) was developed to provide a consistent approach to automate the process of performing a facility siting study. It is a risk-based facility siting analysis tool, which incorporates OSHA requirements [1] and the recommendations provided in API RP-752 [2]. The ultimate objective of this tool is to help manage the risks to on-site personnel from explosive, fire and toxic hazards. The tool uses a database approach to manage facility building data, potential blast sources, hazardous material inventories, and occupation information and thus allows the user to evaluate the siting issues with complete control of data integrity. Figure 1 shows the layout of the tool.

The main feature of FSDAn is its ability to provide management instant access to information for everyday decision-making and a cost effective means to update a facility study, where the impact of changes to the process, occupancy and other items can be instantly assessed and evaluated. For example, a blast analysis could be evaluated instantly to locate safe zones to place temporary trailers that house process staff.

Additionally, FSDAn presents the results graphically through the use of a Geographical Information System (GIS), which provides a highly effective means of internal and regulatory communication. Consequence analysis results can be readily plotted and overlaid on a facility plot plan and/or satellite/aerial imagery map by a few keystrokes. Risk results are provided in several formats, including casualty expectations and, if available, on a facility's risk matrix.

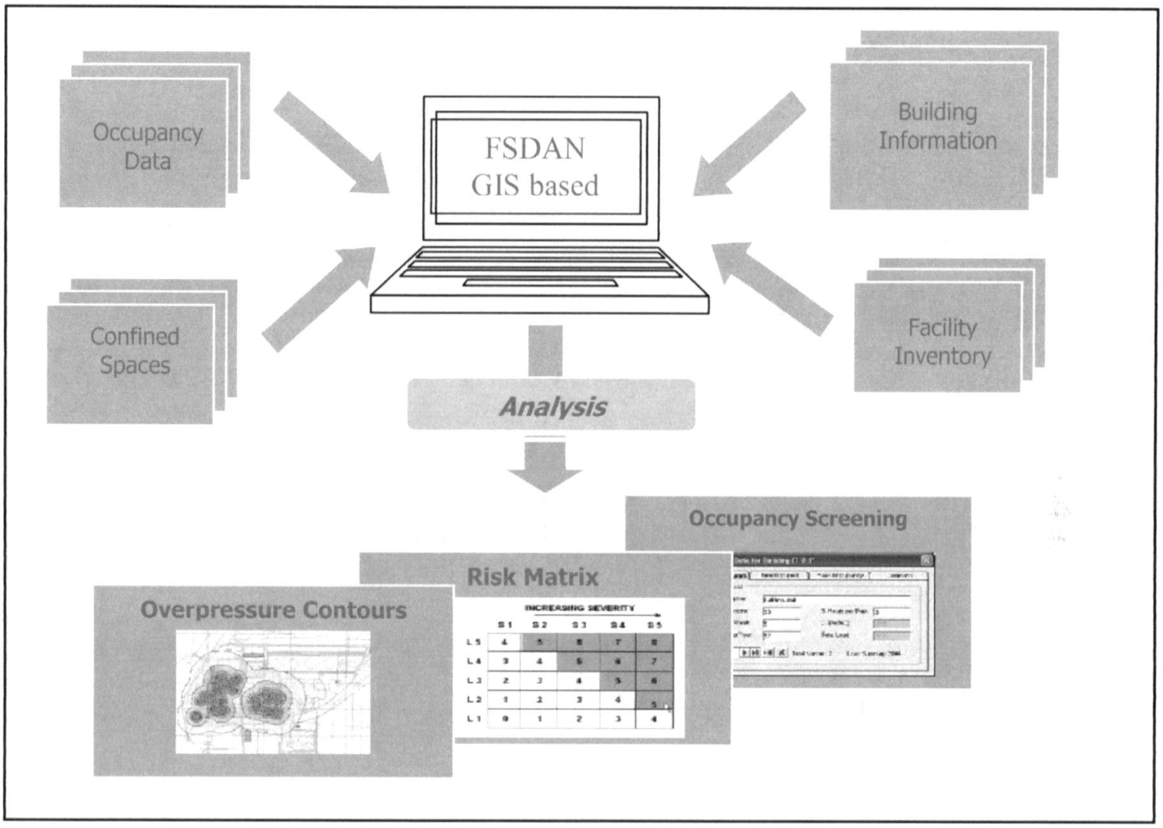

Figure 1. Facility Siting Database and Analysis (FSDAn) Layout

FSDAn prepares a consequence analysis to determine the potential impact on building occupants. The consequence analysis includes (1) computation of the potential overpressure and impulse (the air blast environment), the potential structural fire load, and the potential toxic concentrations interior to each building; and (2) an assessment of the damage to each building as a result of the blast environment, the fire impact to each facility building, or potential toxic exposure to inhabitants inside buildings.

The consequence analysis is based on site-specific scenarios and a facility walkdown. Potential release scenarios are defined, which include vessel rupture, leaks from pipes, flanges, pumps and other equipment. Ruptures and leaks could, if ignited, result in blast and/or fire impact. In addition, some chemicals are toxic. The potential release scenarios can be developed by reviewing the Process Safety Information, P&IDs, and PHAs for each process unit. The data required for analysis includes process unit layout drawings, process flow data, piping and vessel inventories and composition, and process conditions. The following discussion outlines the approach to conducting a refinery facility siting study.

Source terms are defined and dispersion modeling performed to determine the extent of explosions, fires and toxic impacts. For flammable materials, spaces that could potentially confine flammable vapors, in the event of an inadvertent release, are identified. An explosion of refinery material is unlikely to occur unless such materials are confined, become turbulent and are ignited. When a flammable material is released, the vapor will mix with the surrounding air and it may disperse without incident, or quickly ignite to produce a fire. If ignition is delayed, a flash fire or vapor cloud explosion may occur. If a released material is toxic, it could pose a threat to facility employees as it disperses downwind of the release location.

FSDAn uses TNO's confined vapor cloud explosion models [3], and [4] to estimate the overpressure and impulse acting on all sides and the roof of a building. The blast consequence models (in the form of Pressure-Impulse (P-I) diagrams) for estimating building damage and occupant casualties due to an explosion are included in the database for 20 generic building types. They were prepared using the FACEDAP (Facility and Component Explosive Damage Assessment program) [5] component blast models. The code also predicts the potential for window breakage.

Impacted buildings are compared with the consequence results to determine the potential building damage that could lead to occupant injury, or whether building occupants could be affected by exposure to a toxic material. For buildings of continuing concern, the tool allows a risk assessment to be conducted to determine whether the risks are acceptable.

2. Data Input

A plot plan of the facility is required to initiate the setup a database to automate a facility siting study. The plot plan can be in a GIS, CAD, or bitmap format. The purpose of the plan is to identify the location of buildings, vessels, pipelines and other equipment, that contain toxic or flammable materials, and potential blast sources.

A facility survey must be performed to collect all of the required building data and to enable hazard evaluations. The survey includes a thorough examination of the facility to verify building locations, determine building structural information, locate and measure potential blast material confining locations, identify major vessel locations, note ignition sources and mitigating conditions, locate potential fire and toxic release locations, and obtain additional process flow and plot plan information.

2.1 Building Locations and Occupancies

The first step in the data collection process is to identify buildings within the perimeter of the facility on the plot plan.

The next step in the process is the gathering of building occupancy data. The occupancy information, which needs to be collected are:

- Occupancy load - Number of personnel hours that a building is inhabited, based on an annual average
- Percent of time occupied - Percentage of an individual's total time spent in a building based on the criteria of a regular working time of 40 hours a week, or 2000 hrs per year
- Peak occupancy: Number of occupants potentially exposed for a given period

Figures 2 to 4, examples of FSDAn data input screens, show the parameters that should be collected for each occupied building.

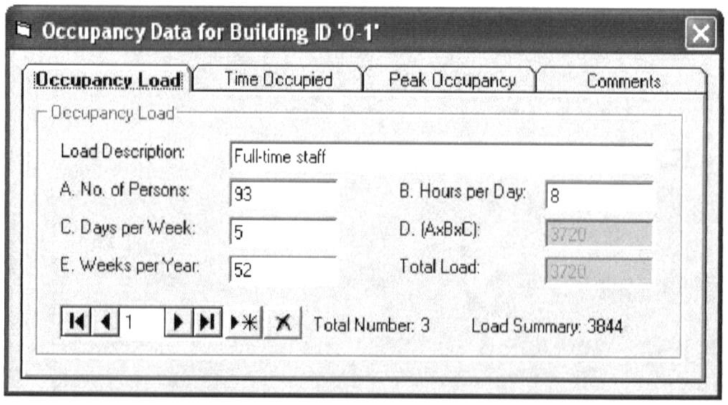

Figure 2. Occupancy Load

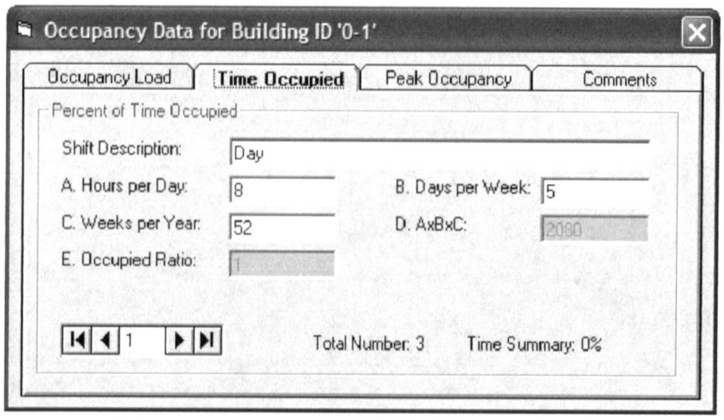

Figure 3. Percent of Time Occupied

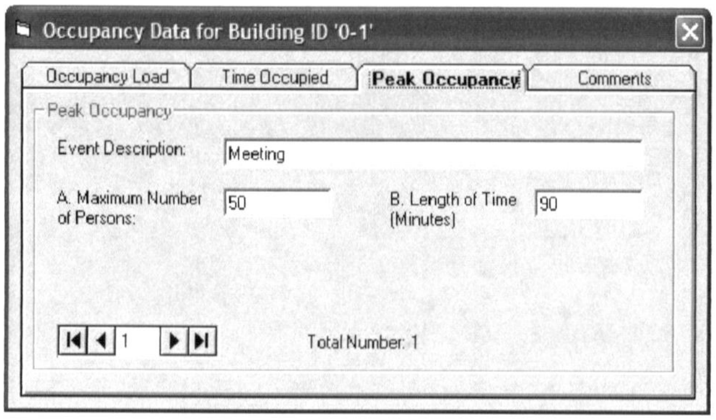

Figure 4. Peak Occupancy

In order to conduct a facility siting study according to API RP-752, occupancy criteria have to be developed. These criteria specify maximum building occupancy levels such that, if one or more of the criteria were exceeded, it will initiate additional study. (Note: API RP-752 states that each company may define its own risk and occupancy criteria. However, in the wake of the March 23, 2005 BP tragedy, the US Chemical Safety and Hazard Investigation Board made an urgent recommendation to the API to issue a new recommended practice to ensure the safe placement of occupied trailers and temporary structures away from the hazardous areas of process plants.)

2.2 Building information

For those buildings that fail the Occupancy Criteria, building construction data must be obtained, which includes structural definitions of walls, roof, frame and windows. In addition, construction drawings may be required for detailed structural evaluation. Figure 5 shows an example of building information as entered to the FSDAn database.

The amount of building data depends on the scope of the evaluation. For an initial analysis, only general information is needed. Additional information is appropriate when a more detailed structural evaluation needs to be performed. That is, when a conservative initial analysis indicates that structural damage is sufficient to yield casualties.

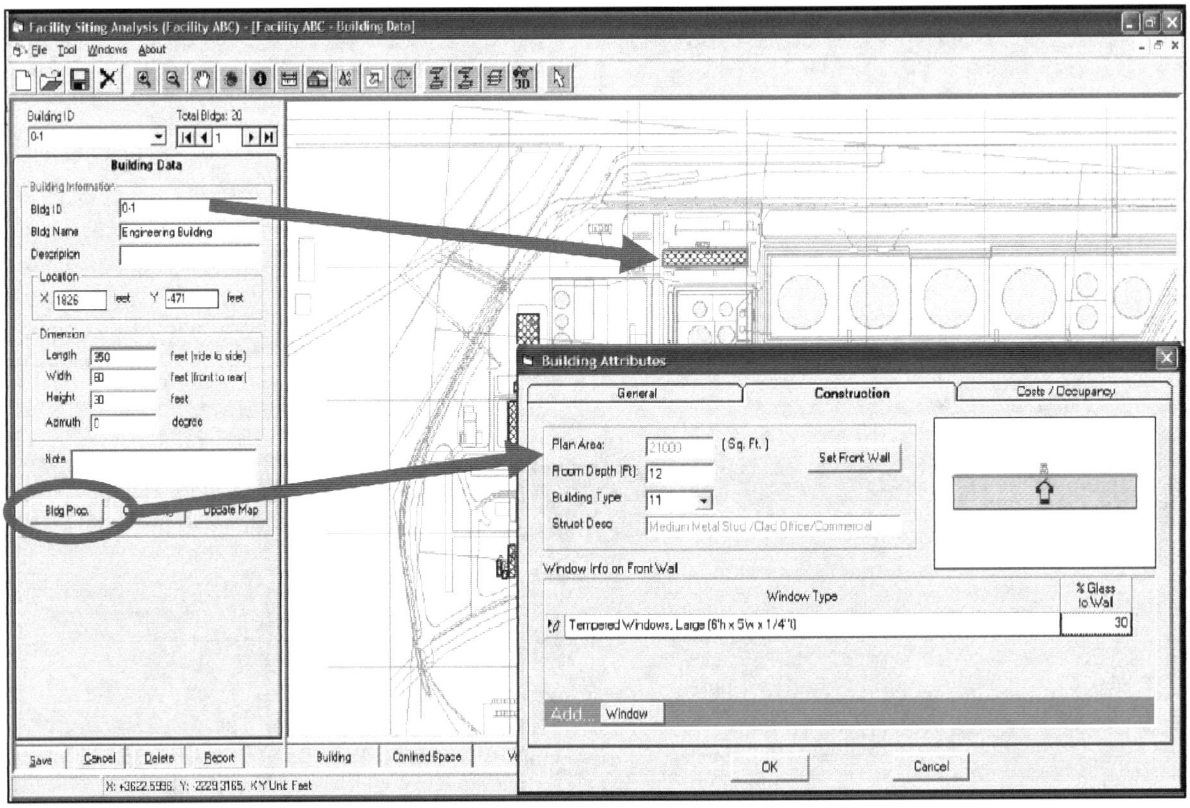

Figure 5. Building Information

2.2 *Confining Spaces Data*

It is highly unlikely that a high energy damaging unconfined vapor cloud explosion will occur in a refinery process due to the nature of the process materials. Refinery vapor cloud explosions are unlikely to occur unless a flammable material is released, then confined, becomes turbulent and is ignited. For this reason, a refinery facility siting study using the approach contained in FSDAn considers only confined vapor cloud explosions. The degree of congestion and the spacing between confining areas is defined during the facility walk down. Confining spaces that could support a blast are specified as follows:

1. A space confined on at least 2 "faces" of a box shape, but more typically on 3 or 4 faces. For a 2-surface confining volume, a small separation between two large surfaces should be considered as confining.
2. Typical confining surfaces are the ground, overhead decks, and pipe racks, side walls of buildings and the surfaces and structural supports of large vessels.
3. Congested pipe areas are also considered as potential confining surfaces. Piping need not occupy a single vertical or horizontal plane but is typically limited to a depth of less than 10 feet. Refinery units have repeated groupings of equipment located side by side with occasional roadway or access paths separating them.

4. Relatively open spaces of 20 or more feet are considered to define boundaries between approximately adjacent semi-confined spaces. For high volume spills or high emission rate leaks it is possible to fill several volumes to define a larger source.
5. Airflow in general tends to be more restricted at ground level. Semi-confined space status is assigned to congested areas on the upper decks of multiple deck units.

Figure 6 shows a typical 3-level confined space and its related parameters as defined for a typical refinery unit.

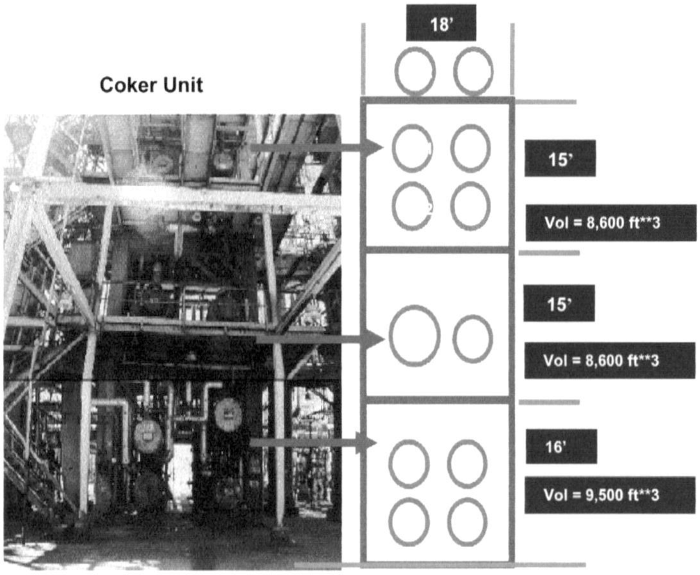

Figure 6. Example of Confined Space

If separate blast sources are located close to one another, a blast may be initiated almost simultaneously in each location. FSDAn has the ability to calculate a blast effects from separate blast sources that are within a specific distance, which is specified by the user.

2.4 Facility Inventory

For each vessel that contains flammable and/or toxic materials, vessel inventory and composition, physical properties and process temperature and pressure, is collected. All materials, including mixtures of materials, are listed. (A substance in the mixture of less than 1% by weight is generally ignored.) The location of vessels, pipes and tanks that contain toxic materials must be identified in order to determine the distances between these sources and buildings that house employees. This information is used in the calculation of building exterior toxic concentrations for each identified toxic material. In addition, during the walk-down, ignition sources are identified.

2.6 Meteorological and Wind Rose Data

Meteorological data is required by the release rate and dispersion models incorporated into FSDAn to calculate material release rates and downwind concentrations of each process material after release to the atmosphere. The relevant weather variables in the vicinity of a facility may be obtained either directly from a local weather service station or through the National Climatic Data Center of the National Oceanic and Atmospheric Administration (NOAA).

A wind rose is often used because it graphically represents changes in wind direction and speed. Weather records are processed and presented in the STAR format, which divides the measurement data according to the Pasquill stability classification system. Typical weather measurements such as wind direction, frequency and magnitude, are summarized on a monthly, seasonal, or annual basis. Figure 7 provides an example of a typical annualized wind rose plot. The movement of a released process material is affected by atmospheric heating and mechanical stirring, low-level inversions, wind speed and direction, ground roughness, and air temperature. Other important meteorological measurements include air density, air temperature and pressure, cloud cover and category, relative humidity and inversion height.

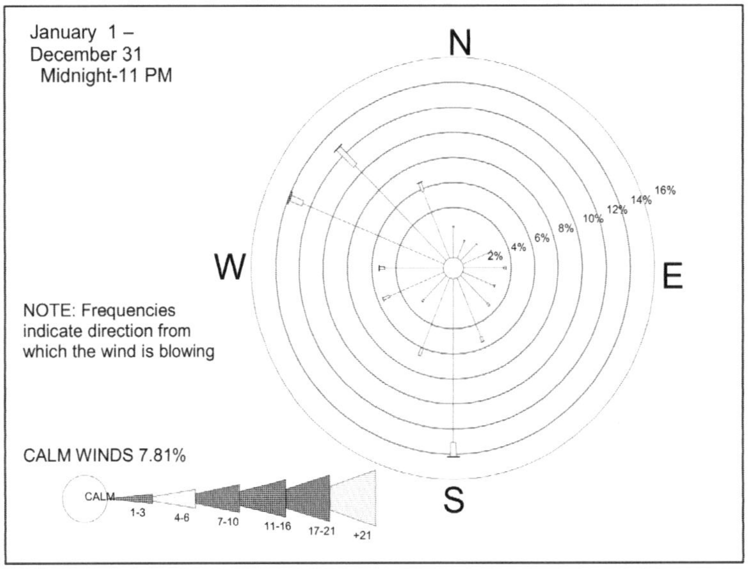

Figure 7. Wind Rose Plot

3. Analytical Procedures

The objective of consequence analysis is to estimate the magnitude of an explosion and fire and evaluate their effects on a building and relate the damage sustained by the building to the degree of potential injury to the occupants or damage to building equipment. For toxic releases, the toxic concentration outside the building is estimated using air dispersion models. Toxic vapor may enter a building in varying amounts of concentration; FSDAn employs a model that determines building interior toxic concentrations, depending on various conditions, including: air exchange rate, wind speed and release duration. The toxic hazard to building occupants, including impairment and/or physiological harm, is computed as a function of the material released, its concentration and exposure duration.

3.1 Blast Analysis Procedure

The analysis process for an explosive hazard first determines the extent of blast loading (in terms of overpressure and duration) for each potential event on each target structure. Explosive sources are identified for the buildings that do not satisfy the occupancy criteria. These occupied buildings are then further evaluated to determine the potential extent of blast damage and the vulnerability of building occupants.

The Multi-Energy (ME) Method [3], and [4], is the procedure incorporated into FSDAn to calculate the overpressure and impulse levels (blast hazard) at each building that does not satisfy a facility's occupancy criteria. (The ME method serves as an alternative to the conventional TNT equivalency approach, which relates the explosive power of a vapor cloud to the weight of some portion of the fuel-air mixture that is considered available to detonate.) In the ME method, the hazard associated with each explosive source is dependent on confined space dimensions and degree of congestion, distance between the source and the building, and the explosive material.

An explosive assessment begins with a screening analysis to eliminate those occupied buildings that are not impacted by an explosion. During the screening process, overpressure and impulse paired values at each building location are calculated for each explosive source assuming the confined spaces identified during the walkthrough can be filled with a mixture of light hydrocarbons in air that result in a stoichiometric mixture. Buildings that are exposed to only low overpressure levels (less than 1 psi incident), and do not have windows, are unlikely to experience significant structural damage or occupant injury and are eliminated from further evaluation.

For buildings that do not pass the initial screening process, the confined spaces that resulted in blast damage are re-examined to determine if they can indeed be filled with explosive materials from nearby vessels. Vessel contents and prevailing wind conditions at the facility are used at this stage of the blast assessment. This process potentially eliminates another set of structures.

The remaining buildings are then subjected to damage evaluations, and those with construction characteristics that provide sufficient strength to protect their occupants from injury and/or death are also eliminated. Finally a risk analysis is performed for only the buildings whose occupants could be potentially injured from an explosive event. If a building does not satisfy facility risk criteria, recommendations are made to bring them into compliance. Recommendations may include suggestions for procedural or operational changes (e.g., reducing occupancy, moving operations) and/or structural/window/door modifications (blast barriers, blast film, blast doors).
The ME method used in the blast assessment is based on gas explosion research performed during the last decade, which indicates that a flammable fuel-air mixture is explosive only under certain conditions; i.e., only when the mixture is partially confined and/or obstructed. In addition, research has found that a turbulence-generating environment is required for development of an explosion. At refineries, the turbulence generating mechanisms are caused by obstructions in the form of piping networks and/or process equipment within the confined vapor cloud. Such obstructions cause the flame front produced by an ignition source to accelerate; if sufficient turbulence is present a shock wave may result.

The energy-scaled distance is determined from the size of the vapor cloud, the combustion energy of the particular fuel-air mixture, and the distance from the center of ignition to the target specified by the user. In addition, the user must estimate the level of explosion that is likely to occur. Weak deflagrations are associated with explosive strength levels from 1 to 5, stronger deflagrations have levels ranging from 6 to 9 and high-energy detonations (TNT-like) are level of 10. The assumption of the explosion level can have a significant effect on the overpressure levels estimated relative to those predicted from TNT models, especially for strong deflagration events at close-in distances and at all distances for weaker deflagration type events. In general, conservative explosion level values are selected for design, but a "realistic value" given a particular vapor cloud incident is, at best, an engineering judgment.

FSDAn can compute a likelihood associated with the full range of explosive intensity levels (1 to 10) for the vapor cloud deflagration of a light hydrocarbon vapor cloud releases into a confined or semi-confined space, [6]. The frequencies, given the occurrence of an event, of each deflagration level are computed and the user can select the most likely occurrence for the given confining conditions. This information is used in subsequent risk computations performed by the code.

3.1.1 Structural damage analysis

The dynamic response of structures and windows subjected to an explosive blast is governed by: (1) the parameters of the air blast (overpressure and duration) and (2) the physical properties (strength and dynamic characteristics) and configuration of the impacted structure. In general, each structure subjected to an explosive blast should be analyzed to determine the appropriate blast environment and response.

In an explosion, the blast wave travels outward from the source with greater velocities in the high pressure regions. In the case of a vapor cloud explosion, the blast wave reaches a limiting velocity that is greater than the speed of sound in the unburned gas. The blast wave profile is such that the pressure rapidly rises to a peak and then gradually drops off. The peak pressure at the blast front falls as the blast wave travels outwards from the center of the explosion. At some distance from the explosion center, the region of positive pressure changes to one of negative pressure. This region of negative pressure is usually quite weak.

Peak Overpressure and Peak Reflected Overpressure:

The peak incident overpressure (or simply peak overpressure) is the peak pressure that exists in a blast wave prior to its impact with a structure and consists of a quasi-static portion (an increase in the pressure above atmospheric) and a dynamic portion (an increase in pressure due to winds caused by the shock wave. When the blast wave encounters a solid surface such as the wall of a building, it will reflect from it and diffract around and over it. The diffraction of the blast around the structure will engulf the target and cause a normal squashing force. There is also a dynamic component as the blast wave pushes the air mass and generates wind loading. The peak overpressure is often referred to as the peak side-on or incident overpressure. When a blast wave impacts a structure the properties of the shock wave are modified, such that the peak reflected overpressure may range from a factor of two to even ten higher than the peak incident overpressure. In addition the properties of the reflected impulse (the area under the overpressure time history) will also be changed depending on the overpressure level and the geometry of the structure it impacts. For lower values of overpressure, peak reflected overpressures are unlikely to be more than a factor of two higher than the peak side-on overpressures.

Angle of Blast Wave Incidence:

The angle at which a blast wave impacts a structure is also important. The degree of modification of the reflected blast wave is lessened as the wave impact angle changes from normal (90° to the surface) through an incidence angle parallel to the structure's surface (0°). For explosions involving refinery materials the peak overpressure typically needs to be modified using factors in the range of one to two, depending on the orientation of a structure in relation to the blast source.

Structural Blast Models

The extent of blast damage to a building depends on a number of parameters such as materials and method of construction, orientation of the building relative to the blast wave, configuration of the building (length to width ratio), and details of reinforcement and connections.

We use and have developed a number of tools to analyze the effects of blast on different types of structures. FACEDAP (Facility and Component Explosive Damage Assessment Program) [5], is one blast damage assessment program that can be used to approximate structural damage due to the effects of overpressure and impulse. The analyst builds a model of the structure from a series of building components (e.g., beams, columns, walls, girders, roof decking, etc.). The procedure then determines the overpressure and impulse loading on each exterior building component (including the effects of reflection) and computes damage based on each component's Pressure-Impulse (PI) damage diagram. An overall structural damage level for a building is determined by using a weighted-average summing procedure over the component damages. The procedure can consider member dependencies, such that if one member fails it may trigger failure of others. The component pressure-impulse (PI) damage diagrams were developed based on structural dynamics (assuming flexural or buckling failures) and calibrated to actual observed behavior (when available) based on test and reported World War II data. The methods are intended to provide a conservative prediction of damage to a structure.

ACTA has developed blast consequences models for estimating building damage as well as occupant casualties and fatalities due to an explosion for several generic building types using the FACEDAP component blast models described above. The combinations of wall, roof and frame characteristics are used to define generic structure types for blast consequences models. Figures 8 to 10 show the damage, casualty and fatality models developed for a small reinforced concrete structure, respectively. The blast models are in the form of Pressure-Impulse (P-I) diagrams. The P-I diagrams consist of hyperbolic contours of constant damage, serious injury and fatality in percent. These P-I diagrams are incorporated into FSDAn and used to compute structural damage and, if necessary, expected casualties within a facility's structures, as a result of various computed blast environments.

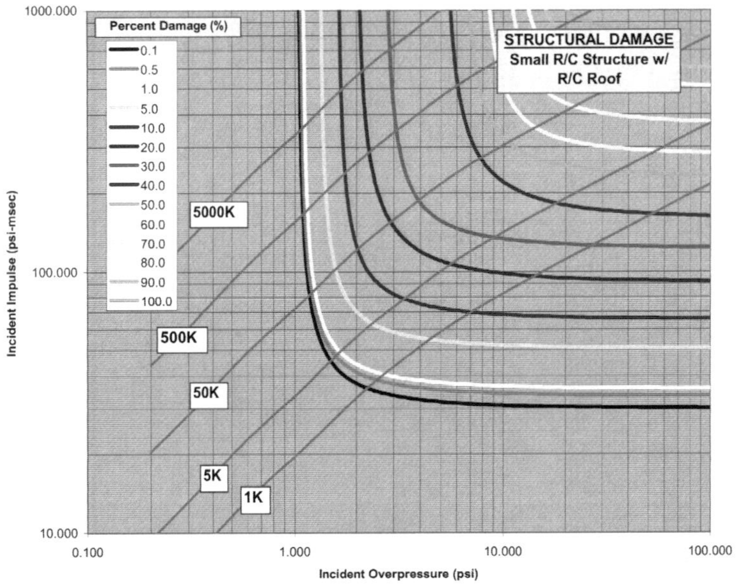

Figure 8. Structural Damage PI Diagram for a Small R/C Office Building (~2500 sq ft) (8" R/C Shearwalls with R/C Roof/Beams)

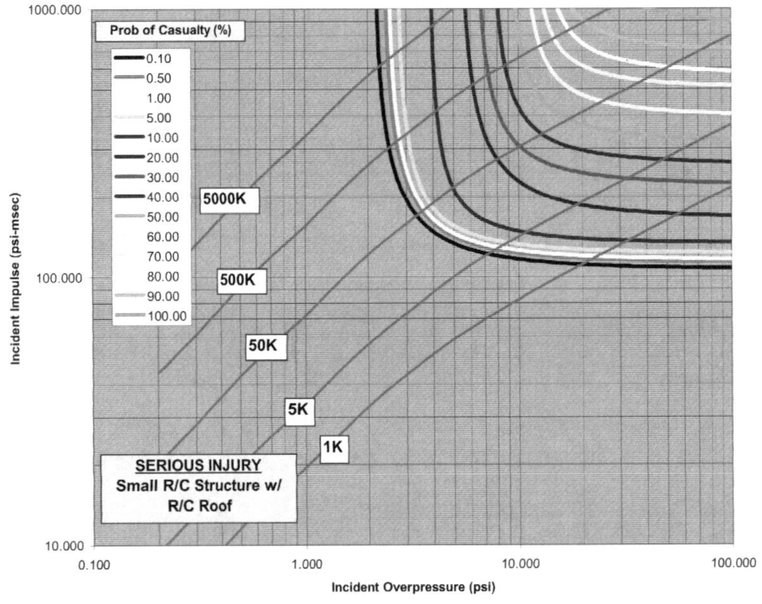

Figure 9. Serious Injury PI Diagram for a Small R/C Office Building (~2500 sq ft)

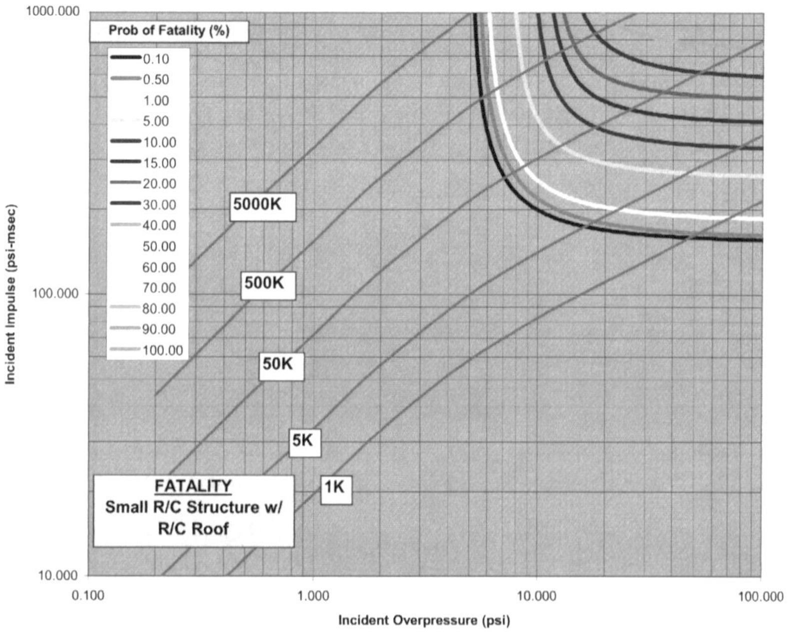

Figure 10. Fatality PI Diagram for a Small R/C Office Building (~2500 sq ft)

3.1.2 Windows damage analysis

ACTA has developed window breakage, serious injury and fatality models for twenty-one generic window types based on HazL breakage algorithms [7]. As for the generic structure types, P-I diagrams were developed to define (a) the probability of window damage (breakage), (b) the probability of serious injuries, and (c) the probability of fatalities as a function of window type and applied loading.

Four window types are included in the computational algorithm incorporated into FSDAn: annealed glass (AG), tempered glass (TTG), insulated glazing unit (IGU) with dual AG panes, and laminated TTG. Three generic window sizes: small, medium and large are defined, although the user can input the actual dimensions of each window and the thickness of glass.

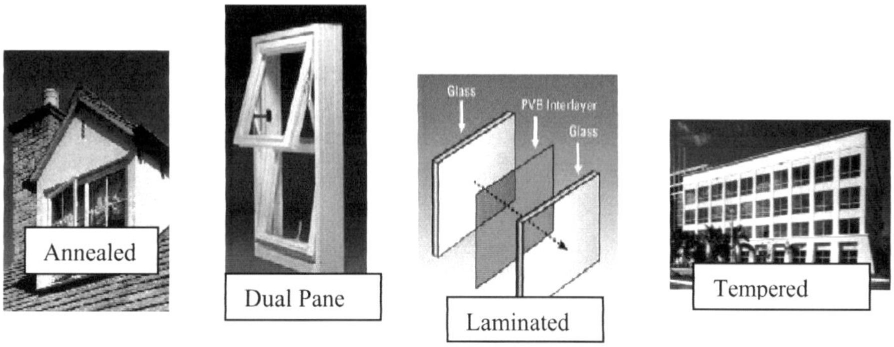

Figure 11. Generic Window Types Supported in FSDAn's Window Breakage Model

3.2 Toxic Analysis

When a toxic material is released to the atmosphere it may disperse harmlessly, or it may collect in a manner that presents a hazard to building occupants. Certain toxic materials can be fatal at levels of some concentration in air. Unlike explosions, however, they develop in such a way that on-site personnel do, in many instances, have sufficient warning to evacuate the area or seek shelter in an appropriate structure.

The dispersion of a toxic material in the atmosphere is affected by many site-specific factors, including the physical properties and release conditions of the material, the quantity released, weather conditions, obstacles, and the direction and source of the release. Toxic vapors may enter a building in varying amounts, depending on a number of factors, including: the degree of leak tightness of the building; the number and location of intake vents; and the difference in temperature and pressure between the inside and outside of the structure.

Hazards to building occupants include impairment and/or physiological harm, depending on the material released and its concentration and exposure duration. FSDAn computes the threat to a building's occupants through an initial determination of the exterior toxic gas concentration and subsequently the amount of toxic material that could potentially enter the building.

Screening analyses are employed to identify the buildings in the facility, and their construction features, in proximity to processes or vessels from which acutely toxic materials could be released. Only those buildings that exceed one or more of the Occupancy Criteria are considered for toxic hazard analysis.

3.2.1 Sheltering Model

The dispersion model is used to provide results of peak concentrations averaged over a period of exposure time at the exterior of the buildings of concern. In order to determine the potential impact to building occupants, the code determines the potential building interior concentrations.

Many buildings offer a good degree of protection to employees. ACTA has developed toxic gas sheltering models for several types of structures. After computing the potential concentrations of toxic gases at the exterior of a building, these models are used to predict building interior concentrations. A number of authors provide equations that define interior concentrations of a gas from external and internal sources, e.g., Wadden and Scheff, [8] and Kelnhofer, et. al. [9]. The equations describing outdoor and indoor concentrations for a tracer gas as a function of time, t, are:

$$RV = Q \tag{3.1}$$

$$dC_i / dt = R(C_o - C_i) \tag{3.2}$$

where,

- R = the air exchange rate (/hr)
- C_i = indoor air concentration (assumed well mixed) (ppm)

C_o	= exterior concentration (assumed constant),
V	= volume of the building (m^3), and
Q	= infiltration, or volumetric flow rate through the building (m^3/hr)

The abrupt arrival of an external toxic gas plume of sufficient duration is regarded as a sudden change from a concentration of zero to the average toxic concentration of the plume. The inside toxic concentration is

$$C_i = C_o [1 - e^{-RT}] \tag{3.3}$$

The inside exposure is the time integral of the inside air concentration

$$I_E = C_o [T - (1/R) + (e^{-RT})/R] \tag{3.4}$$

and the outdoor exposure is the time integral of the outdoor concentration

$$O_E = C_o T \tag{3.5}$$

The ratio of the inside and outside exposure is the Dose Reduction Factor (DRF),

$$DRF = 1 - (1/RT)(1 - e^{-RT}) \tag{3.6}$$

where T is measured from the time of the cloud arrival and does not exceed the length of time that the plume envelopes the building. As T increases, the DRF approaches unity, i.e., no interior dose reduction.

The Air Force Safety Office conducted experiments to determine air exchange rates for several types of buildings at an Air Force facility. Based on analysis of the Air Force's experimental results, it is reasonable to assume that an exchange rate (R) of 0.4/hr may be assigned to the office buildings and trailers of a refinery. For blockhouses and control rooms, an R value of 0.2/hr may be assumed.

Once the code computes the DRF, an estimate of the peak or time-history of toxic impact to building inhabitants is made. This information is used in conjunction with an exposure-response function to determine the probability of a specified health effect to building occupants.

3.2.2 Exposure Response Functions

ACTA works with toxicologists to develop exposure-response functions (ERFs), which provide the probability of a specific effect to an individual subjected to a specified concentration of a toxin [11]. An ERF is a function that provides an estimate of the probability of a specified health effect for different population groups for a specified concentration of toxic material. Figure 12 is an example of the form of an Exposure Response Function (ERF) as used in FSDAn.

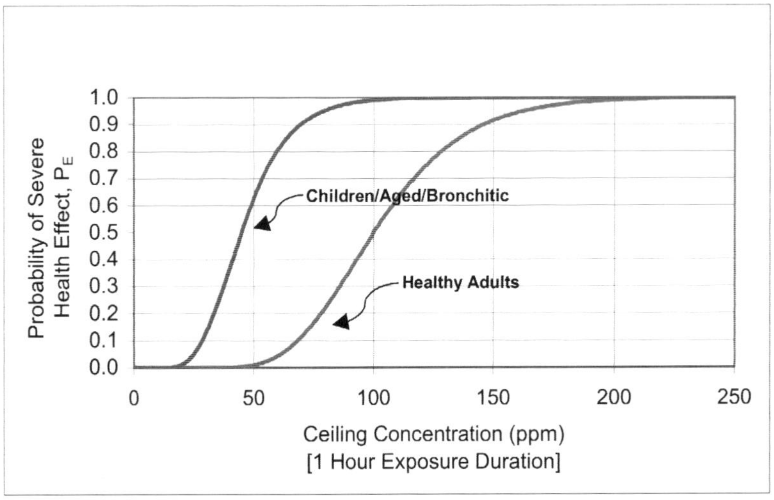

Figure 12. Example of Exposure-Response Function (ERF)

3.4 Fire Analysis

When a flammable material is released to the atmosphere, it may be ignited and burn, and/or it may cause an explosion. Facility fires can result in considerable damage to plant and equipment. Unlike explosions, however, they develop in such a way that on-site personnel do, in many instances, have sufficient warning to evacuate the area.

The evaluation of fire hazards to building occupants considers the occurrence and effects of potential fires at the facility. Fires may result from, or cause, explosions and can be classified as: pool fire, fireball, jet fire, BLEVE, or flash fire.

For a typical facility such as refinery, there are more than 1,000 vessels/tanks that store flammable materials, which, if released, may result in a fire. A screening analysis is performed to determine which vessels/tanks and pipes potentially hazard occupied buildings. The selection of fire sources and the fire type (jet fire, fireball/BLEVE, and pool fire) depend on the following conditions:

- storage operating conditions;
- physical state of the flammable materials in storage;
- type and quantity of the flammable material; and
- distance to buildings.

The models incorporated in FSDAn to estimate the impact of various types of fires are based on equations presented in the AIChE publication, *Guidelines for Evaluating the Characteristic of Vapor Cloud Explosions, Flash Fires, and BLEVEs* [10].

3.3.1 Fire Types

A flash fire results from the ignition of a released flammable cloud in which there is essentially no increase in combustion rate. The principle dangers of a flash fire are radiation and direct flame contact. The size of the flammable cloud determines the area of possible direct flame contact effects. The cloud size depends on dispersion and release conditions.

The cloud dimensions and concentrations obtained from the gas dispersion models are used to estimate the area affected by a fire. It is assumed that a flash fire spreads throughout the part of the vapor cloud in which the material concentration in air is between the material's flammability limits. FSDAn incorporates air dispersion models, e.g., SLAB [12], which is used to establish the UFL (upper flammable limit) and LFL (lower flammable limit) contours.

Pool fires usually result from the burning of a spilled flammable or combustible liquid in the diked area of a storage tank.

Jet fires occur at the source of release of a pressurized flammable material, such as a release from a ruptured pipe, or pipe flange opening.

A fireball burns quickly and usually results from a large release of flammable materials in a short period of time. A BLEVE (Boiling Liquid Expanding Vapor Explosion) involves an external heat input to the outer surface of a vessel containing a pressurized flammable liquid. The source of this heat input is often a leak from the impacted vessel, which runs down the outside of the vessel and catches fire. A BLEVE can potentially impact a building in three ways: (1) thermal radiation, (2) blast effects and (3) high energy debris (pieces of the ruptured vessel).

3.3.2 Fire Impact

The extent of fire damage to a building depends mostly on the thermal resistance properties of the structure's outer walls. Thermal fluxes due to a specific type of fire can be correlated with radial distance from the source center. Many relationships have been established to predict fire damage to various structures. Fire damage is time dependent; sustained building damage increases with the level of heat flux and the duration of exposure.

The API RP-752 does not indicate the need in a facility siting study to determine the effects of fires on structures. However, this is a necessary input to such an analysis. Methods described in an AIChE document, [10] and in the TNO document, [13], are used within the code to determine fire effects to structures and building inhabitants.

4. Risk Analysis

To determine whether risk reduction is required, it is necessary to have a standard to compute whether computed risks are acceptable. A facility's risk matrix, or other means of communicating risk such as a risk profile or casualty expectation, may be used to determine whether risk mitigation measures are required. Once established, such risk matrix or other approach is incorporated into FSDAn. The risk results computed by the code (e.g., frequency versus severity for various events) are then compared to the specified standard to determine which events require mitigation measures to be adopted. Figure 13 shows an example of a risk matrix.

Risk reduction measures must be considered if the assessment shows that the risk to an occupied building that fails the Occupancy Criteria are not acceptable according to the facility's risk acceptance standard. In general, risk reduction measures may be specified to reduce event probability or event consequence.

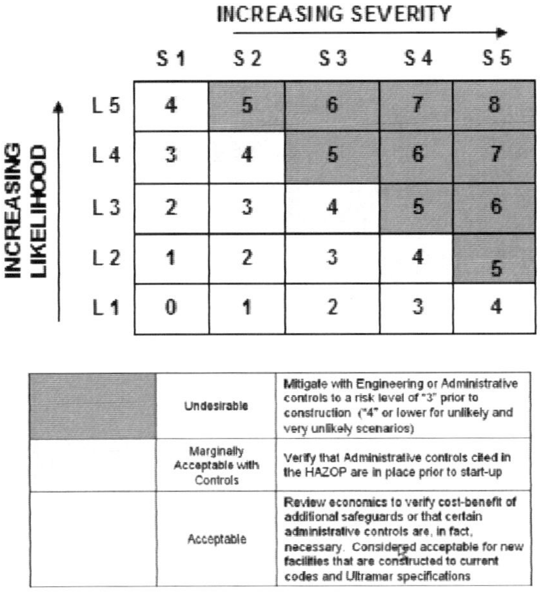

Figure 13. Example of a Risk Matrix

4.1 Blast Event Probability

The likelihood component of risk for each blast event is computed using the following equation:

$$P_{BD} = P_{(IE)} \cdot P_{(VCF|IE)} \cdot P_{(Expl|Ign)} \cdot P_{(Ign|VCF)} \tag{4.1}$$

where,

P_{BD} = Probability of building damage due to an explosion

$P_{(IE)}$ = Initiating event probability

$P_{(VCF \mid IE)}$	= Probability that a vapor cloud forms in a confined volume (a function of the annual wind directional frequency), given the initiating event
$P_{(Ign \mid VCF)}$	= Probability of an ignition given an ignition source is present and a vapor cloud forms
$P_{(Expl \mid Ign)}$	= Probability of an explosion (and not a fire), given an ignition

4.2 Toxic Event Probability

The likelihood component of risk due each toxic release is computed using the following equation:

$$P_T = P_{(IE)} \cdot P_{(W/S)} \tag{4.2}$$

Where,

P_T	= Probability of Toxic Exposure (exterior to a building)
$P_{(IE)}$	= Probability of initiating event
$P_{(W/S)}$	= Annual wind probability in a 22.5° sector in the direction of the building from the toxic source for a given atmospheric stability class (from wind rose data)

4.3 Fire Event Probability

Fireball/BLEVE:

The building damage probability estimate due to a fireball event is computed using the following equation:

$$P_{FD} = P_{(IE)} \cdot P_{(Ign|Qty)} \cdot P_{(E|Ign)} \tag{4.3}$$

where,

P_{FD}	= Building damage probability due to a fireball
$P_{(IE)}$	= Initiating event probability
$P_{(Ign|Qty)}$	= Probability of an ignition given an ignition source is present and for a given quantity of material
$P_{(E|Ign)}$	= Probability of a fireball/BLEVE, given an ignition

<u>Jet Fires</u>

The estimate of the likelihood of a jet fire event is computed using the following equation:

$$P_{JFD} = P_{(IE)} \cdot P_{(Ign|Qty)} \cdot P_{(E|Ign)} \tag{4.4}$$

where,

P_{JFD}	= Building damage probability due to a jet fire

P_{IE} = Initiating event probability

$P_{(Ign|Qty)}$ = Probability of an ignition given an ignition source is present and for a given quantity of material

$P_{(E|Ign)}$ = Probability of a jet fire, given an ignition

Pool Fire:

The estimate of the likelihood of a pool fire is computed using the following equation:

$$P_{PFD} = P_{(IE)} \cdot P_{(Ign|Qty)} \cdot P_{(E|Ign)} \tag{4.5}$$

where,

P_{PFD} = Building damage probability due to a jet fire

P_{IE} = Initiating event probability

$P_{(Ign|Qty)}$ = Probability of an ignition given an ignition source is present and for a given quantity of material

$P_{(E|Ign)}$ = Probability of a jet fire, given an ignition

Flash fire:

Building damage probability estimates due to a flash fire event is computed using the following equation:

$$P_{PFD} = P_{(IE)} \cdot P_{(W/S)} \cdot P_{(Ign|Qty)} \cdot P_{(E|Ign)} \tag{4.6}$$

where,

P_{FD} = Building damage probability due to a flash fire

$P_{(IE)}$ = Initiating event probability

$P_{(W/S)}$ = Annual wind probability in a 45° sector in the direction of the building from the flammable source for atmospheric stability class F (from wind rose data)

$P_{(Ign|Qty)}$ = Probability of an ignition given an ignition source is present and for a given quantity of material

$P_{(E|Ign)}$ = Probability of a flash fire, given an ignition

5. Presentation of Results

FSDAn outputs overpressure contour plots, structural damage and window breakage from explosions. It provides radiant heat flux impact to each structure and toxic concentrations at each building. Risk results are provided in several formats, including casualty expectations and, if available, on a facility's risk matrix. Figure 14 is an example of the type of results provided by FSDAn. It shows overpressures contours plotted and overlaid on a facility plot plan and aerial imagery map. Figure 15 provides an example of a casualty risk profiles for the occupants of a building.

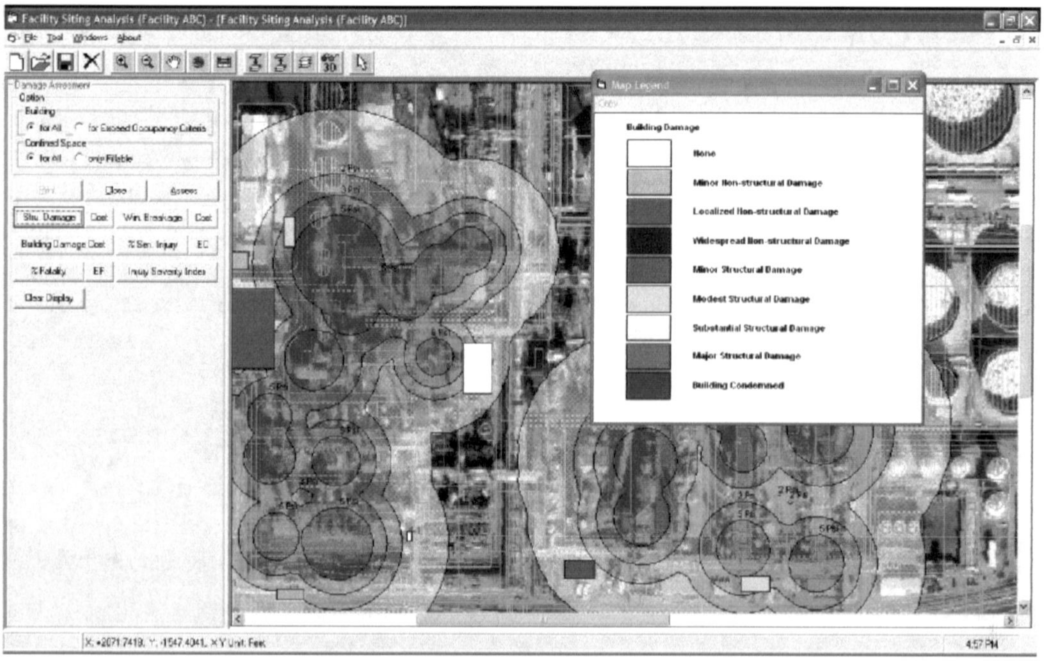

Figure 14. FSDAn's Sample Output

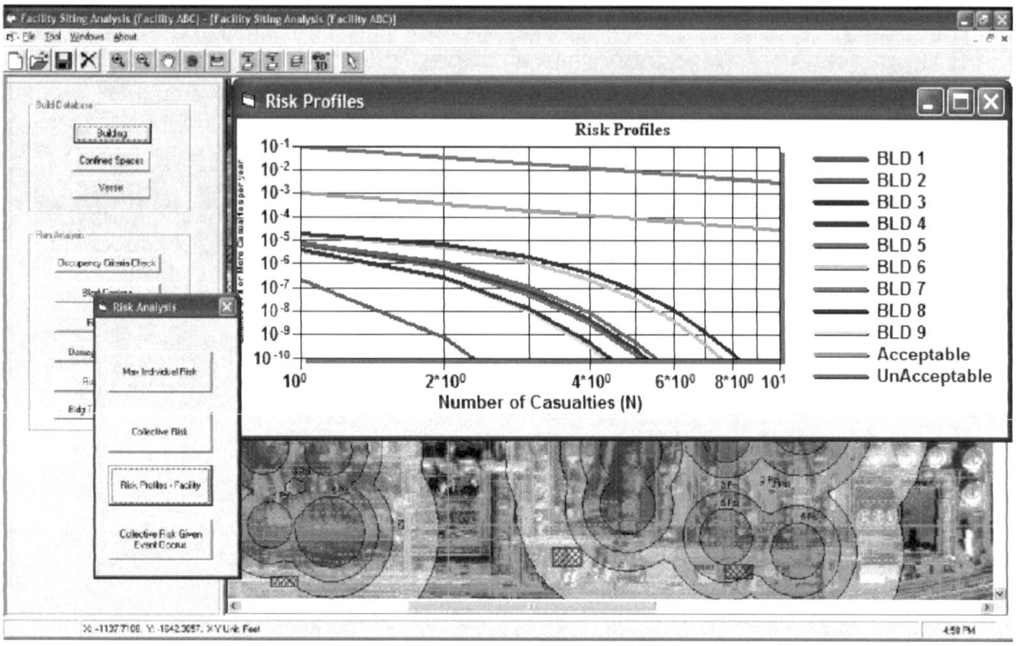

Figure 15. Risk Profiles Presentation

6. Summary/Concluding Remarks

A facility must consider all potential hazard sources in the placement of structures, permanent and temporary. Unless all potential sources are considered, it is not possible to determine whether the selected location is suitable with respect to hazardous impact. The use of a database approach to manage building data, potential blast sources and hazardous material inventories allows a facility siting study be conducted with complete control of data integrity and to provide results that are consistent and in the best interests of the company on-site personnel. In addition, the database approach allows a facility siting analysis to be easily updated and the impact of changes determined within a short period of time.

In light of the March 23rd 2005 incident at the BP Texas City Refinery, the U.S. Chemical Safety and Hazard Investigation Board recommend that API RP752 be revised to ensure the safe placement of occupied trailers and similar temporary structures away from hazardous areas of process plants (CSB news release on October 25, 2005). The Board also issued the following new recommended practice:

1. Protect occupants from accident hazards such as heat, blast overpressure, and projectiles;
2. Establishes minimum safe distances for trailers and similar temporary structures away from hazardous areas of process plants;
3. Evaluate the siting of trailers under a separate methodology from permanent structures, since trailers are more susceptible to damage, are more readily relocated, and likely do not need to be placed near hazardous areas.

The FSDAn software described in this paper is being used to enable facilities to meet the requirements and more of the API. In addition it is a highly suitable tool to meet the suggested requirements of the CSB.

6. References

[1] Process Safety Management of Highly Hazardous Chemicals; Explosives and Blasting Agents; Final Rule," U.S. Federal Register, Pat II Department of Labor, Occupational Safety and health administration, 29 CFR Part 1910, (February 24, 1992).

[2] Management of Hazards Associated with Location of Process Plant Buildings," API Recommendation Practice 752, American Petroleum Institute, Washington D.C., 1995.

[3] Van den Berg, A.C. and A. Lannoy, Methods for Vapour Cloud Explosion Blast Modeling, Journal of Hazardous Materials, 34 (1993) 151-171.

[4] A.C. van den Berg, et. al., "Vapor Cloud Explosion Blast Modeling", Proceedings of the International Conference and Workshop on Modeling and Mitigating the Consequences of Accidental Releases of Hazardous Materials, May 1991

[5] Oswald, C.J., and Skerhut, D., "FACEDAP User's Manual," Contract No. DACA 45-91-D-0019, US Army Corps of Engineers, Omaha District, April 1993.

[6] Van den Berg, A.C. and Eggen, J.B.M.M., "GAME - Guidance for the Application of the Multi-Energy Method", The 2nd International Specialist Meeting on Fuel-Air Explosions, Bergen, Norway, June 1996.

[7] Hazl (Window Fragment Hazard Level Analysis) Version 1, U.S. Army Corps of Engineer Research and Development Center, June 2000.

[8] Wadden, R.A., Scheff, P.A. Indoor Air Pollution, Characterization, Prediction and Control. 213 pp., John Wiley & Sons, New York, 1983.

[9] Kelnhofer, W. J., Hunt, C. M., Didion, D. A., Determination of combined air infiltration and ventilation rates in a nine-story office building. Proceedings of the Conference on Improving Efficiency and Performance of HVAC Equipment and Systems for Commercial and Industrial Buildings. Purdue University, pps 322-328, April 12-14, 1976.

[10] Guidelines for Evaluating the Characteristics of Vapor Cloud Explosions, Flash Fires, and BLEVEs, Center for Chemical Process Safety of American Institute of Chemical Engineers, New York, 1994.

[11] "An Expert Elicitation of Estimates of Exposure Limits for Space and Missile Launch Toxicants," Report No. 99-400/11.2, Contract No, FO4684-97-C-0001, Task 11, ACTA Inc, Torrance, September 1999.

[12] Ermak, D.L., "User's Manual for the SLAB Model, An Atmospheric Dispersion Model for Denser-Than-Air Releases," Lawrence Livermore Laboratory, 1989.

[13] TNO, Methods for the Determination of Possible damage to People and Objects Resulting from Releases of Hazardous materials ("The Green Book"), The Netherlands, 1992.

Additional references not cited:

[14] Guidelines for Chemical Process Quantitative Risk Analysis, Center for Chemical Process Safety of American Institute of Chemical Engineers, New York, 1989.

[15] Guidelines for Use of Vapor Cloud Dispersion Models, Second Edition, Center for Chemical Process Safety of American Institute of Chemical Engineers, New York, 1996.

[16] Chrostowski, J.D., et. al., " Blast Damage, Serious Injury and Fatality Models for Structures (Rev. 1)", ACTA Technical Report No.02-487/32.1, September 2002.

[17] "Risk Management Program for Chemical Accidental Release Prevention", U.S. Environmental Protection Agency, 40 CFR Part 68, October 20, 1993.

[18] Spicer, T. And Havens, J.A., "User's Guide for the DEGADIS 2.1 Dense Gas Dispersion Model," EPA-450/4-89-019, United States Environmental Protection Agency, Research Traingle Park, NC, November, 1989.

Risk Abatement Provided by Safety Instrumented Systems May Cause Remote Hazards with Higher Risks

Adrian L. Sepeda, PE
AIChE—Center for Chemical Process Safety

Prepared for Presentation at the
8th Process Plant Safety Symposium
AIChE 2006 Spring National Meeting
Orlando, Florida
April 23-26, 2006

"No good deed goes unpunished."
- Clare Booth Luce, *in H. Faber, The Book of Laws, 1980*

Abstract
Section: Safety Instrumented Systems and Safety Critical Devices

A basic law of physics states that for every action there is an equal and opposite reaction. Translating that law into the world of risk analysis, we might state that for every action (even a risk mitigation action) there are consequences—and not all of those consequences are 1) clearly identified, 2) fully understood, and/or 3) good. Even when the initial actions are an effort to mitigate a perceived risk, the range of consequences (i.e. reactions) can sometimes create another set of hazards and possibly larger risks than initially addressed.

The complexity of safety instrumented systems facilitates the creation of at least four types of new hazards that should be addressed. First, some consequences that may result from the proper functioning of a single safety system may not be desirable. Second, the accumulative and interactive effects when multiple safeguards function properly may cause unexpected and unwanted consequences. Third, unanticipated and sometimes unwanted consequences at seemingly remote locations may occur. And fourth, because of interactions with other control components, safety instrumented systems can easily create several layers of risk abatement actions cascading from the initiating event making it virtually impossible for operations personnel to troubleshoot and correct the original anomaly in a timely manner—adding yet another hazard to the operations.

This paper addresses the potential of creating "downstream" or remote hazards by the installation of safety instrumented systems and then discusses how and when consequence analyses should be expanded to identify these potential new hazards. This is done using the life cycle approach for the design and maintenance of safety instrumented systems discussed in the newly published CCPS book, "Guidelines for Safe and Reliable Instrumented Protective Systems."

....

....

Risk Abatement Provided by Safety Instrumented Systems May Cause Remote Hazards with Higher Risks

Adrian L. Sepeda, PE

INTRODUCTION

Over the years safety systems have evolved, solving some problems but creating others. Initially, local/field indications were read and acted on by "outside" operators. Later, with the help of technological improvements, these field readings were compiled by instruments and transmitted to a central control system to be displayed and then analyzed and acted on by a "control room" operator. In today's world we now have the capability for these field readings to be compiled and transmitted to a central control system for analysis. They are then acted on by the central control system itself and sometimes simply monitored by the "control room" operator.

With this, we moved forward in safe operations of our facilities primarily due to the speed and programmed analytical expertise of the controlling system, but we gave up some "hands-on" control. And, along the way, we began to lose tools, human ability and expertise to analyze and respond to various anomalies as they occur in the process partly because of lack of use and partly because of the complexity of the instrumented systems. A single safety system can cascade into multiple other safety systems which can cascade and influence even other safety systems causing an impact on the manufacturing operations itself. The end manifestation of a process upset may be physically remote from the initiating event due to these cascading safety effects. The "Cause-Consequence" relationship normally used to troubleshoot problems is easily hidden by transferring "the problem" from one safety system to another.[(3)(4)] Computerized systems can keep track of all that is done and why, but it may be beyond human capabilities—at least within the time allotted for decision making. Operational response to system upsets that demand human interventions may be delayed, misdirected, or non existent because the initiating cause was so well-hidden by layers of safety instrumented systems (SISs).

In some (usually most) cases, SISs handle the "problem" without significant human intervention.[(5)(6)] However, this success brings its own new hazard. Near-misses may go unnoticed leaving the initiating hazard unrecognized and unaddressed unless every safety instrumented system (SIS) activation is analyzed for cause.

CURRENT CONSIDERATIONS

To offset this reduction in human control, we have implemented strict design parameters and management of change systems. We systematically analyze each SIS for potential failure modes and build in additional safeguards if we are uncomfortable with the level of security the original system offers. We use safety standards, such as IEC 61511 and ISA S84, to define the level of safeguards needed and to calculate the efficiency and availability of each.[5][6] We design each system so that we are comfortable with the percentage of time the safety device will function as desired when needed. SISs designed and maintained throughout their life cycle as recommended in the CCPS book, "Guidelines for Safe and Reliable Instrumented Protective Systems,"[6] provide the basis for proper failure analysis.

After designing around potential failures in the system, we sometimes take a quick look at the initial and immediate needs when the SIS functions properly. When doing this, we usually look to assure ourselves that the immediate downstream receptor (the system into which the "problem" is transferred) is there and available. Often our downstream analysis goes no further and rarely does it include the accumulative effects of activating multiple safeguards.

THE PROBLEM

Although our efforts to assure successful functioning of SISs are usually more than adequate, we seldom analyze all of the consequences of that successful operation. In other words, we fail to fully and accurately determine all downstream consequences when the safety systems function as we desire.[1][2]

To make this analysis we must look at new specific sets of hazards that consider:

- all (not just the desired) consequences of a single safeguard functioning properly;
- the accumulative effects and interaction consequences when multiple safeguards function properly sequentially or at approximately the same time;
- the transformation of hazards (into a different state or mode) by the actions of a safeguard;
- the transfer of hazards downstream or to remote locations by properly functioning safeguards; and
- the potential for hiding "near-misses" when safeguards function as desired.

Consider all Consequences of Successful SIS operation...

Each time a single safeguard functions properly multiple consequences occur. We quickly take credit for the one we want to happen and tend to overlook those that do not solve the operational problem.[2] In most cases the safety system produces the desired consequence or action. And, also in most cases, the other unanticipated consequences are of little or no concern. However, occasionally they may have a significant impact on operations. Following the adage that for every action there is an equal and opposite reaction, an analysis of all possible consequences of safety system activation is needed to assure that those "opposite reactions" are properly addressed and controlled.[1][7] Since by their very nature these consequences are unanticipated, a semi-structured hazard analysis technique such as What-If? may be the appropriate tool for discovery.[3]

Consider Accumulative Effects and Interaction Consequence...

Most systems are designed to address a single malfunction. However, safety systems have an influence on other safety systems, sometimes causing hazards because of the interaction and accumulative effects. When more than one malfunction occurs, either caused independently of the other, caused by the other malfunction, or caused by the other's safety system, the result of interaction between the responding safety systems may not have been included in the hazard analysis. This new hazard presents itself when multiple safeguards function properly, either sequentially or at approximately the same time. To address these integrated accumulative issues, the design analysis, operating procedures, all subsequent management of change analyses, and any accident investigations should consider scenarios that include the activation of more than one SIS and the resultant (accumulative) consequences.[8]

Consider Transferring the "Problem" Downstream Consequences...

These interactions and layering of SISs tend to transfer the "problem" from one area to the next, eventually hiding the original initiating event. This requires operations personnel to back-track through the SISs and their successful actions to find the original or initiating "problem." In the end, it delays or misdirects correction efforts of the humans responding to the anomaly. It also moves the original "problem," or however it manifested itself, downstream into another part of the manufacturing process. That part of the process may or may not be able to successfully and safely handle the "problem" or how the "problem" presents itself in the new environment. In some cases, the "problem" is not resolved but simply transformed into another issue requiring another safety system to react.[9]

Consider the Issue of Hiding Near-misses...

When SISs function properly, they can also hide the initiating events and underlying hazards. If the SIS successfully handles the "problem" and there are no other adverse consequences, we call this a near-miss and unless there is a diligent management system to recognize and investigate these, they will be lost or hidden from view. Lack of knowledge about near-misses leaves a major hole in risk management efforts. Basic process control systems (BPCSs) normally operate and control the plant manufacturing operations. However, activation of a SIS is beyond basic process control and should be treated as such — requiring an analysis to determine cause of activation. Only by determining a SIS's cause of activation can the underlying hazard be exposed and then addressed.

A SIMPLE EXAMPLE—A One-Function Safety Instrumented System

A simple one-function SIS is easy to analyze and is a good example of how successful activation can cause new hazards—or at least raise questions that should be answered. Consider the example shown in Figure 1. A jacketed reactor's temperature is controlled by cooling water flow through the jacket. A temperature controller is part of the basic process control system (BPCS) and senses high temperature. It increases cooling water flow—eventually to its maximum rate in an effort to control the temperature. However, the temperature continues to rise and a SIS is automatically activated. It switches the reactor from cooling water to chilled water for emergency cooling. (Let us assume that this SIS functioned as designed.) The temperature of the run away reaction in the reactor is now being reduced. Everything should be okay. However, this action does not go unnoticed by the rest of the process.

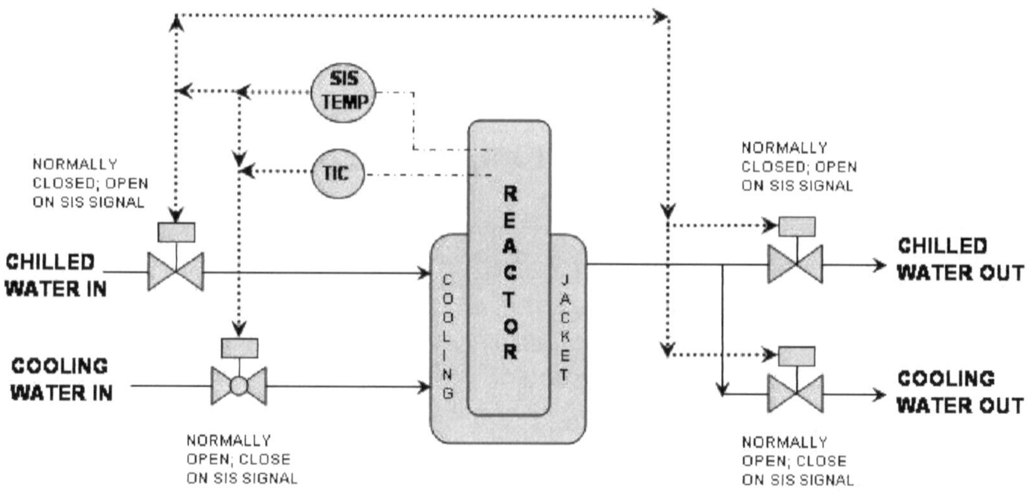

FIGURE 1

The quick change from cooling water to chilled water might upset both water systems causing swings in the flow rates. Those swings might cause the temperatures in other reactors to start to swing. If the temperature gets too low in any of the reactors, the reactions die and the product instantly becomes scrap. Responding to these swings then demands time and attention from the central control system and the control room operator. Because of the sudden increase in chill water demand, other systems that were on chilled water suddenly experience a drop in pressure and/or flow, causing their flow control valves to open quickly. These systems might begin to experience some swings as well, causing more demands on the central control system and operator. If the water flow swings become large enough, other reactors might require emergency cooling water flow compounding the problem. Other not adjacent systems might be impacted by the emergency in the reactor area and begin to experience unusual conditions. Eventually, the problem of one runaway reaction was handed off to several other systems and hidden as the initiating cause behind multiple upsets in the plant. If you were an operator in another area of the facility, would you be able to troubleshoot why your unit suddenly began to experience problems? If you were the control room operator, would you be able to piece all of these safety system actions together and realize that the initiating cause was the switch from cooling water to chilled water on one reactor—and then be able to control all downstream consequences?

To stop this scenario from occurring, some questions should have been addressed during the design phase of the plant. These questions include: Do we know that quickly maximizing cooling water flow to this reactor does not negatively impact other reactors in the chain? Are the cooling water pumps capable of providing the new flow rate without tripping off due to high amps—an event that might cause the entire facility to require an emergency shutdown? Will supplying extra cooling water to this reactor negatively impact other cooling water users in the plant? Will adding chilled water to this reactor on an emergency basis negatively impact any other operations at the site? Have all instrumented systems been "tuned" properly to account for sudden activation by any SIS? Also, how will the operator know if the consequences of increased cooling water flow or providing emergency chilled water to this reactor caused another plant problem in a different unit?

A COMPLEX EXAMPLE WITH MORE SEVERE CONSEQUENCES

Consider as an example a facility that manufactures chlorine, caustic, and ethylene dichloride (EDC). This facility has two gas-fired boilers that produce steam for the entire complex. Ethylene and chlorine are reacted in the presence of a catalyst in one large reactor. The liquid level in the EDC reactor is controlled at 75% by varying the proportional flow of chlorine and ethylene. Natural gas is used to "sweep" the vapor space of the EDC reactor ensuring a fuel rich (out of the explosive range) environment. This "sweep" gas, laden with unreacted ethylene and chlorine, is scrubbed and then sent to the boilers for use as fuel. Excess "sweep" gas is directed to the onsite flare system via a pressure sensitive diverter/proportioning valve. See Figure 2 for a simplified flow diagram.

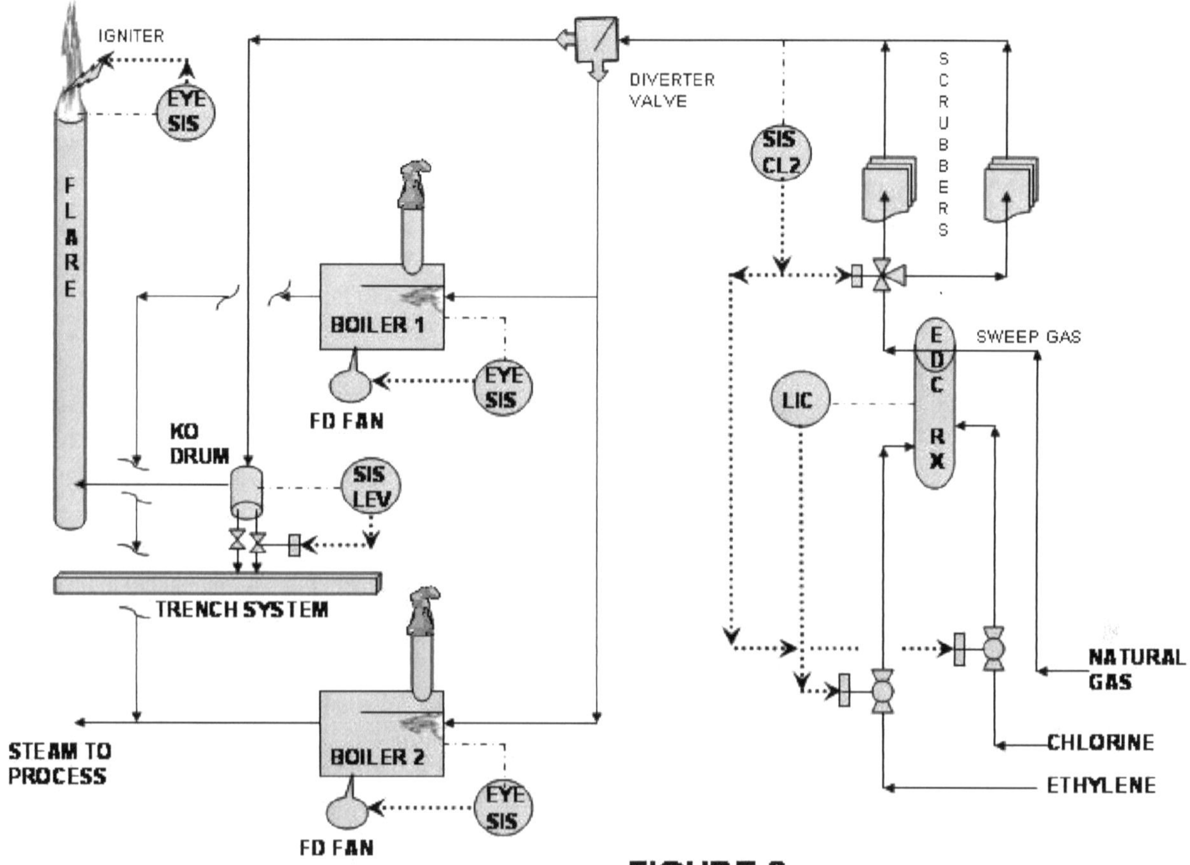

FIGURE 2

Suddenly, the level controller of the EDC reactor fails "low" indicating a false low level. The BPCS sensing this low level tries to maintain the proper level by increasing the ethylene and chlorine flows substantially. Because the level is not really low in the reactor, liquid EDC, ethylene, and chlorine begin to overflow into and through the scrubber. This liquid travels through the piping system, overwhelms the boiler fuel liquid knock out drums, and feeds directly into the gas-fired boilers, extinguishing the flames. The heat in the fire boxes vaporizes the liquid quickly and causes the fire boxes of both boilers to almost instantly become fuel rich to the point that they are out of the explosive range. Both boilers cease producing steam and all manufacturing operations dependent on steam begin an emergency shutdown activated by their BPCSs and SISs.

Excess liquid flow from the EDC reactor is diverted to the flare system and quickly overwhelms its liquid knock out drum. The liquid knock out drum has a level sensor that sends a signal to the control room when it reaches 50%. An operator is usually sent to manually drain the drum. There is also a SIS that dumps the liquid from the drum if the level reaches 90%. This feature was installed to stop any liquid from burping into the flare. Unfortunately, this emergency dump system is inadequate to keep up with this large of a flow. A portion of the volatile liquid is dumped into the trench system, but most of the liquid travels up the flare and extinguishes the flame. Liquid rains down on the surrounding area.

Several SISs have already begun to do their jobs successfully. The scrubber for gas flow out of the EDC reactor has sensed a chlorine breakthrough at its outlet and has switched flow to the online spare scrubber. When the second scrubber experiences a chlorine breakthrough, the reactor's SIS starts an emergency shut down.

Safety systems are activated in the boiler and flare areas as well. The fire eye sensors in the gas-fired boilers detect no flame so they start a programmed relight procedure that includes increasing the forced draft fans to maximum flow to clear the fire boxes of combustion products and excess fuel. Only when the fire boxes are clear will the SIS grant permission to relight the boilers.

The flare fire eye also senses a flame out and starts the timer for relight. Originally, a signal was to be sent to the control room operator that the flare had gone out and a manual restart would be required. The automatic relight system was put in during the construction stage as an environmental measure to minimize the amount of chemicals leaving the flare not thermally destructed. It automatically relights the flare after 30 seconds.

Everything seems to be taken care of. The plant is down and no one is happy about that, but all SISs are functioning as designed and it looks like the plant is in a safe condition. Instead, however, an explosion rocks the boiler area. In less than a minute there is another explosion in the flare area. Fire is reported in almost every trench in the plant and a chlorine odor is everywhere. What went wrong? Every SIS functioned as designed! But a careful exhaustive examination of all of the consequences of successful operation and accumulated effects had not been done.

Here is why the successful operation of some SISs in conjunction with some BPCSs caused this disaster. First, consider the EDC reactor area. The BPCS sensed a low level and did what it was programmed to do—increase flow to reestablish the level. Because the consequence and downstream effect of excess flow was not considered in the hazard review of the system, flow was continually increased, eventually to the point of overflow. The "problem" was then transferred downstream to another system. A better hazard review of that first system might have considered the possibility that the low level signal could be false. If the potential for a false low level signal had been recognized and then evaluated, a second level controller or a limit on the time or volume of flow increases might have been put in place. That same evaluation would have also considered the possible consequences of excess flow into and out of the reactor.

At the scrubbers, the SISs functioned as desired—they shut down the EDC reactor because of chlorine breakthrough. This system was designed properly with considerations for the consequences of activation. Flows were shut off and the reactors were isolated from all systems except the excess pressure relief system. Production was interrupted, but the reactor area went to a safe mode and did not transfer problems elsewhere.

Incomplete consideration of all consequences of SIS activation in the boiler area was the major contributor to the boiler explosions. When the liquid surge from the EDC area extinguished the flames in both boilers, the SIS recognized that there was no flame. It then started a programmed relight procedure based on failure to sustain a flame due to low fuel or improper fuel to air ratio. This included instructing the forced draft fans to go to maximum flow to clear the firebox of any combustion products and unburned fuel. Unfortunately, low fuel was not the reason for flame failure this time. A better hazard analysis might have considered the possible consequences of flame failure due to liquid in the line or excess fuel instead of too little. By having the forced draft fans quickly add air to the fire box, the fuel rich environment was diluted, bringing the fuel to air ratio into the explosive range. Because this happened so quickly, surfaces in the fire box were still hot enough to ignite the mixture, causing an explosion that severely damaged both boilers. This explosion was caused because the hazard analysis did not consider all potential activation scenarios and the related consequences.

Activation of the flare liquid knock out drum emergency dump system provided a path for flammable liquids and chlorine to be dumped into the plant trenches. The expectation was that liquid trapped in the knock out drum would be water. No consideration was given in the original hazard analysis to the probability that it might be flammable or toxic. Manually dumping of liquid from the drum would have detected the flammables or chlorine, but the SIS automatically dumped it when no one was around. As soon as an ignition source was found, the flammables in the trench systems ignited.

The explosion at the flare is a typical example of how last-minute changes or additions can defeat originally well-designed safety systems. The automatic pilot relight system was put in at the last minute to ensure there would be no environmental exceedences due to the flare not being lit. Little or no consideration was given as to why the pilot might go out or what bad things might happen if it were immediately relit. The objective was to keep it lit. When liquid squirted out of the top of the flare, it extinguished the flame and pilot. When the EDC reactor shut down completely, all process generated flow to the flare stopped. Only pilot gas flow remained. After 30 seconds, the automatic relight system sparked at the pilot. It ignited. In fact, since the liquid had come out of the top of the flare and rained down to the ground below, the flames followed that liquid path and traveled the full length of the flare and into the trench system. The trenches already had combustible liquid in them because the flare knock out drum had been dumping into them. The fire in the trench followed the flow of liquid all the way to the waste water pond on the other side of the plant.

The ill-conceived "good deed" of installing an automatic relight had caused an explosion at the flare and spread fire throughout the plant trench system. All of the potential downstream consequences were not evaluated before the automatic relight system was installed. Neither were all potential initiating causes of flame failure evaluated. If they had been, the flare explosion might not have happened. But because they were not evaluated, the result was that a "problem" was transformed into something else and then transferred into another system incapable of handling it.

In the end, proper functioning of the following process controls and SISs contributed to the disaster.

- EDC reactor area: All of the potential consequences of the BPCS functioning properly were not evaluated. It did not trigger alarms or any SISs designed to handle low level emergencies. This inaction transformed the initiating "problem" (a malfunctioning level controller) into a different "problem," liquid overflow from the reactor, and then transferred it into the next system. The downstream systems were not available to handle the transformed "problem," and the central computerized control system and its human operator did not recognize that a "problem" had been transformed and transferred.

- Flare knock out drum level controller: The BPCS for the flare knock out drum was backed up by a SIS. The "problem" that volatile liquids could be in the piping system feeding the flare was never recognized. The SIS's response simply ignored that possibility and transferred the "problem" downstream into the trench system which was incapable of handling it. Because the actual "problem" was never recognized, the SIS's action also helped hide it from other safety systems.
- Flare automatic relight system: The prime objective, keep the flare lit, overrode considerations of possible consequences of that demand. Therefore, the transformed "problem" that was transferred into the SIS was not expected or within its ability to abate. In fact, the SIS's programmed response actually caused an explosion remote from the initiating event — a malfunctioning level controller on the EDC reactor. This, at least in part, was a result of accumulative responses to several upstream events — malfunctioning EDC reactor level controller and flare pilot automatic relight system.
- Boiler relight system: The boiler relight SIS had not been programmed to deal with consequences of this transformed "problem" — combustible liquids in the gas feed lines. Therefore, this unrecognizable "problem" was not addressed and the SIS went about starting the relight sequence which included purging the fire box. The accumulative responses to high level in the EDC reactor plus boiler flame out led to disastrous results.

MANAGING THE HAZARDS

Use the tools that are available, and if they are inadequate, develop new ones.

The way to manage hazards caused by the success of SISs is multifaceted. First, the Cause-Consequence relationship must be firmly established as the analysis framework.[3][4] Every time a safety system is activated it is due to some Cause and that activation results in some Consequence. The key is to discover each potential Cause and follow through on each potential Consequence that is applicable in every stage of a facility's life cycle.

Thorough analyses must be done during the design phase so that all the activation scenarios and the downstream and accumulative consequences are identified. Particular care should be taken to not simply pass the "problem" on into the next system without fully evaluating the downstream system's ability to handle it. Further, do not transform the type or mode of the "problem" without fully understanding the consequences of that transformation.

Recognize that the activation of a SIS is usually due to a failure in the BPCS and is, therefore, a change in normal operations that should be evaluated using MOC or accident investigation tools.[6] Results of these analyses should be programmed into the computer control/distributive control system so that it can react or supply enough information to the control room operator to react appropriately without compounding or simply passing the "problem" on.

For the future, we also need to develop new tools. Logs of alarm sequence may no longer be adequate for on-line problem identification. We need to fully implement alarm prioritization and logging prioritization. From that we should develop a system that looks back at all of the signals that may have cascaded into causing the last alarm. And we need a system that predicts potential downstream consequences based on the alarms that have already been activated. Perhaps we can program into the computer system a "troubleshoot" button the control room operator can push that graphically displays the cascading of all signals that might have caused this last alarm to sound and then displays the potential consequences of the last alarm. The process hazard analysis methodology, Cause-Consequence Analysis,[3] seems an appropriate frame work for such a computer-generated graphic.

Disclaimer: Although the information in this paper is believed to be factual and accurate, no warranty or representation, expressed or implied, is made with respect to any or all of the content thereof and no legal responsibility is assumed therefore. The examples used are for illustrative purposes only and do not necessarily represent actual events.

REFERENCES

(1) Dowell, A. M., III and Hendershot, D. C. (1997). "No Good Deed Goes Unpunished: Case Studies of Incidents and Potential Incidents Caused by Protective Systems." *Process Safety Progress* 16,3 (Fall, 1997), 132-139

(2) Gilovich, Thomas (1991). *How We Know What Isn't So,* 49-87. The Free Press.

(3) *Guidelines for Hazard Evaluation Procedures, Second Edition with Worked Examples*, Center for Chemical Process Safety, AIChE, New York, 2006

(4) *Guidelines for Investigating Chemical Process Incidents, Second Edition*, Center for Chemical Process Safety, AIChE, New York, 2003

(5) *Guidelines for Safe Automation of Chemical Processes*, Center for Chemical Process Safety, AIChE, New York, 1993

(6) *Guidelines for Safe and Reliable Instrumented Protective Systems*, Center for Chemical Process Safety, AIChE, New York, 2006

(7) Kletz, Trevor (2003). *Still Going Wrong! Case Histories of Process Plant Disasters and How They Could Have Been Avoided*, 149-151. Gulf Professional Publishing.

(8) *Layer of Protection Analysis*, Center for Chemical Process Safety, AIChE, New York, 2001

(9) Sanders, Roy E. (1999). *Chemical Process Safety. Learning from Case Histories*, 45-46. Butterworth-Heinemann

Failure Conundrum

Michela Gentile, Ph.D.
Angela E. Summers, Ph.D., P.E.
SIS-TECH Solutions, LP
Houston, TX 77034 USA
Phone: 281-922-8324

Prepared for Presentation at
American Institute of Chemical Engineers
2006 Spring National Meeting
8th Process Plant Safety Symposium
Orlando, Florida
April 24-26, 2006

February 2006

UNPUBLISHED

Failure Conundrum

Michela Gentile, Ph.D.
Angela E. Summers, Ph.D., P.E
SIS-TECH Solutions, LP
Houston, Texas 77034

Abstract

Safety instrumented system (SIS) failures are caused by random, systematic, and common cause faults. IEC 61511 (ANSI/ISA 84.00.01-2004) stresses the importance of minimizing the propagation of faults into failures through design, operating, and maintenance practices. To fully understand the lifecycle requirements, it is first necessary to understand the types of failures and their potential effects on the SIS. Although several technical standards and other specialized literature address the topic, it is still a "fuzzy" matter, subject to misunderstanding and discussion.

IEC 61511 Clause 11.9 requires that the SIL be verified using quantitative analysis, such as reliability block diagrams, fault tree analysis, and Markov modeling. This analysis includes only those dangerous failures that are random in nature. Common cause faults may or may not be included in the verification calculation dependent on whether they exhibit random or systematic behavior. Any personnel assigned responsibility for verifying the SIL should understand each failure type and the strategies that can be used against it. Consequently, this paper provides an overview of random, systematic, and common cause failures and clarifies the differences in their management within IEC 61511.

1. Introduction

Modern control systems are designed to keep the process within specified parameters considered acceptable for normal and safe operation. However, when the process excursions are larger than the allowed range of variation, hazardous conditions may occur. If the situation is not addressed, it can degenerate to a hazardous situation with potential consequence to human life, environment, and plant assets. The failure of the control system is often listed in hazard and risk analysis findings as an initiating event for hazard scenarios. The risk associated with such as process excursion may require mitigation using a safety instrumented system (SIS).

During an H&RA study, risk reduction is allocated to each safety function used to mitigate process risk. When a safety instrumented function (SIF) is identified, a safety integrity level (SIL) is also assigned that is related to the SIF's required risk reduction. The risk reduction and SIL establish a benchmark for the design and management practices used throughout the SIF life. This benchmark serves two purposes: 1) it defines the minimum performance that should be achieved with regard to random failure and 2) it determines the rigor of the functional safety management system required to reduce the potential for systematic errors to a sufficiently low probability.

For many years, the chemical process industry has addressed SIS integrity and reliability issues through the use of prescriptive design and maintenance practices that are sufficiently conservative to assure that the risk reduction can be met in a wide variety of process applications. These internal practices are frequently criticized for what is perceived as "over-design," because the practices seem to go beyond the minimum requirements of IEC 61511[1] or one of the other industrial standards. However, it should be recognized that any industrial standard documents general practices that are acceptable to participating stakeholders. A specific application may necessitate more rigor, including fault tolerance, diagnostics, testing, etc.

Project and plant personnel are under pressure to "optimize" anything and everything, while adhering to the requirements of industrial standards. This "optimization" is generally intended to make the design as simple and cost effective as possible. However, when taken to extreme, optimization can reduce the design safety margin to a minimum and can result in the heavy reliance on devices whose failure modes are not well known. As less and less safety margin is built into the design, greater importance is placed on the precision of the predictive failure calculation. Thus, the device's failure rate can become a critical design parameter rather than simply a performance benchmark. While the nature of process hazard itself does not change, excessive focus on minimal cost can result in the SIF being designed and operated with less tolerance for failure.

Consequently, while this paper presents a discussion of how to classify and manage failures, it must be recognized that the calculations are not a means to an end. Safe operation is achieved by designing and implementing SISs that take into account a wide variety of site specific criteria. The primary advantage of internal practices is that, rather than turning every SIS design into an exercise in optimization, the risk management philosophy is clearly laid out with a sufficient safety margin to address uncertainty. The SIL is then verified using dangerous failure rates estimated for each device. The calculations are therefore used to verify, not design the SISs.

2. A "Bathtub" Viewpoint

The failure behavior of a population of hypothetical devices or components over its lifecycle (i.e., from production to disposal) is commonly represented by the "bathtub" curve, as shown in Figure 1A. This plot shows how the overall failure rate of a device changes with time. The initial failure rate of the hypothetical device is driven by its "burn-in" or "infant mortality" rate, which declines rapidly. The middle flat section of the curve represents the "useful life" of the device and is typically characterized by a constant failure rate. The SIF performance calculation is based on this middle section. The last part represents the "wear out" or "end-of-life" failure rate and is characterized by an increasing failure frequency.

As indicated by Smith[2], the Bathtub Curve is in reality composed by three overlapping curves, one for each of the three sections, Figure 1B.

Figure 1. Overall Bathtub Curve (A) And Components Of The Bathtub Curve (B).

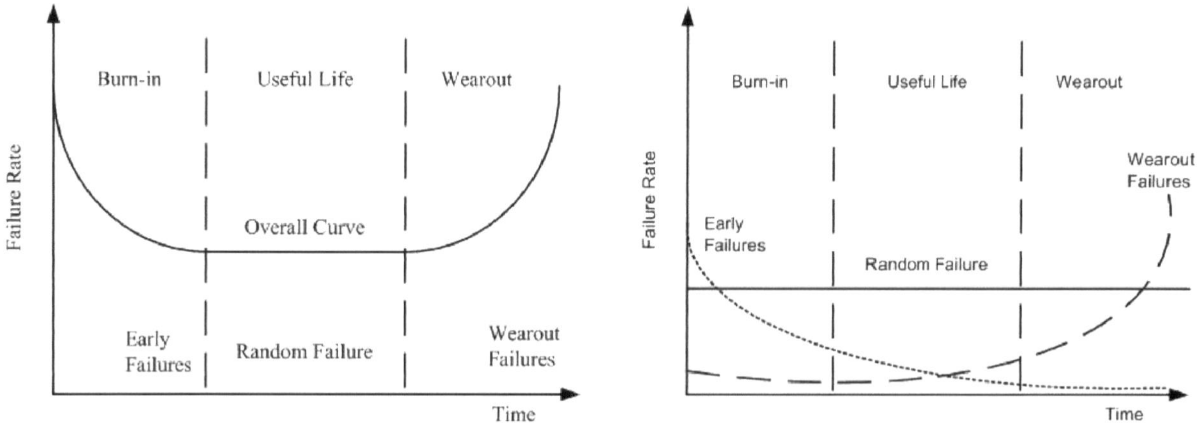

Failures that occur during the burn-in time are predominantly caused by manufacturing flaws in the components that comprise the device. Manufacturers affect the shape of this curve when they perform burn-in and function testing prior to release from production. However, not all flawed components fail immediately, and some cases of "infant mortality" occur during the useful life of the device. This is indicated by the extension of the burn-in curve to the useful life and wear out region.

The transition between the burn-in and useful life curves (Figure 1A) occurs when the failure rate becomes constant. Failures during the useful life of the device occur mainly due to random events that increase the stress on the device beyond its physical limits. When the wear-out failure rate becomes dominant over the useful life failure rate, the overall failure rate increases ending the useful life of the device.

The wear out period is characterized by an increased slope that is related to the device type. Programmable devices tend to have a very sharp increase in the failure rate due to the large number of aging components. Electromechanical devices tend to have a more gradual increase in failure rate as they age, and the slope of their failure rate curve is heavily affected by how well their mechanical components are maintained. Preventive maintenance has been shown to extend the useful life of electromechanical devices. As with infant mortality, not all wear-out failures occur after the end of the useful life; these failures can happen during any stage of the lifecycle of the device and as time passes its rate of occurrence increases.

A device should undergo burn-in and function testing to detect as many infant mortalities as reasonably possible. The device should also be replaced when maintenance records demonstrate that it has reached its wear-out phase. The middle section (i.e., the useful life) is the only part of the "bathtub curve" that is then relevant to the analysis of an SIF device. In this region the device's failure rate is considered constant (Figure 1A).

The constant failure rate assumption is a fundamental one in the analysis of the SIF performance; however, it has been demonstrated that this region does not exist for all devices[3]. For example, some devices, such as block valves, are characterized by dominant failure mechanisms which are a function of time (i.e., lower failure rate at early stages of useful life, high failure rate at wear-out stages). In this case, specific statistical distributions, such as Weibull should be considered, as well as the random failure models.

The random failure rate is characteristic to the device when it is operated in accordance with its specification. The estimated failure rate is based on specific operational conditions implicit in the statistical data set used to estimate it. Operating the device outside of the manufacturer's specification for the device may damage or stress it, thereby causing early failure. Unexpected process application impact, such as chemical attack, corrosion, and deposition, or external environmental impact, such as electromagnetic interference, vibration, and heat, can also shorten its useful life.

3. Systematic, Random, and Common Cause Failures

An SIF uses many different devices to execute the safety function that mitigates a specific process hazard. The performance of an SIF depends on the characteristics of each SIF device (i.e., individual failure rates), the properties of the system itself, and the interactions among its components (i.e., voting architecture, common cause failures). An implicit assumption made during SIF design and its performance verification is that the devices are in their "useful life" period and that the SIF is designed and managed according to its design and operating basis

The SIF probability of failure is a function of the random failure rate of its devices (i.e., field sensors, final elements, logic solvers, and other events) and the design basis parameters, such as redundancy, test interval, and diagnostic coverage. Other systematic, non-random contributions affect the observed performance of the SIF. Systematic and random faults can cause the failure of an individual device or the simultaneous failure of multiple devices. The individual device failure is easily assessed using probabilistic techniques that are described in many publications, such as Smith[(1)], Lees[(2)], and ISA TR84.00.02 [(5)]. While the failure rate of most user-approved devices is low, there is ample failure rate data available to support quantitative modeling.

Independent failures of devices are tracked and analyzed as part of the functional safety management system. In contrast, simultaneous failure of multiple devices is more difficult to assess quantitatively. The goal of failure tracking program is to identify root causes and data trends, which does include the identification of systematic and common cause failure. However, these failures happen very infrequently and, when one occurs, it tends to be considered subsystem specific, so mapping trends in dependent failures is problematic.

Faults are managed within the IEC 61511 work process using different strategies depending on whether they are random or systematic. Consequently, the following sections briefly discuss random, systematic, and common cause failures, so that the strategies for their management can be better understood.

3.1 Random failures

Random failures are physical failures brought on by excessive stress on the device. Random failures can occur at any time during the device life and are generally well understood by the time devices achieve user-approved status. The occurrence of these failures does not follow a pattern, but instead occur randomly in time. Stress factors are the result of a variety of causes, such as abnormal process conditions, the presence of an adverse microenvironment (e.g., corrosion produced by chemical impurities or metallurgy), or normal device wear and tear. Low frequency atmospheric events (e.g. snow in Houston, Texas USA) can also be considered random events.

The bathtub curve for an individual device is developed by counting the failures that occur in a population of identical (or sufficiently similar) devices over a certain period of time. When the devices are in their useful life period, the collected data set is used to calculate the failure rate using statistical techniques. Collected data typically includes some systematic failures, e.g., it often takes multiple failure reports before it is recognized that there is an issue with instrument air quality. Identified systematic failures contribute to the observed failure rate. Figure 2A and 2B illustrate how the observed failure rate is higher due to the presence of systematic events that are often perceived as random. These failures are illustrated in Figure 2B using additional failure rate lines. The intent of the analysis is to predict SIF performance, so systematic failures should be tracked and their inherent presence is generally considered when estimating the random failure rate. As more information is collected, trends can be identified and used to minimize random and systematic failures in new or modified designs.

Figure 2. Hypothetical device's random failure rate is constant (a), however other non-random failure sources affect the observed failure rate of the device (b).

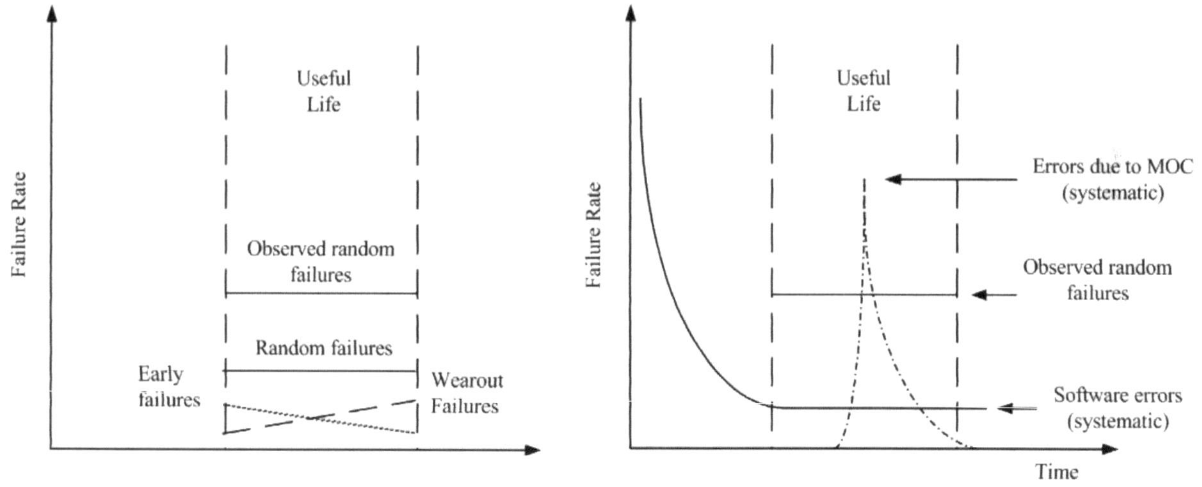

Random failures can be detected by internal diagnostics, externally-configured diagnostics, and full function tests. Redundant, fault tolerant subsystems are often used to reduce the probability that a single failure will cause the SIF to fail. Redundant subsystems also provide the potential for diagnostic coverage, where a diagnostic algorithm is executed at a specified interval to detect certain device failure modes. The test interval is generally chosen based on maintenance history, manufacturer's recommendations, industrial standard's requirements, insurance requirements, regulatory requirements, and what is necessary to achieve the required risk reduction. As the testing interval gets longer, there is an increasing probability that multiple devices within a subsystem have malfunctioned prior to fault detection.

3.2 *Systematic failures*

While random failures are caused mainly by physical factors, systematic failures are the direct consequence of device and SIF complexity. Every hardware and software component in a device is subject to failure due to design, specification, operation, maintenance, and implementation errors. These mistakes immediately put the devices on the path to failure. As the complexity of the process or SIF increases, the potential for systematic errors increases exponentially due to the combination of failure modes. As complexity increases, the probability of detecting these errors decreases. Each device has many known opportunities for systematic error. With any new process application, there is the potential for many unknown (or as yet unidentified) failures. When issues associated with interconnectivity, communication, and utility quality are added to the analysis, there are generally a large number of potential systematic failures. Due to the complex and non-random nature of systematic failures, it is difficult to predict or statistically analyze them.

Further, only a limited number of device failure modes and failure paths can be tested. When the failure patterns are not detected by the limited testing that is practically achievable, systematic failures can happen every time the specific set of conditions occurs. These conditions become an intrinsic part of the device and thus the SIF. Systematic errors are a major source of common cause failure, because of the potential to disable redundant devices.

Systematic failures include many types of errors, such as:

- Manufacturing defects, e.g., software and hardware errors built into the device by the manufacture.
- Specification mistakes, e.g. incorrect design basis and inaccurate software specification.
- Implementation errors, e.g., improper installation, incorrect programming, interface problems, and not following the safety manual for the SIS devices.
- Operation and maintenance, e.g., improper testing and improper bypassing.

Improper management of change applied during manufacture, specification, implementation, operation and maintenance may generate systematic failures. For example, the modification of vendor software in a transmitter may change how the device is configured to fail. If the maintenance procedure is not changed, errors may occur during commissioning or maintenance. For this reason, the functional safety management system incorporates configuration management, as a specialized management of change procedure.

As an example of a systematic error, consider a specification mistake that occurs during the engineering, installation, commissioning and validation phase. This error may be as simple as not properly specifying the process application on the hardware specification form, leading to the purchase of a device with inappropriate materials of construction. While the device itself may fulfill all of the requirements necessary to gain user-approved status, the incorrect specification results in the premature failure of the device. Redundancy and voting do not address the specification error, because it is typical practice to purchase both devices using the same specification.

To illustrate further, let's assume that there is a subsystem composed of 2oo3 voting temperature transmitters installed in the same thermowell. The designer feels that this is acceptable, because the specification indicates that the process is clean. However, in reality, the process contains polymer which rapidly fouls the thermowells, insulating them from the actual process. None of the three transmitters detects the actual process condition. This mistake may result in the transmitters not being able to detect high temperature within the process safety time.

According to the data from a sample of 34 incidents published by HSE[4] that there is no single lifecycle phase to blame for SIS non-performance. Mistakes are made throughout the lifecycle. Consequently, systematic failures must be addressed throughout SIF lifecycle. The proper implementation of a management system decreases systematic errors and enhances safe and reliable operation.

Systematic errors related to manufacturing defects can be reduced through the use of diverse redundancy. Diversity typically involves the use of different technologies or different manufacturers. While device manufacturer errors can be addressed by diversity, this increases the SIF complexity. Incorrect specification, implementation, operation, and maintenance constitute root causes that are not solved by diversity and can actually increase if unnecessary complexity is introduced into the SIS design. The need for diversity should be balanced with consideration for the greater difficulty in long-term operation and maintenance.

Further, the perceived improvements gained by diversity are based on the assumption that different devices have different failure rates and modes. Consequently, it is less probable for all of them to fail simultaneously, if they are different. However, as discussed in the next section, diversity only reduces the potential for common mode failures. Many common cause failures are not addressed by diversity.

Systematic errors are best addressed through the implementation of the functional safety management system of IEC 61511[1], which overlays a quality management system with a project development process. Each verification, functional safety assessment, audit, and validation required by the standard is aimed at reducing the probability of a systematic failure to a sufficiently low level.

3.3 Common cause failures (CCF)

All common cause failures have the potential to reduce the SIF performance; however, they must be addressed in different ways depending on the nature of the failure (e.g., systematic or random). Throughout the IEC 61511[1] lifecycle, it is recommended that the devices, systems, or protection layers be assessed for independence and the potential for common cause failure. The concepts of independence and common cause are interrelated. A lack of independence means that there is a potential for a common cause failure. Likewise, an identified common cause indicates a lack of independence and therefore some level of dependency.

Common cause failure is a term that is used to describe random and systematic events that cause multiple devices, systems, or layers to fail simultaneously. Another term is "common mode" failure, which describes the simultaneous failure of two devices due to the same failure mode. Common mode failure is related to the use of identical devices in the redundant subsystem. For example, two redundant differential pressure sensors on a pipeline can be simultaneously disabled due to loss of signal (common mode failure) originated from diaphragm damage (failure cause) caused by water hammer. Common mode failure is a subset of common cause failure.

Diversity is often suggested as a means to eliminate common cause failure. However, it is important to understand that common cause failure can impact identical and diverse devices. For example, process application or external environmental conditions can affect different technologies simultaneously when the conditions trigger each device's failure mechanisms. These devices may eventually fail due to different reasons, but the abnormal process condition is
root cause that started the failure propagation. The use of diverse technologies (i.e., diversity) does reduce the potential for common mode failure. In summary, diversity reduces the potential for dependent failure by minimizing common mode failure, but does not eliminate the potential for common cause failure.

The approach taken to manage common cause failure (CCF) is specific to the nature of the failure. Within the standard(1), two types of CCF are addressed: 1) single points of failure where one malfunctioning devices causes an SIF failure and 2) single events that lead to multiple failures in a redundant subsystem. Figure 3 and Table 1 refer to these as "single point of failure" and "redundant," respectively.

Single points of failure were discussed in random and systematic sections above. As shown in Figure 3 and Table 1, the errors can be systematic (B) or random (A). Systematic failures (B) occur when human errors result in the violation or invalidation of design and operating basis assumptions (e.g., process assumed to be clean but in reality is not). Random failures (A) can occur throughout the useful life of a device. These failures are managed using redundancy, diagnostics, and function testing.

As with single points of failure, redundant subsystems can fail due to systematic errors (C) in the device manufacture, specification, design, installation, and maintenance. These errors typically happen due to lack of knowledge, information, and training and are generally unknown to engineering personnel. Test procedures may not identify these errors, since they are not expected. Systematic errors are difficult to test even if easily identified. For example, it is possible that the valve actuator is incorrectly specified, so how do you test to determine that the valve actuator will close under all emergency process conditions? Specification errors must be caught during the design and engineering phases using verification. Installation and commissioning errors are reduced by comparing the delivered equipment specification to the device specification sheet.

Table 1. Methods Used To Address The Different Types Of Common Cause Failures And Examples (See Figure 3)

CCF Type	METHOD	COMMENTS	EXAMPLES
SINGLE POINT FAILURE			
A	Analytical (i.e., Single event in fault tree)	Can affect several part of the process.	Power loss, instrument air loss due to abnormal or transient operation. Normal random device failure.
B	Lifecycle management	Caused by mistakes during design and puts devices immediately of failure path. Assumptions are violated and invalidated. Not enough technical knowledge is available. Not possible to detect these errors.	Insufficient supply of instrument air to the valves actuators during normal operation due to wrong process specification of plant utilities.
REDUNDANT ARCHITECTURE			
C	Lifecycle management	As B.	Wrong device design: software bug, defective device components due to manufacturer error Wrong device specification due to engineering error, therefore: device plugs because used in dirty environment believed to be clean. Runaway reaction caused by well-known process chemistry.
D	Analytical (i.e., Single event in fault tree)	Caused by the process, but user induced thru equipment connection and installation. Explicit failure data available.	Environmental stress on the device (e.g., abnormal corrosivity, vibration, pressure, temperature, solids, erosion, etc.) due to transient or abnormal operation. Gasket failure due to unlikely process chemistry in typical applications for which prior-use data exist.
E	Analytical (i.e., β-factor)	Caused by the process, but user induced thru equipment connection and installation. No explicit failure data available therefore β-factor estimated on prior-use information and experience for similar devices and systems.	Environmental stress on the device (e.g., abnormal corrosivity, vibration, pressure, temperature, solids, erosion, etc.) due to transient or abnormal operation. Gasket failure due to process chemistry combined with temperature higher than the specified due to fouling of upstream heat exchanger.
F	Analytical (i.e., β-factor)	Inherent part of the design of the device. β-factor estimated on prior-use information and experience.	Device failure due to manufacturing errors such as defective raw materials or production processes.

Figure 3. Taxonomy Of Common Cause Factors (CCF) And Methods For Their Analysis (See Table 1).

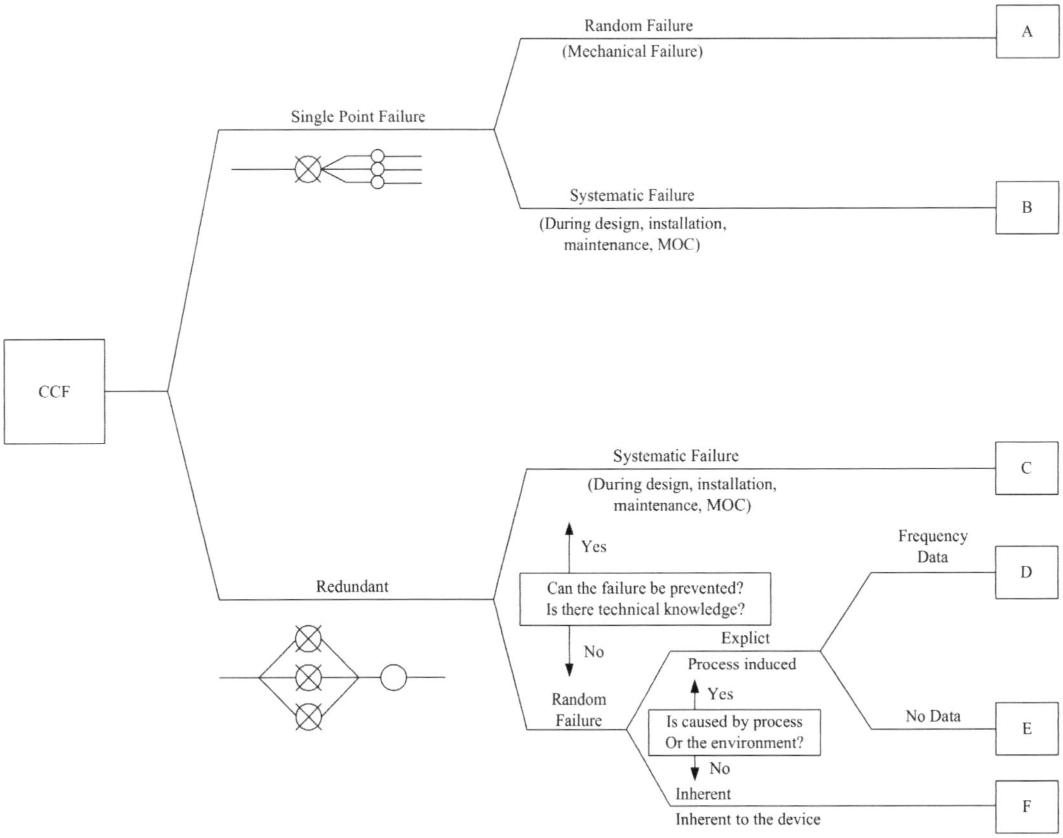

Random errors of redundant subsystems (D, E, and F) can be caused either by conditions that are inherent to the device (F) or inherent to the system (D and E). Random failures are caused by conditions that are inherent to the device that are introduced by defective manufacturing (F). Sources external to the device are induced by the application environment, the device installation, and the devices interconnection to other systems (D and E).

This type of system-induced random errors can be divided into two categories depending on the availability of failure frequency data. If data is available (D) the failure can be modeled explicitly as an event. For example, if fault tree analysis is the selected analytical methodology, this type of CCF is treated as a basic event with its own failure rate. If data is not available, CCF can be addressed using the beta factor method (E). In many cases, the beta factor is simply accounting for random events inherent to the device (F) in the process application. The value of the beta-factor is selected based on prior-use data for the specific device or a sufficiently similar one in the application environment. Many owner/operators use a beta factor between 0.1 and 5.0% when good engineering practices are applied in the design and installation to minimize CCF. The beta factor can be substantially higher if good engineering practices are not followed. Refer to ISA TR84.00.04[5] for more discussion of CCF and the beta-factor.

Checklists can be used to identify CCF during the design and engineering stage. The list of questions guides the engineer through the design aspects examining opportunities for CCF.

Analytical methods for CCF include fault tree analysis and Markov models. In terms of mathematical complexity, Markov models are the most difficult to apply but offers high flexibility and the opportunity to model time dependent systems. Fault tree analysis is based on simpler mathematics and offers flexibility but has limited application for the modeling of failures derived from sequences.

4. Conclusions

Device failures can be classified as random, systematic, and common cause failures. Random failures occur due to different types of random events and can be assumed statistically to occur at a constant failure rate. Random failures are easily modeled using probabilistic math, allowing the SIF performance to be determined. Systematic errors are caused, or indirectly induced, by human error or unforeseeable complex conditions. Systematic failures are not random events and are addressed by the lifecycle management system. Since they are not easily analyzed, most systematic errors are not included in the SIL verification calculations.

Both, random and systematic failures can induce common cause failures (CCF) that can be in the form of single points of failures or failure of redundant devices. The nature of the CCF determines the methodology used to address.

It is fundamental for process and control engineers to understand the different type of failure a device can suffer in order to identify the optimal solutions for each situation. Depending on the type of hazard to be controlled and the type of process, different architectures based on voting and diversity can be used to achieve the required balance between reliability and performance.

5. References

[1] IEC 61511, International Electrotechnical Commission, Geneva, Switzerland (2003).

[2] Smith D.J, “Reliability Maintainability and Risk”, 5th edition, Butterworth-Heinemann Ltd., Oxford (1997).

[3] Lees F.P., “Loss prevention in the Process Industries”, 2nd Edition, Butterworth-Heinemann Ltd, Oxford, UK (1995).

[4] HSE,“Out of control: why control systems go wrong and how to prevent failure”, 2nd edition (1999).

[5] ISA 84.00.04 “Guidelines on the implementation of ANSI/ISA 84.00.01-2004 (ISA 61511 Mod.)”, 1st edition (Expected publication 2005).

NOTE: This paper is adapted from a draft Center for Chemical Process Safety (CCPS) book “Guidelines for Safety and Reliable Instrumented Protective Systems“, American Institute of Chemical Engineers, New York, New York (Expected publication 2006).

190c Using Layer of Protection Analysis to Define Safety Integrity Level Requirements

Randy Freeman, ABS Consulting, 16800 Greenspoint, Houston, TX 77060

The publication of ISA standard S84.01 – 2004 defines the life safety cycle for the design and installation of safety instrumented functions (SIFs) using safety instrumented systems (SISs). However, the determination of the required safety integrity level (SIL) is an activity that the process designers need to complete prior to detailed design. This paper outlines an approach for the determination of the required SIL using the layer of protection analysis (LOPA) method. An example is presented along with a suggested format for documenting the LOPA, SIL, and SIF requirements in a manner that can be transferred to the detailed design team for implementation.

Use of Layers of Protection Analysis (LOPA) To Determine Protective System Requirements

W. Kent Goddard, PE
Solutia Inc.
Process Safety & Property Protection Lead

Prepared for Presentation at
American Institute of Chemical Engineers
2006 Spring National Meeting
8th Process Plant Safety Symposium
Orlando, Florida
April 24-26, 2006

UNPUBLISHED

Use of Layers of Protection Analysis (LOPA) To Determine Protective System Requirements

W. Kent Goddard, PE
Solutia Inc.
Process Safety & Property Protection Lead

Abstract

Risks to safety, health, and the environment cannot be totally eliminated from process industries. However, tools such as Layers of Protection Analysis (LOPA) can be used to better understand and manage risks related to process safety.

Properly designed and maintained Safety Instrumented Systems (SIS) are important safeguards for reducing process risks to company-tolerable levels when other safeguards (i.e., other layers of protection) or inherently safer designs are not feasible. Before these protective systems can be designed, the proper number and types of Safety Instrumented Functions (SIF) must be determined. Determining the appropriate Safety Integrity Level (SIL) of each SIF is important to avoid under-protection (i.e., inadequate risk reduction) or over-protection (i.e., excessive cost). LOPA is an effective tool for making these determinations.

This paper provides an overview of LOPA methodology and how it may be used to (i) evaluate various safeguards that reduce the likelihood and severity of process safety incidents, (ii) determine if SIFs are required, and (iii) determine what SILs are necessary to reduce risks to company-tolerable levels. This paper also addresses the major factors a company should consider when developing a LOPA procedure specific to its needs.

1.0 Managing Process Safety Risks

Process safety risks are inherent to the operation of chemical manufacturing plants and refineries. The nature of the raw materials and final products (reactivity, flammability, corrosivity, etc.) and the way in which these materials are handled (high temperatures, high pressures, etc.) present challenges to safe operations. These types of risks may be categorized in different ways. For example, there are risks related to:

- Safety and health of on-site personnel and the surrounding community
- Environment (air, ground, and water)
- Financial (property damage, business interruption, and corporate image)

In general terms, the level of process safety risk is proportional to the severity of the potential consequence and the likelihood of that consequence occurring. It's important to recognize that risk can be reduced but it cannot be totally eliminated. Hence, to reduce risk we must reduce (i) the likelihood of an event, (ii) the severity of the consequence if an event does occur, or (iii) both likelihood and severity of an event.

Each company must determine the level of risk they deem tolerable. Risks at or below the tolerable level require no action while risks above the tolerable level require action to reduce them to a tolerable level.

The first thing to consider to reduce risk, especially for new facilities, is inherently safer designs. For example, consideration should be given to reducing batch sizes or the on-site quantities of highly hazardous chemicals to limit the severity of a potential release. Another example is constructing vessels to contain the pressure of a deflagration or other high-internal-pressure event. These approaches have the advantage of being passive; i.e., they do not rely on the proper functioning of an active safeguard. Such approaches, however, often have tradeoffs, such as lower production capacities or higher installed costs. Additionally, inherently safer designs are often difficult to implement in existing processes. Hence, there will always be a need for adequate safeguards.

Safeguards are used to reduce the likelihood of an event occurring. Examples of different types of safeguards include:

- **Basic Process Control System (BPCS) IPL** – While a failure of the BPCS (distributed control system, programmable logic controller, or other system) may initiate a hazard scenario, alarms and normally-operating control loops are often implemented in the BPCS to respond to process excursions. Additionally, BPCS IPLs are special systems that are independent of the initiating cause and are managed to provide protection against identified hazards.
- **Standard Operating Procedures (SOP)** – Trained operators can respond to process excursions using well documented procedures to correct out-of-control conditions.
- **Alarms and operator actions** – Alarms with operators trained to take specific documented actions to correct out-of-control conditions.
- **Safety Instrumented Systems (SIS)** – Often called “interlocks”, SIS are independent of the BPCS and consist of a sensing element (such as a pressure transmitter) sending a signal to a logic solver that takes the process to a safe state using a final element (such as a shut-off valve).

Consequence Mitigation Systems (CMS) do not prevent an upset condition, but instead reduce the severity of the resulting consequence. Examples of CMS include:

- **Pressure Relief Valves and Rupture Disks** – These devices prevent the catastrophic overpressure of equipment or piping systems but may have lesser, yet unacceptable, secondary consequences such as the release of hazardous chemicals to the environment.
- **Fire suppression and mist suppression systems** – These systems do not prevent a release of hazardous material, but reduce the level of damage following the release.
- **Emergency ventilation systems** – These systems do not prevent a release, but they can be used, for example, to keep the concentration of flammable vapors inside a process building below the explosive limit.

The first step is to understand the risks associated with a particular process and then determine which risks exceed the company’s risk tolerance. There are various methods for analyzing risks and evaluating the adequacy of safeguards, each having unique benefits and shortcomings. The following are examples:

- **Qualitative methods** - Qualitative methods include Process Hazard Analysis (PHA) techniques such as HAZOP, What-If, Checklist, etc. These techniques are designed to identify various hazard scenarios by considering different initiating causes (deviations from normal safe operations) that can propagate into process safety incidents. The severity of the worst credible consequence is ranked using a company-defined table often comprised of four to six levels of severity. The various safeguards that could stop the propagation of the hazard scenario are listed. These safeguards, plus the frequency of the initiating cause, are qualitatively "weighed" by the PHA team to estimate the frequency of occurrence of the worst credible consequence. Like for severity, this frequency (or likelihood) is ranked using a company-defined table often comprised of four to six levels of frequency. The risk is then determined using a severity-by-likelihood matrix. See Figures 1-3 for example tables and matrix used for PHAs. These methods are good at identifying the broad spectrum of upset conditions and resulting process safety hazards. One shortcoming, however, is the difficulty most teams have at estimating the likelihood of high-consequence scenarios occurring, especially when these scenarios already have multiple associated safeguards. This difficulty is due partly to the human mind's limited ability to qualitatively judge whether something is expected to happen once every 10,000 years, say, or once every 100,000 years.
- **Quantitative Risk Analysis (QRA)** – QRA methods such as Fault Tree and Event Tree use numerical frequencies of initiating causes, probabilities of enabling events such as available ignition sources or unfavorable meteorological conditions, and probabilities of safeguard failures. The resulting estimated frequency of a particular process event with a defined severity is compared with the company's tolerable risk criteria to determine if additional or improved safeguards are needed. These methods have the benefit of reducing the subjectivity of qualitative methods, providing substantiated and more-consistent results when reliable values for various probabilities are available as inputs to the QRA. One drawback to QRAs is the large amount of time and cost associated with this method.
- **Semi-Quantitative Layers of Protection Analysis (LOPA) method** - LOPA "splits the difference" between qualitative and quantitative methods. It removes the subjectivity associated with estimating frequencies using qualitative methods and it is quicker and easier to use than quantitative methods.

The rest of this paper is devoted to a discussion of the LOPA method.

3.0 LOPA Overview

LOPA is a semi-quantitative (order of magnitude) method for evaluating the level of risk associated with potential process safety hazard scenarios. It can be used for various purposes, such as:

- Identifying the need for Independent Protection Layers (IPLs) to reduce process safety risks to tolerable levels;

- Classifying the required SIL of SIFs for compliance with "Functional Safety: Safety Instrumented Systems for the Process Sector," ANSI/ISA 84.00.01-2004; and
- Providing a consistent means for identifying the equipment used in IPLs that should be included in a company's mechanical integrity program.

Each LOPA study is performed on a particular hazard scenario. The specific scenarios normally are identified by hazard identification techniques such as PHAs, incident or near-miss investigations, management of change hazard evaluations, or when it is necessary to define the SIL requirements for a SIF.

Following are the generic steps used by many LOPA procedures. An example is included to aid understanding.

- LOPA evaluations are normally team efforts; the first step includes forming a team and gathering the process information relative to the scenario(s) being studied.

- Begin the study by precisely describing a particular scenario, including the initiating cause, scenario propagation, and ultimate consequence assuming that all safeguards fail.

  Example: A failure of the Reactant B control loop could causes too much Reactant B to be added to the reactor. This could result in a runaway reaction that could over-pressure and rupture the reactor. This could cause injuries and/or fatalities within the department due to the catastrophic failure of the reactor and/or the subsequent toxic release.

- List the initiating cause(s) and determine its frequency(s) using a table specific to the company's LOPA procedure. Because this is an order-of-magnitude method, initiating cause frequencies are usually listed as once/year, once/10 years, or once/100 years.

  Example: *Failure of BPCS control loop: Once/10 years (or* $1x10^{-1}$*) per the table.*

- Rank the severity of the ultimate consequence using a table specific to the company's LOPA procedure. This table often is the same table that the company uses to rank severity during PHA studies. In addition to safety, the consequence severity may also be ranked with respect to environmental impact and/or economic loss. The tolerable frequency for each severity level is determined from the company-approved table (see Figure 4 for an example tolerable frequency table).

  Example: *For the sake of simplicity of this example, we will consider the safety severity only. The company's table lists this as a Level 2 event with a tolerable frequency of once/10,000 years (or* $1x10^{-4}$*).*

- Use simple math to calculate the risk reduction required to reduce the risk to the company-tolerable level.

  Example: *10,000/10 = 1000 (or* $1x10^{3}$*).*

- Evaluate the existing safeguards to determine which ones can be given credit as IPLs. To be considered an IPL, a safeguard must be:
  - Effective at stopping the propagation of the hazard scenario;
  - Independent of the initiating cause;
  - Independent of other safeguards (to avoid the potential for common cause failures); and
  - Auditable, so the assumed effectiveness and reliability can be validated in some manner.

  <u>Example:</u> The PHA for this department listed an interlock in the BPCS to shut off the flow of Reactant B if the totalized flow exceeds 105% of the set point. IPL credit could not be taken for this safeguard because it is not independent of the initiating cause. The PHA also listed a Safety Integrity Level 2 (SIL 2) SIS to add additional cooling to the reactor shell and overhead condenser if the temperature in the reactor exceeded 200° C. This SIS was assigned a Risk Reduction Factor (RRF) of 100 per the company's table of RRFs. The PHA also listed a shortstop system that would kill the reaction and prevent the overpressure condition if the temperature in the reactor exceeded 200° C. IPL credit could not be taken for the shortstop system because it is not independent of the SIL 2 interlock – both rely on the reactor temperature probes. Hence the total RRF from current safety systems is 100 (or $1x10^2$).

- List and evaluate the CMS.

  <u>Example:</u> The PHA listed a relief valve as a safeguard. But the relief valve could not be considered an IPL because it is sized for a scenario less severe than this runaway reaction.

- Compare the actual risk reduction to the risk reduction required to reduce the frequency to the company-tolerable level.

  <u>Example:</u> RRF Available: 100 (or $1x10^2$). RRF Required: 1000 (or $1x10^3$). Gap: 1000/100 = 10 (or $1x10^1$).

- If the company risk tolerance criteria are not met, then the LOPA team makes recommendations accordingly.

  <u>Example:</u> The LOPA team recommends modifying the existing shortstop system to make it independent of the SIL 2 SIS. The modified system needs to meet a SIL of 1. One approach would be to trigger the shortstop system based on pressure in the reactor.

- If there is a CMS, such as a relief valve, the secondary consequence from proper actuation of the CMS needs to be ranked using the company's severity ranking table. Calculate the frequency that the CMS is expected to activate based on the probability of failure of the primary IPLs. If the resulting risk (severity and frequency) of the CMS activation is above the company's tolerable limit then a recommendation is necessary.

Example: While this reactor does have a relief valve it is not evaluated as a part of this LOPA scenario. It would, however, be evaluated as part of a less-severe overpressure scenario where it is relied upon as a CMS.

4.0 Factors to Consider When Developing a Company-Specific LOPA Procedure

The book Layer of Protection Analysis – Simplified Process Risk Assessment published by AIChE and the Center for Chemical Process Safety is an excellent source of information about the LOPA methodology. It is not, however, an off-the-shelf LOPA procedure. Each company that chooses to use LOPA needs its own specific procedure. This section describes some of the considerations and decisions a company will need to make in developing its procedure.

Each company must decide when it will use the LOPA procedure. A company may allow PHA teams to determine for which scenarios LOPA is needed, require a LOPA for all high-severity consequences identified by PHA teams, or choose to fully integrate LOPA into the PHA procedure, essentially conducting a LOPA analysis for all PHA scenarios. Other uses of LOPA are listed above in the **LOPA Overview** section. The company should also establish the minimum requirements for LOPA team composition and training for LOPA facilitators.

A significant decision each company must make is whether to include enabling conditions, time at risk, and/or other frequency adjustments in the procedure. These adjustments reduce the expected frequency of a scenario propagating to its worst credible case scenario. Examples may include "factoring down" for the number of hours per year that a batch reactor is at the high-risk point of the batch cycle or for the probability of a person being "in the wrong place at the wrong time" when a hazard scenario occurs. These adjustments add another layer of complexity that may underestimate the risk or may reduce the consistency of results between similar risks evaluated by different LOPA teams. Very well-defined rules must be established if these adjustments are made part of the LOPA procedure. Such adjustments move LOPA closer to a Quantified Risk Assessment and negate some of the value derived from LOPA being a simplified technique.

Corporate management must quantify the corporate risk tolerance. This decision should be made after the decision about use (or non-use) of enabling conditions, time at risk and/or other frequency adjustments is made.

The LOPA procedure must have clear rules with which to evaluate safeguards to determine if they qualify as IPLs. Many of these rules are available in the CCPS LOPA book, including requirements for effectiveness, independence, and auditability. A company also should define the minimum time-available-to-respond requirements for IPLs relying on operator responses to alarms and upset conditions.

The LOPA procedure must include tables for initiating cause frequencies and Risk Reduction Factors (RRF) for various types of IPLs. The CCPS book contains such tables with the frequencies expressed as ranges. A company's LOPA procedure should settle on a specific frequency for each initiating cause and a specific RRF for each specific type of IPL and each type of CMS.

The LOPA procedure cannot be finalized until the specific software is chosen to document the LOPA studies. Options included spreadsheets and databases either developed in-house or available from various consultants that facilitate LOPA studies. There are also software companies that market LOPA products. If many of a company's LOPA scenarios will be generated by PHAs, consideration should be given to a well-integrated PHA/LOPA software package.

5.0 Summary and Conclusion

LOPA is a useful method for evaluating the adequacy of safeguards protecting against process safety hazards. It is more powerful than qualitative methods (such as PHAs), making it especially valuable for evaluating relatively complex scenarios and scenarios with relatively severe consequences. A more rigorous Quantitative Risk Analysis may be more appropriate for extremely complex scenarios and scenarios involving very severe consequences.

If a company chooses to use LOPA, it should develop its own LOPA procedure specific to its needs. Factors discussed above will help a company develop an appropriate LOPA procedure to meet its desired risk tolerance levels.

Figure 1: Example of an Incident Severity Ranking Table[1]

Incident Class	Negligible (V)	Moderate (IV)	Serious (III)	Critical (II)	Catastrophic (I)
SAFETY: **Injuries and Illnesses**	Non-injury or first aid	Recordable injuries (e.g., cuts, burns, etc.)	Lost Work Day Injury (e.g., serious cuts and burns, broken bones, etc.) OR multiple OSHA recordable injuries	In-patient hospitalization of three or more personnel OR one or more fatalities within a unit or local area of the plant	Multiple fatalities across a plant site OR one or more injuries or fatalities to the public (e.g. large toxic release, vapor cloud explosion, etc.)
SAFETY: **Health**	No or negligible health effects to employees, contractors or visitors	Slight health effect (e.g. irritation or discomfort) to personnel in a section of the plant and the health effect would be reversible	Health effect to personnel on-site and the health effects would be considered reversible after medical treatment or only minor irreversible (e.g. respiratory problems or illnesses)	Health effects to personnel across a plant site and the effects would be considered irreversible	Threat to the lives of personnel across a plant site OR one or more injuries or fatalities to the public
ENVIRONMENTAL	No or slight impact that can be easily contained and cleaned up by onsite personnel	Moderate impact that can be handled by regular emergency crews OR any permit or regulatory deviation that doesn't exceed a Federal, state or local agency Reportable Quantity (RQ)	An impact on in-plant property requiring a costly clean-up effort OR any release that exceeds a Federal, state or local agency Reportable Quantity (RQ)	Major impact to the environment off-site (e.g., substantial harm to wildlife)	Environmental release that harms the public
ASSET PROTECTION: **Property Loss and Business Interruption**	Damage to a piece of equipment that is easily replaced or spared with an expected total loss of less than $25,000	Isolated property damage or minimal business interruption with an expected total loss between $25,000 and $500,000	Property damage or business interruption with an expected total loss between $500,000 and $2,000,000	Property damage or business interruption with an expected total loss between $2,000,000 and $10,000,000.	Property damage or business interruption with an expected total loss greater than $10,000,000 OR substantial damage to homes and buildings within the community
ASSET PROTECTION: **Impact On Public Image and Community Relations**	No or negligible impact	Public relations efforts needed due to questioning by media or others, but no press coverage; ACC-reportable process safety event	Small articles released by local media in the newspaper or coverage by the local news	Concerns expressed by local community officials; headline coverage by local newspaper or local news	Significant impact causing public relation difficulties; headline news coverage by national media

[1] Each company should develop its own table as appropriate to its own circumstances.

Figure 2: Example of a Frequency Ranking Table[2]

Rating Of Frequency (Likelihood)		
Class	**Rating**	**Frequency (Likelihood)**
A	Frequent	Likely to occur often
B	Occasional	Likely to occur several times during the life of the process
C	Rare	Likely to occur once during the life of the process
D	Unlikely	Not expected to occur during the life of the process but it's possible
E	Highly Unlikely	So improbable that it's assumed it will never occur; Engineering and administrative controls meet or exceed company-adopted "recognized industry practices"

Figure 3: Example of a Risk Matrix[3]

	Negligible (V)	**Marginal (1V)**	**Serious (III)**	**Critical (II)**	**Catastrophic (I)**
Frequent (A)	Team's Discretion (5)	3	2	1	1
Occasional (B)	Acceptable Risk	4	3	2	1
Rare (C)	Acceptable Risk	Team's Discretion (5)	4	3	2
Unlikely (D)	Acceptable Risk	Acceptable Risk	Team's Discretion (5)	4	3
Highly Unlikely (E)	Acceptable Risk	Acceptable Risk	Acceptable Risk	Acceptable Risk	Acceptable Risk

Note: The numbers 1 to 5 represent the Priority of the risk related to a PHA team finding/recommendation.

---

2 Each company should develop its own table as appropriate to its own circumstances.

3 Each company should develop its own matrix as appropriate to its own circumstances.

Figure 4: Example of a Tolerable Frequencies Table[4]

Consequence:	**Negligible (V)**	**Marginal (IV)**	**Serious (III)**	**Critical (II)**	**Catastrophic (I)**
Tolerable Frequency:	1 / 10 years	1 / 100 years	1 / 1,000 years	1 / 10,000 years	1 / 100,000 years
Tolerable Frequency (per year):	.1	.01	.001	.0001	.00001
Likelihood in Any Given Year:	10%	1%	0.1%	0.01%	0.001%

[4] Each company should develop its own table as appropriate to its own circumstances.

"Asset Protection, Applying Safety Life Cycle Methods"

Richard R. Dunn
William H. Johnson
Victor J. Maggioli
Stanley A. Urbanik

Prepared for Presentation at
American Institute of Chemical Engineers
2006 Spring National Meeting
8th Process Plant Safety Symposium
Orlando, Florida
April 24-26, 2006

February 2006

UNPUBLISHED

"Asset Protection, Applying Safety Life Cycle Methods"

Richard R. Dunn
William H. Johnson
Victor J. Maggioli
Stanley A. Urbanik

Abstract

Today's programmable electronic technology allows automation of chemical processes to an ever increasing level of complexity. Benefits are being achieved in product manufacturing, throughput, and quality. However increased automation can result in complex functional safety issues (i.e., safety-related process hazards). Functional safety issues are addressed by American Institute of Chemical Engineers (AIChE) - Center for Chemical Process Safety (CCPS) (1)(2)(8), Instrumentation Systems and Automation Society (ISA) (3)(6)(7), and International Electro-technical Commission (IEC) (4)(5).

These resources [(1) through (8) inclusive] each utilize the functional safety life cycle as a tool to ensure all aspects of functional safety are addressed. Many safety issues have an asset dimension (e.g., loss of production, equipment damage) that has to be considered. This paper addresses non-safety related asset issues (i.e., process hazards resulting in financial loss only). This paper is based on how one chemical company addresses asset issues using the safety life cycle as an asset protection life cycle (see Figure 1).

1. Key Terms:

Approved Independent Backup (AIB)—a non-electrical/instrument safety layer, such as a relief valve that will prevent overpressure in a vessel. The AIB must prevent an initiating event from resulting in a hazardous event. There are a variety of devices, barriers and materials that can be used as an AIB if selected and used properly. The choice of a safety interlock implementation based on the existence of an AIB assumes that the process is only operational when the AIB is present and functional.

The backup should be approved (see above definition) and have the following features:

- Uniqueness:
- The backup should be designed to prevent specific hazardous events. It should not have other functions in the normal operation of the facility.
- Independence:
- The backup should be independent of other safety features such as safety interlocks.
- Dependability:
- The backup should perform with a high degree of reliability, have high availability and should not require human action. Some examples are dikes, relief valves, containment vessels, barricades, and restricted access.
- Auditability:
- The backup should be designed to facilitate validation/testing. It should be tested or inspected on a periodic basis.

Asset Protection (AP)-mitigates or prevents significant damage to equipment and or significant loss of production. AP has three classifications (see Figure 2 and definitions below):

Basic Process Control System—a system which responds to input signals from the process, its associated equipment, other programmable systems and/or an operator. It generates output signals causing the process and its associated equipment to operate in the desired manner. It can be suitable for performing AP-1, -2, and -3 protection functions when properly designed.

Class AP-1 event—a hazardous event that results in significant damage (e.g., >1 million dollars) to equipment and/or significant loss of production. It does not result in serious injury to personnel; it does not have significant environmental impact. In the judgment of the classification team, the event merits protection by two independent process interlocks, typically implemented in the BPCS.

Class AP-2 event—the same as a Class AP-1 event except that, in the judgement of the classification team, the event is adequately prevented by one AIB and merits protection by a single process interlock, typically implemented in the BPCS.

Class AP-3 event—an event which represents a low-risk business loss, or an AP-1 event which in the judgment of the PHA team is adequately mitigated by two or more AIBs. Class AP-3 events are adequately mitigated by existing corporate Electrical and Instrument (E&I) design practices that are added to, or part of, the Basic Process Control System (BPCS).

Hazardous event—an undesired, dangerous release of materials or energy (e.g., toxic/corrosive discharges, fires, and explosions) with potential for causing serious injury to people and/or significant property or environmental damage. It may result from a single unplanned event or sequence of events.

Layer of Protection Analysis (LOPA)–a simplified form of risk assessment. See the LOPA text book (8).

Process hazard—a condition that has the potential to result in a release of, or exposure to, a hazardous substance, or in the release of energy (mechanical or chemical), which in turn can cause serious injury to personnel, significant property damage or significant environmental harm.

Process Interlock—a system or function that detects an out-of-limits (abnormal) condition or improper sequence and either halts further action or takes corrective action. A process interlock operates automatically; no operator action is involved. It consists of a sensing function, a control function, and a final control element. A process interlock is generally part of the BPCS and is carried out with the same equipment and the same level of security as normal sequencing and other start/stop functions in the BPCS. Process interlock is synonymous with non-safety interlock. Failure of a process interlock will not result in serious injury to personnel or have significant environmental impact.

NOTE: A process interlock provided as mitigation for a Class AP-1, AP-2, or AP-3 event requires periodic tests at a frequency appropriate to the required risk reduction, to ensure reliability is maintained.

Safety Interlock—a system or function that detects an out-of-limits (abnormal) condition or improper sequence and brings it to a safe condition. A safety interlock operates automatically; no operator action is involved. It consists of a sensing element, a control function, and a final control element. The safety interlock control function (logic solver) must be separate from the logic solver performing the BPCS control functions. A safety interlock deals with Safety Integrity Levels (SIL) 1, 2, and 3 only, which is a quantitative measure of probability to fail dangerous. An accepted synonym is safety instrumented function (SIF). This paper does not address safety interlocks.

2. Abbreviations:

- AIB = Approved Independent Backup
- AP = asset protection
- BPCS = Basic Process Control System
- FTA = Fault Tree Analysis
- LOPA = Layer of Protection Analysis
- MTTFd = Mean Tine to Fail (dangerous)
- PFD = Probability of Failure on Demand
- PHA = Process Hazardous Analysis (consistent with definition in OSHA 29CFR 1910.119)
- SIF = Safety Instrumented Function
- SIL = Safety Integrity Level

3. Introduction - The asset protection life cycle (see Figure 1)

The asset protection life cycle uses many of the safety life cycle tools with some modification. These include:

- Extension of the functional safety classification method to encompass asset protection,
- Expansion of the screening method for hazardous event classification to include asset protection interlock classification (see Figure 2)
- Implementation of asset protection interlocks (process interlocks) utilizing architectures that were previously developed to address safety interlocks
- Use of instrumentation devices' quantitative reliability data (available as a result of the development of prior use, certified, and compliant safety devices) to develop mean time to failure (dangerous and spurious) calculations (6)
- Integrating asset protection terminology with the comparable terminology found in the safety life cycle; see forthcoming "Guidelines for Safe and Reliable instrumented Protective Systems"(2).

While process asset protection issues are typically the domain of each company, it is hoped that this paper may initiate comments, ideas, critique, and constructive alternatives to the methods presented in a way that the process sector can benefit from the resulting dialogue.

3.1 Classification of process hazardous events - Extension of the functional safety classification method to encompass asset protection

The company's PSM guideline states, "The classification of events involving injuries and environmental impact where a safety interlock is present or being considered shall be part of the process hazards analysis process. Events that involve financial loss only may be classified at this time or delegated to another team for further analysis and final classification. The PHA team should determine that the impact of the event is limited to financial loss. Classification of events of concern and the appropriate interlocks should be done by a multi-disciplined team that includes engineering, operations, and maintenance input. For events involving injuries and environmental impact a PHA team is frequently the classification team."

In most cases the classification for financial loss events is initiated in the PHA and finalized by multi-disciplined teams. With the advent of classifications (safety, injury, environmental) as part of the safety life cycle (3) the company developed an extended asset protection life cycle classification scheme for addressing significant equipment damage/production loss.

3.2 Asset Event Classifications - Expansion of the screening method for hazardous event identification to asset protection interlock classification

The Company utilizes an internally-developed screening process known as the AIB method to classify hazardous events to establish the design basis for interlocks. The AIB classification method features a straightforward procedure integrated into the PHA process. The AIB method considers only the credibility of an event and the effectiveness of AIBs in determining the reliability requirements of the interlock system.

The AIB method is intended to be conservative. That is, the classification (interlock requirements) under this method may be more stringent than the classification determined by a risk-based method (e.g., LOPA, FTA).

This paper discusses application of the AIB method to asset loss hazards only.

Each asset event is classified based on whether it has no AIB (AP-1), a single AIB (AP-2), multiple AIBs or whether the event consequence meets the definition for significant financial loss (AP-3). Figure 2 illustrates the classification scheme.

The classification is defined in a way that establishes the minimum design approach. This prescriptive method mandates 2 channel (i.e., sensor through logic solver to final element (synonym: 2 process interlocks)) asset protection interlock architecture for AP-1 classifications. For AP-2 classifications a single channel is adequate. For AP-3 classifications the team determines if an interlock is required.

Classification of AP-1, AP-2, and AP-3 events is documented and distributed to the business sector and the E&I design team. However, asset protection interlock information is kept separate from the safety interlock information.

Note: For extremely high consequence or complicated hazard scenarios, the classification team may choose to use more quantitative risk-based methods (e.g., LOPA) to define the asset interlock risk reduction required. Whichever method was used, the classification team documents the findings.

The following seven steps (see Figure 2) provide an overview of the AIB classification method:

1. Establish site ground rules for significant asset risk-related events prior to classification of events. These ground rules apply to the AIB method only.

   Ground rules for assessing significant asset loss and acceptable AIBs are established by a PHA team and/or the management of each site and, in some cases, for individual areas of a site. The following are examples of ground rules established for past projects:

   - Significant asset loss—property or business loss greater than $2MM; or property loss greater than $1MM; or outage lasting more than two weeks; or more than five days' lost production. A cost-benefit analysis may be required to understand the value of risk reduction provided by interlocks.

   - Potential AIBs (all are application-specific)— relief valves, dikes, check valves, weak roof to shell seam in storage tanks, flow restrictors, pump recirculation lines, retention ponds.

2. Select a hazardous event that was identified in a PHA.
3. Determine if an interlock will provide effective protection against the event being considered.
4. Determine if the event consequence has safety or environmental impact, or if it affects only asset losses. Events that pose safety or environmental risks are not discussed further in this paper.
5. Determine if the event results in significant asset loss. If it does the initial classification is AP-1. If not, the initial event classification is AP-3.
6. Identify all AIBs for the event being considered. If the event classification team cannot reach consensus on whether a protective device can be considered as an AIB, then a risk-based method (e.g., LOPA) should be used for the event being considered.
7. Determine final classification for design purposes, based on the number of AIBs- which apply. For any events classified as an "AP-1" event, consider using LOPA to further refine the requirements for interlock design.

The results should be documented (see Figure 4 for an example of the documentation). Sensors and final elements used to prevent the event should be identified in the documentation.

4. Design – Application of SIF design techniques to design of asset protection interlocks

The design of Safety Instrumented Functions (SIF) for SIL 1, SIL 2, and SIL 3 performance requires the use of voting sensors, logic solvers and final elements. This results in various single and multi-channel architectures. Skills necessary to design these safety interlock systems include an understanding and implementation of:

- Features to mitigate common mode problems
- Fail-to-safe application software
- Embedded software security techniques
- Fault tolerance techniques
- Systematic error reduction techniques
- Quantification of the reliability of the design architectures (i.e., PFD, SIL, and spurious trip rate)

The skills noted above are used in the development of process interlock architectures to provide effective interlock designs for asset protection.

5. Equipment Design - Use of instrumentation device(s) quantitative reliability data (available as a result of the development of prior use, certified, and compliant safety devices) to develop mean time to failure spurious calculations (6)

A major contribution realized from the quantification of SIF reliability was the need to quantify the reliability of sensors, logic solvers, final elements, networks, and related devices for their use in safety interlocks. A number of methods to accomplish this emerged. They included:

- Manufacturers developed new products to meet global functional safety standard's requirements (4).
- Manufacturers quantified the reliability of their off-the-shelf products to determine if they could immediately compete for a share of the safety market (e.g., does their off-the-shelf transmitter meet the requirements for a SIL 1, 2, or 3 claim limit?).
- Users and manufacturers combined their performance experience to quantify device(s) PFD using "prior use" data (3).

Today user's, integrators, manufacturers etc have access to reliability data in a way never before so readily available. The result is that asset protection interlock design can be readily quantified so as to reduce dangerous failure (e.g., leading to equipment damage) or to reduce spurious trips (e.g., causing loss of production).

Available sources of reliability data include:

- Corporate reliability databases
- Vendor data
- Instrument Reliability Handbook (ISA)
- OREDA (SINTEF)
- Process Equipment Reliability Database (CCPS)

The use of reliability data requires certain considerations, including:

- The device has been applied according to the manufacturer's specifications and installation instructions.
- Determine if the reliability data includes process effects (e.g., impulse line pluggage)
- Determine if the reliability data for valves includes fails-safe action (e.g., a pneumatic actuator with a spring to drive the valve to its safe position).
- Determine if the reliability data is based on de-energize-to-trip or energize-to-trip application approach.
- Determine the architectural constraints assumed in the reliability data (e.g., logic solver is equipped with redundant power supplies, communications, and main processor).

Using reliability data from any external source requires analysis to determine whether the data is valid for the application, since the data is very application-specific. Plant experience may dictate higher or lower device reliability numbers be used.

The Corporate interlock standards include example process interlock designs for AP-1, AP-2, and AP-3 events. The PFD and/or spurious trip rate is calculated for each example design.

6. Conclusions

The ability to use the techniques associated with the safety life cycle and the resulting contributions made by manufacturers, integrators, and users has allowed asset protection capability to be enhanced for everyone. The users must decide if their application warrants the considerations noted in this paper.

It is also very important to understand that the safety life cycle is a global tool (4) (5) thus making this approach potentially appropriate for all international applications.

7. References

(1) "Guidelines for the Safe Automation of Chemical Processes", Center for Chemical Process Safety of the American Institute of Chemical Engineers, New York, 2000, ISBN No: 0-8169-0703-X.
(2) "Guidelines for Safe and Reliable instrumented Protective Systems", Center for Chemical Process Safety of the American Institute of Chemical Engineers, New York, 2000, ISBN No: Projected Publication 2006.
(3) "ANSI/ISA 84.00.01-2004 (IEC 61511 mod), Functional Safety: Safety Instrumented Systems for the process industry sector" Parts 1, 2 and 3.
(4) "IEC 61508 - Functional Safety of electrical/electronic/programmable electronic safety-related systems", Parts 1, 2, 3, 4, 5, 6, and 7.
(5) "IEC 61511 Functional Safety: Safety Instrumented Systems for the process industry sector" Parts 1, 2 and 3.
(6) "ANSI/ISA TR84.00.02-2002 Part 2, Safety Instrumented Functions (SIF)—Safety Integrity Levels (SIL) Evaluation Techniques

(7) ISA TR84.00.02-2005 Part 2, Example Implementation of ANSI?ISA 84.00.01-2004 (IEC 61511 mod).
(8) "Layer of Protection Analysis-Simplified Process Risk Analysis" Center for Chemical Process Safety of the American Institute of Chemical Engineers, New York, 2000, ISBN No:0-8169-0811-7.

Figure 1. Asset Protection Life Cycle

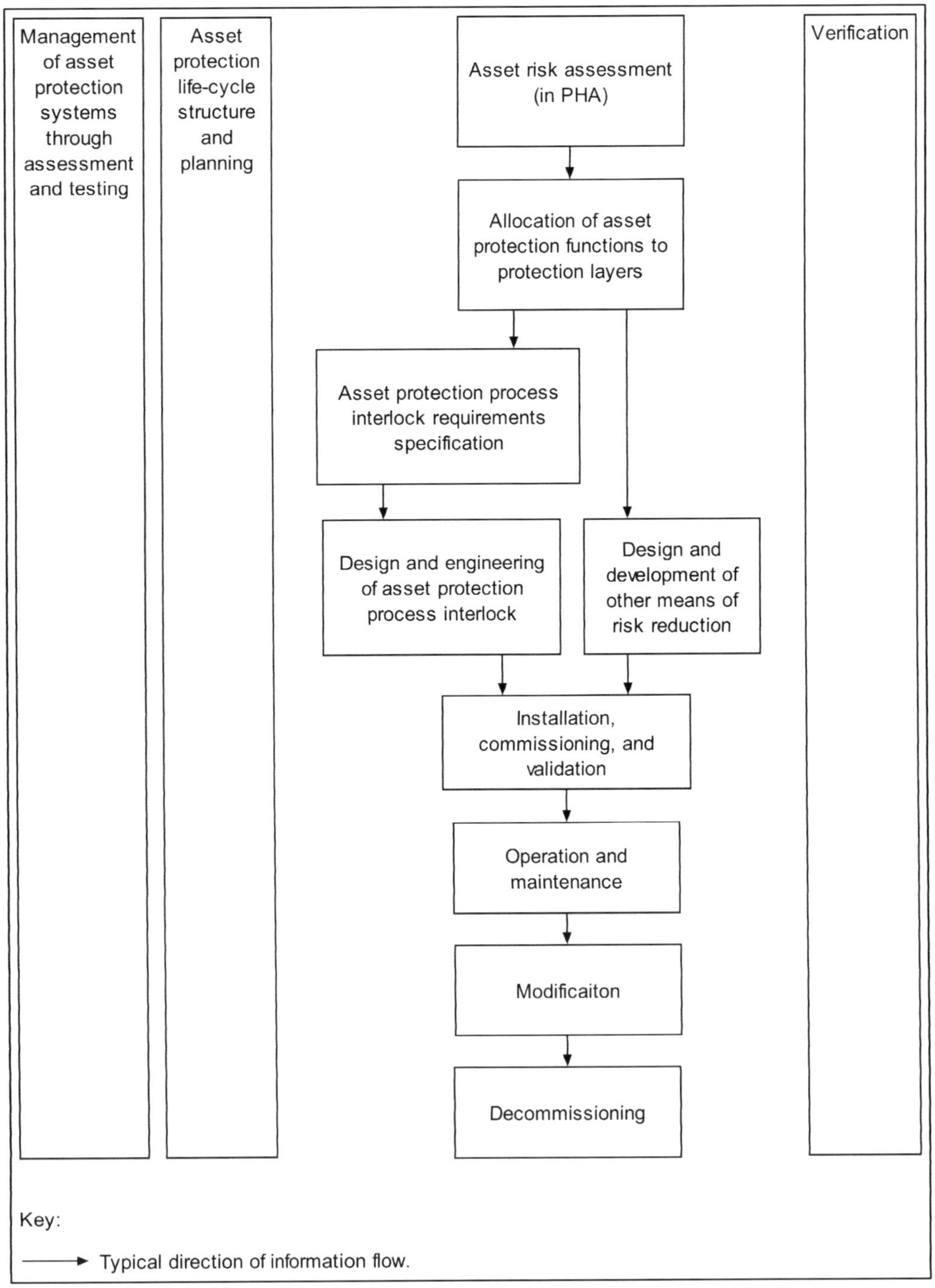

Figure 2. Decision Tree For Asset Protection Interlock Classification

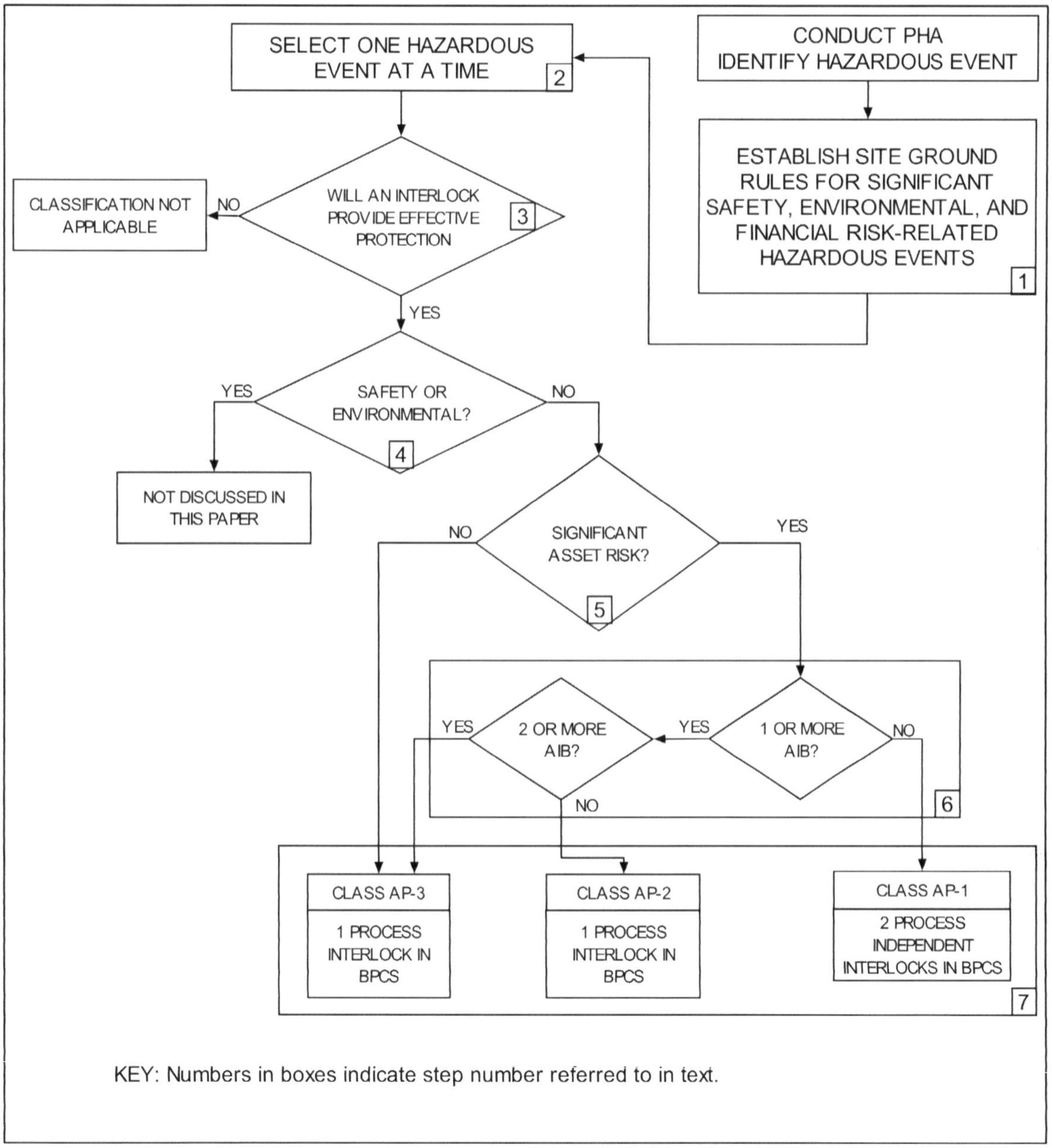

Figure 3. Layers of Protection

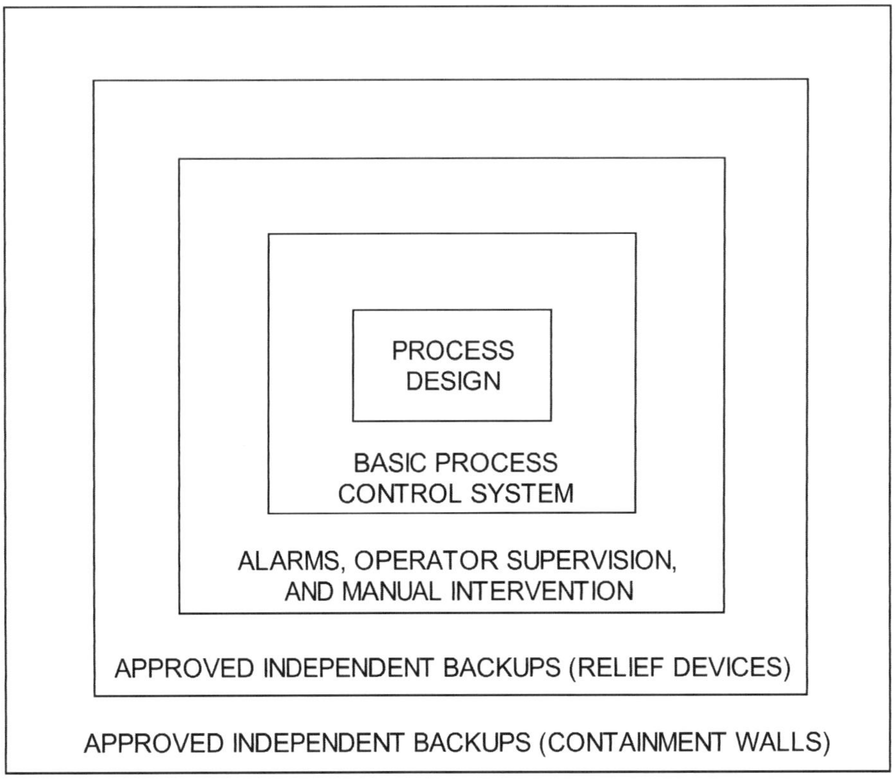

Figure 4. Asset Protection Event Classification Example Documentation

Equipment piece	Event	Event class	Causes considered	Classif- method	AIB	Design	Preventive Instrument tag #	Notes
Tank, pump, compressor, etc.	Overfill, over-temp, overpress, etc.	AP-1	Failure of control loop component, operator error, etc.	Prescriptive or LOPA	RV, dike other	2 channel	450 LT, 630 LV, etc.	
Example storage tank 1507-2033-01	Overfill	AP-2	Level control loop failure	Prescriptive	Dike	1 channel	5708 LT, 5708 LV, 5708 SV,	Vapors from fluid contained in dike are not hazardous.

190f Panel Discussion - Safety Instrumented Systems and Safety Critical Devices

Harry West, Texas A&M University, College Station, TX

The panel discussion, moderated by Harry West, will focus on Safety Instrumented Systems and Safety Critical Devices. In an interactive format, session presenters will address questions raised by attendees, discuss topical issues, and provide valuable insights in this field.

A Case Study of a TFE Explosion in a PTFE Manufacturing Facility

Ali Reza, P.E. and Erik Christiansen, Ph.D., P.E.
Exponent, Failure Analysis Associates
5401 McConnell Ave.
Los Angeles, CA 90066
(310) 754-2700
echristiansen@exponent.com

Abstract

A 1999 explosion in the purification area of a fluorinated polymer manufacturing plant in the United States caused fatal injuries to three employees. An additional employee was severely injured. The affected plant personnel were attempting to locate a suspected polymer plug in lines containing Tetrafluoroethylene (TFE) monomer upstream of the polymerization plant. In an attempt to dislodge a possible plug in an elbow leading off a pressure vessel, one of the employees rapidly actuated a manually operated ball valve. High-pressure TFE upstream of the valve rushed into the downstream piping, which likely contained air at sub-atmospheric pressure. Adiabatic compression against a blind flange created temperatures sufficient to ignite a TFE-air mixture. The initial deflagration provided sufficient energy for a self-sustaining decomposition reaction to propagate within the TFE line into the upstream pressure vessel. Liquid TFE had been collecting in the pressure vessel due to a deliberate increase in the flow of cooling brine associated with a separate process. This allowed the decomposition to continue, causing a rapid increase in pressure inside the vessel, sufficient to fail the bolts on the upper flange. The four employees were exposed to burning TFE sprayed from the pressure vessel and were also injured by shrapnel created by the explosion.

Exponent Failure Analysis Associates was retained to conduct an engineering investigation of the incident, including a technical analysis of the conditions leading to the explosion. We also reviewed the HAZOP performed by the employer that failed to note critical parameters that contributed to the incident. This paper discusses our findings as well as lessons on handling TFE that can be learned from the incident.

1. Introduction

1.1 Background

Tetrafluoroethylene (TFE) is manufactured for the purpose of creating poly-tetrafluoroethylene (PTFE) polymer, which has applications as a coating in non-stick cookware and other items. TFE vapor can form flammable mixtures in air, with a lower explosion limit of 10% at standard temperature and pressure. [1] The upper flammability limit at one atmosphere has been measured to be anywhere between 40% and 100%, depending on the size of the apparatus and the strength of the ignition source used. The TFE/air combustion reaction is [1]:

$$C_2F_4 + O_2 \rightarrow 2COF_2 \qquad \Delta H = -150 \text{ kCal/mol}$$

Pure TFE can undergo a self-sustaining exothermic decomposition reaction under certain conditions if the ignition source is strong enough.

$$C_2F_4 \rightarrow CF_4 + C \qquad \Delta H = -66 \text{ kCal/mol}$$

The reaction front can propagate through TFE vapor in a manner similar to a deflagration. TFE decomposition fronts have also been shown to transition to detonation in long pipelines when the initial pressure is greater than 65 psig. [1]

1.2 TFE Purification Process

After production, TFE monomer is typically stored with terpene added as an inhibitor to prevent premature polymerization. [2] However, any additives must be removed before the polymerization reactors. In the subject process, the majority of the terpene was removed in a distillation tower followed by an absorption tower packed with silica gel to remove any remaining additive.

The plant has two parallel trains to handle the TFE flow as shown in Figure 1. Each train includes a distillation tower and two silica gel towers in parallel to allow continuous throughput during silica gel replacement. Flow through each train is controlled by a valve downstream of the silica gel towers. High pressure nitrogen, vacuum and low pressure TFE recovery connections attached to the top of the silica gel towers are used to dilute and purge the lines before they are opened.

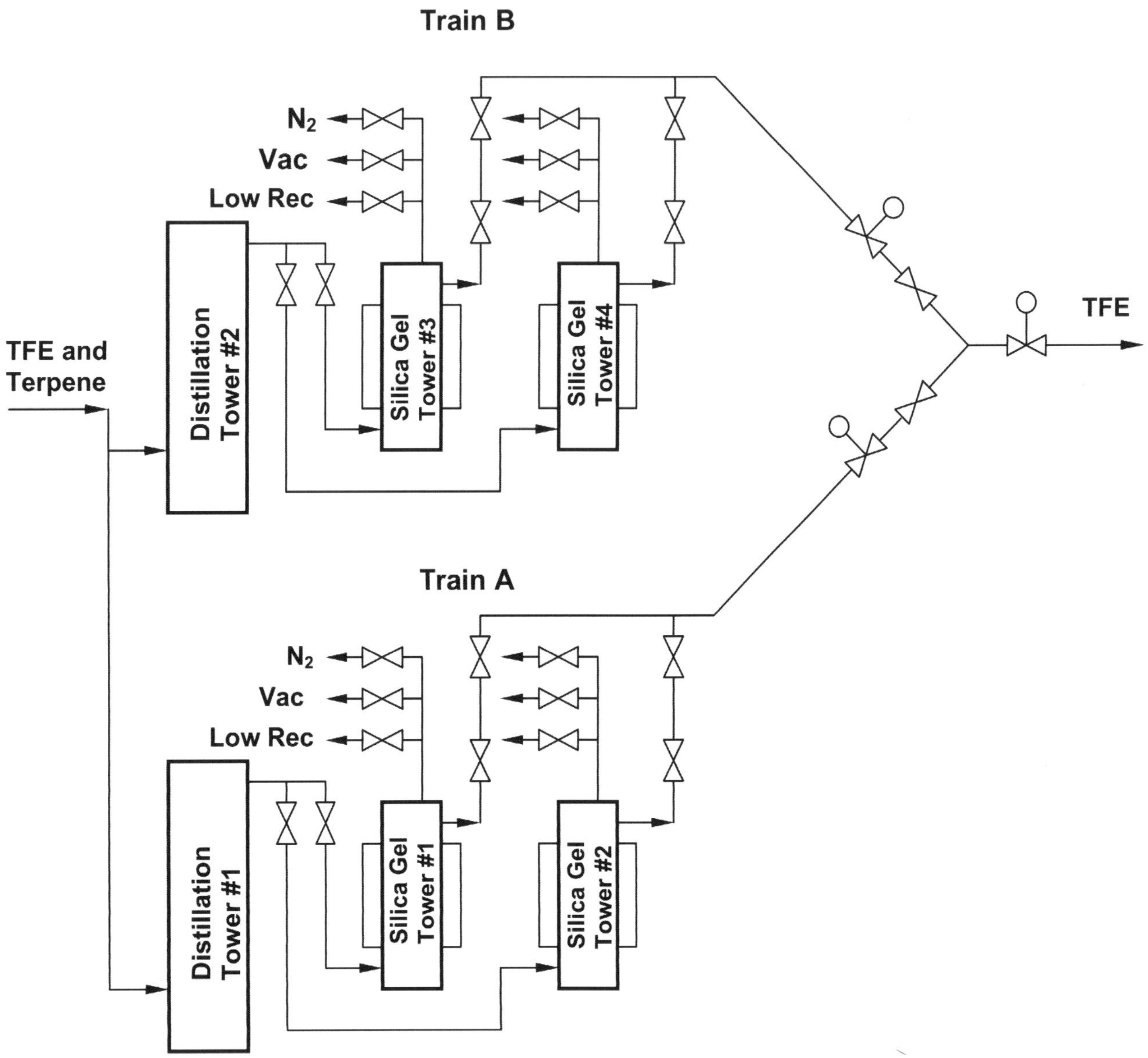

Figure 1. Schematic of Terpene removal system (illustrative only).

The silica gel towers are ASME Section VIII, Division 1 pressure vessels, 10 feet tall and 16 - inches in diameter. They are constructed from ⅜-inch thick stainless steel pipe and rated for a maximum allowable working pressure of 300 psig at a temperature of –40° F. The top and bottom of each tower towers consists of flat flanges held on with 20 ¾-inch bolts. Each tower is surrounded by a 20-inch diameter brine coolant jacket that allows operating temperature of 15 to 25 ° F in order to prohibit polymerization and runaway reaction.

Each tower holds approximately 225 pounds of silica gel, sufficient to process approximately 500,000 pounds of TFE before it has to be replaced. Since TFE reacts exothermically with silica gel, the tower is isolated and the replacement media is slowly saturated with TFE to prevent overheating or a runaway reaction. The saturation process typically takes 24 hours to complete.

1.2 Incident Description

Workers at the plant noticed that the flow of TFE through one of the trains was very low. This was a known occurrence in the uninhibited TFE lines and typically occurred due to excess polymer build-up. Accordingly, they shut the train down to check for restrictions in the lines. They locked out and tagged out the ball valve at the exit of the online silica gel tower (#4), the ball valve upstream of the offline silica gel tower (#3), and the ball valve just ahead of the "Y" connection where the two TFE trains converge. They used the offline silica gel tower (#3) to vent the high pressure TFE from the lines to the low-pressure recovery system and then pulled a vacuum of 25 to 27 inches of mercury before pressurizing with nitrogen up to 10 to 30 psig. The workers repeated the nitrogen replacement and vacuum on the line several times before disassembling the line between the locked valve at the end of the train upstream of the "Y" connection and the control valve. No restrictions were found and the control valve was found to operate properly when manually stroked. The line was reassembled and nitrogen replacement was performed. The line was left under vacuum and the employees unlocked the ball valve between the control valve and the "Y" connection. An employee cracked open the ball valve to allow TFE to flow backwards into Train B from Train A. The employees heard TFE flowing, and thus confirmed that there was no blockage in the ball valve.

The ball valve was closed and locked out again. The employees then focused their attention on the line at the exit of silica gel tower #4. They performed the vacuum and nitrogen purge procedure once again and then removed 4-foot spool section of 2-inch diameter pipe immediately downstream of the valve at the exit of silica gel tower #4. No polymer plug was found, and they reassembled the line and once again performed a purge by introducing nitrogen at 30 psi and pulling a vacuum. The purge process was repeated several times.

Suspecting that the plug might be in the short elbow connecting the silica gel tower to the exit ball valve, the workers decided to open the ball valve rapidly with the line under vacuum in an attempt to dislodge the plug. They removed the lock-out/tag-out locks and rapidly opened the valve about halfway when the explosion occurred. Three workers in the vicinity of the silica gel tower were killed and one was severely inured when the top flange of silica gel tower #4 failed, releasing burning TFE vapor and shrapnel. The one surviving worker reported hearing a "pop" and seeing a flash within one or two seconds of the valve being open. He saw flames come out of the valve and was knocked down by the overpressure.

2. Accident Investigation

After the incident, the upper flange on silica gel tower #4 was found to have blown off due to the force of the explosion, with all 20 ¾-inch diameter retaining bolts failed in tension. Evidence of soot was found on the underside of the flange. The tower itself was pushed downward through the expanded metal floor by the force of the blast.

Disassembly of the ball valve that was being actuated at the time of the accident also established evidence of soot deposition in the interior of the valve, as shown in Figure 2. The PTFE seal removed from the valve had soot deposits and evidence of blow-by of hot gases, as shown in

Figure 3. Disassembly of all of the lines found no evidence of a polymer plug. The control valve upstream of the "Y" connection was disassembled and found to have relatively less soot inside of it. A cross-section of this valve is shown in Figure 4. The ball valve located between the control valve and the "Y" connection was also disassembled, as shown in Figure 5. The downstream side of the ball is clean, indicating the valve was closed at the time of the incident and did not leak downstream. The PTFE gasket removed from the valve shows small soot stains consistent with the edges of the ball, as shown in Figure 6.

A vent line with a rupture disc was attached to the top of each of the silica gel towers. The rupture discs were 1 inch in diameter and rated for 280 psig. The rupture disc attached to silica gel tower #4 was found to have failed, as had the rupture disc on an unused silica gel tower that was connected to the same vent line. The connected tower did not fail, but did show evidence of soot in the interior.

Figure 2. Ball valve immediately downstream of silica gel tower #4.

Figure 3. PTFE seal removed from ball valve downstream of silica gel tower #4.

Figure 4. Cross-section of control valve upstream of "Y" connection.

Figure 5. Ball valve located between control valve and "Y" connection.

Figure 6. PTFE gasket removed from ball valve between control valve and "Y" connection.

3. Discussion

3.1 Liquefaction of TFE in the Silica Gel Tower

Review of internal documents indicated that an operator had increased the rate of cooling brine flow through the silica gel towers in order to speed the saturation of silica gel that had recently been replaced in silica gel tower #1. The cooling brine loops through all of the silica gel towers without individual controls. The increase in cooling brine flow through the towers dropped the temperature in silica gel tower #4 sufficient to liquefy TFE and cause it to accumulate in the tower. The silica gel towers were originally designed to operate at a pressure of 157 psig and a temperature range of 15 to 25° F. In 1998, a process change was implemented to increase the operating pressure to 167 psig in order to increase throughput from the trains. A review of the process change management documentation indicates that no consideration of liquefaction was made. Liquefaction of TFE in the silica gel towers was also not considered during the original process hazard analysis. Consequently, workers were not aware of the possibility of liquid TFE accumulation.

During the period preceding the incident, the temperature within silica gel tower #4 was found to be 10° F, which corresponds to the intersection of the vapor dome with the operating pressure of 167 psig (Figure 7). This verifies that the observed reduced throughput through Train B was a result of liquid accumulation as opposed to a plug or blockage. The brine flow rate had been increased approximately 24 hours before the accident occurred, allowing enough liquid accumulation to nearly fill the silica gel tower to the exit opening by the time the incident occurred. The absence of liquid level indicators in the tower, or other controls such as temperature and pressure alarms that would indicate the presence of liquid TFE, allowed this condition to go unnoticed.

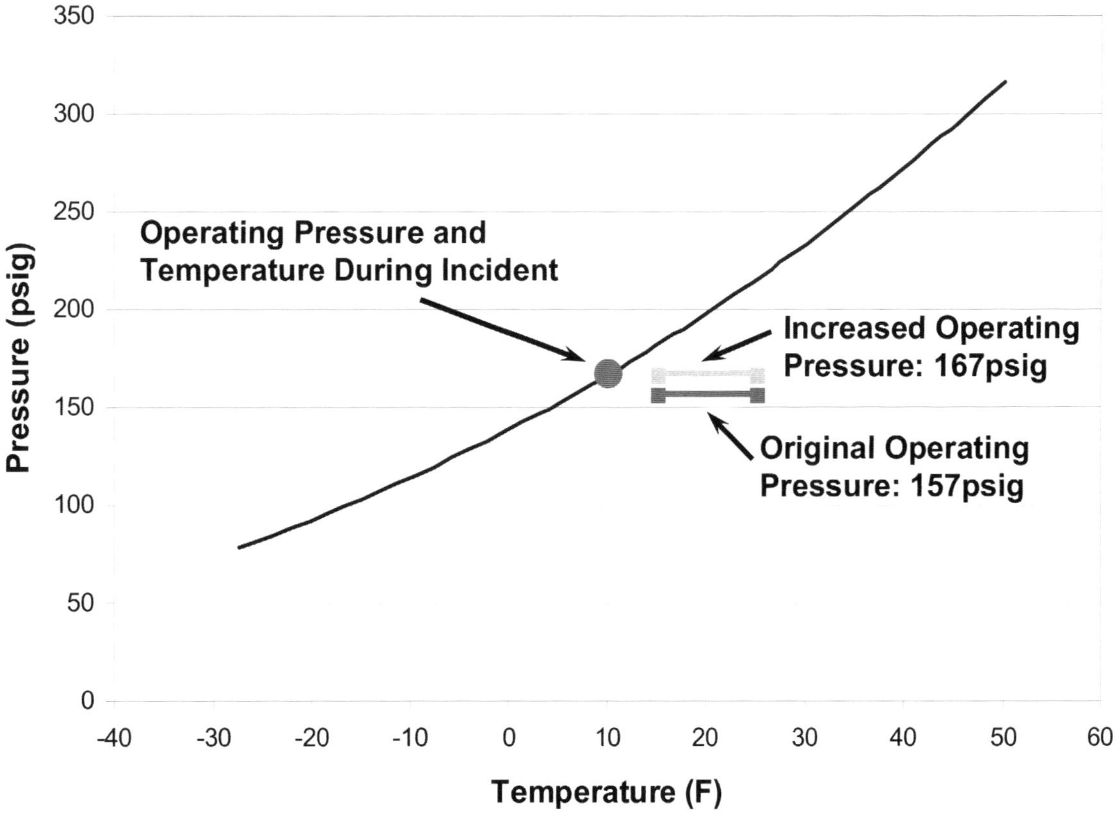

Figure 7. Vapor pressure of TFE with silica gel tower operating conditions.

3.2 Ignition Mechanism

Eyewitnesses indicate that the initiating event for the subject explosion was the rapid opening of a ball valve at the exit of silica gel tower #4. The partially open ball valve allowed high pressure (167 psig) TFE vapor to rapidly expand into piping that had been evacuated to a vacuum of 25 to 27 inches of mercury. Although the piping had previously been purged with nitrogen, witnesses report that a leak check was not performed and gaskets may have been reused. In addition, a final leak check had not been performed on silica gel tower #3, which was used to perform the vacuum and nitrogen purge procedure on the subject piping. Therefore, it is likely that air had been drawn from the ambient into the piping while it was under vacuum.

The high pressure TFE compressed the air inside the piping, causing an increase in temperature. Calculations by Exponent confirmed that, at this compression ratio, the air temperature can reach 1360° F, which exceeds both the autoignition temperature of TFE/air mixtures and the temperature required to initiate the self-decomposition reaction in pure TFE. The process is sufficiently rapid that heat transfer effects are negligible and therefore the compression phenomenon can be considered adiabatic until ignition is achieved. The calculated temperature profile at the interface between the TFE and air is shown in Figure 8 for time increments following compression. The results indicate that a TFE/air deflagration can be initiated at the interface. The deflagration can initiate a self-decomposition reaction front in the pure TFE that will propagate through the piping and into the silica gel tower. Previous work by Couture [1] has demonstrated that rapidly compressing air in 3-inch pipes could create temperatures that are sufficient to ignite TFE. This mechanism is also consistent with the witnesses observing a flash of flame at the valve 1 to 2 seconds after the operator opened it.

The soot deposition on the interior of the piping, valves and the silica gel tower is consistent with this ignition mechanism. Sooting at the ball valve just downstream of the control valve, the likely location of hot, compressed air necessary for ignition, is light compared to piping closer to the tower and the tower itself. TFE decomposes into fluorine and carbon, so it is expected that there will be large amounts of soot produced by pure decomposition and less by TFE/air deflagrations.

Other ignition mechanisms were considered but ruled out as unlikely. For example, TFE can form explosive levels of peroxide. [3] However, peroxide formation is associated with exposure to heat, light and, most importantly, oxygen. There is no evidence that there was any oxygen in silica gel tower #4, as a successful leak check had been performed. No evidence of peroxide or internal damage to the tower associated with a peroxide explosion was found after the incident. Additionally, the increase in cooling of the tower would tend to inhibit peroxide formation. Finally, long induction times can be necessary for peroxide formation, [4] which is not consistent with the timeline of events before the explosion. Spontaneous polymerization was also considered and discounted for similar reasons. The timing of the event with the opening of the ball valve makes it highly unlikely that the polymerization reaction coincidentally experienced runaway at the moment the valve was opened. The low temperature and lack of any evidence of polymerization further discount this possibility.

3.3 Silica Gel Tower Failure

Calculations were performed to determine the internal pressure necessary to fail the 20 ¾-inch bolts that secured the upper flange to the tower. The bolts were constructed of SA-320-L7 steel, with a yield strength to 105 ksi. [5] The pressure inside the silica gel tower required to fail the bolts was calculated to be 3,840 psig. The pressure required to fail the vessel itself, using a thin-walled shell assumption, was calculated to be approximately 1500 psig. The vessel is surrounded by the brine jacket, which itself has a failure pressure of approximately 1300 psig. Due to inertial effects, the combined strength of the coupled inner shell and outer jacket allowed the vessel to sustain pressures in excess of the failure pressure for the top flange.

The pressure required to fail the bolts is much greater than the maximum pressure produced by a TFE vapor decomposition, which in an unvented vessel will reach a final pressure of 1,770 psig. In actuality, TFE vapor decomposition inside the subject silica gel tower will produce a lower pressure due to venting through the vent line after the rupture disc bursts. Calculations were performed using the method of Epstein et al. [6] to determine the maximum pressure in the vessel while modeling venting through the ¾-inch line connected to the rupture disc. The reduced pressure was calculated to be 1,560 psig.

The failure of the bolts attaching the top flange to the silica gel tower can only be reconciled in the context of liquid TFE accumulation in the tower. When the decomposition reaction front propagated back into the silica gel tower, it created an overpressure in the headspace above the liquid TFE of approximately 1,500 psig. However, the decomposition also produced a large amount of heat that boiled off liquid TFE, with the resulting vapor decomposing in the head space. The liquid TFE essentially acted as a source of fuel for the reaction, leading to super-adiabatic pressures. This pressure-piling effect is a documented hazard of liquid TFE storage,

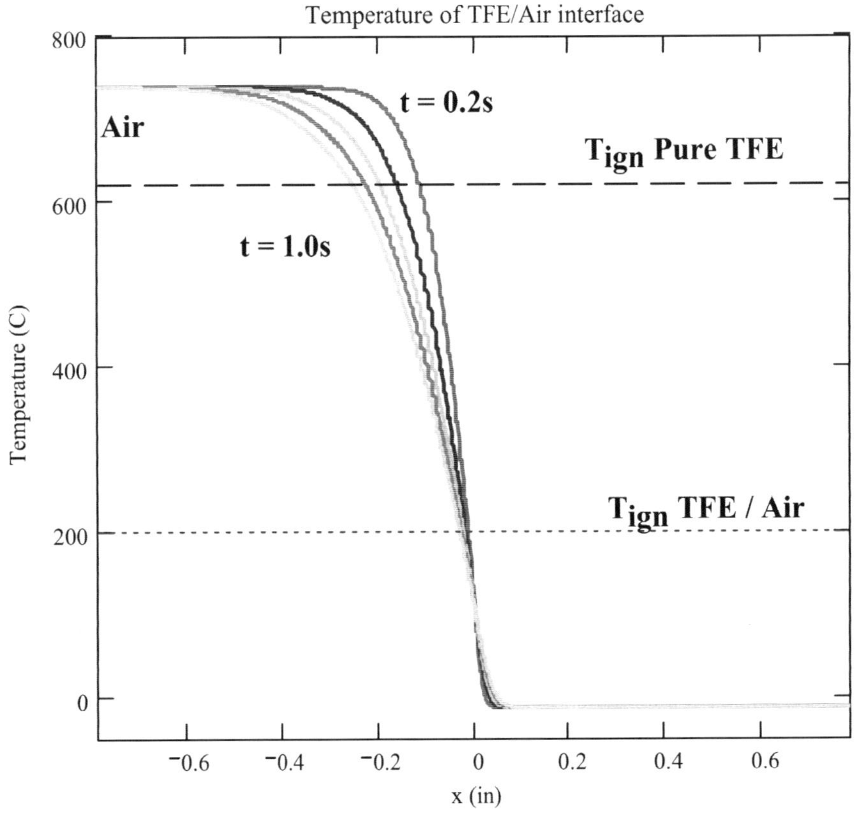

Figure 8. Transient temperature profiles of TFE/Air interface.

and explosion vents have been shown to be ineffective in mitigating overpressure due to this phenomenon. [7] Due to the extreme difference in densities between liquid and vapor TFE, very large overpressures can be achieved from vessels containing liquid TFE. Presence of soot found in the unused tower that shared a vent line with silica gel tower #4 is an indication that the TFE decomposition spread to this tower. However, this unused tower did not fail because it had no liquid TFE present to produce super-adiabatic pressures.

4. Conclusions

1. The original process hazard analysis did not address the hazards of liquid TFE accumulation in the TFE purification area. The accumulation of liquid TFE in the silica gel tower was the primary cause of this incident.
2. The Management of Process Change review when the operating pressure was increased from 157 psig to 167 psig was defective. While this change moved the thermodynamic state of TFE in the silica gel towers closer to the vapor dome and created additional

constraints on the temperature control, the hazards associated with liquid TFE accumulation were not explicitly identified or addressed.

3. There were no liquid level indicators or temperature and pressure alarms or controls installed on any of the silica gel towers. Accordingly, workers had no way of determining that liquid had accumulated within the towers.
4. Using a vacuum purge procedure allowed for the possibility of air infiltrating the lines if a proper seal was not achieved. The workers failed to perform an adequate leak check or replace all gaskets after breaking the lines. This likely allowed air to leak into the lines while they were under vacuum. Oxygen sensors were not used to check the lines for air before introducing TFE. Using a purge process that did not involve drawing a vacuum inside the lines could have potentially avoided this problem.
5. Improper operation of a manual ball valve (rapid opening) allowed adiabatic compression of the air that had leaked into the piping to occur, resulting in temperatures high enough to ignite a TFE/air mixture at the interface. The TFE/air deflagration initiated a self-decomposition reaction front that propagated through the piping and back into silica gel tower #4.
6. Accumulation of liquid TFE in the silica gel tower caused a pressure-piling effect to occur, producing super-adiabatic pressures inside the vessel. The resulting pressure exceeded 3,840 psig, resulting in the failure of the 20 ¾-inch bolts securing the upper flange of the silica gel tower. Without the presence of liquid in the vessel, the silica gel tower would not have failed.
7. The rupture disc and vent connected to the silica gel tower was insufficient to vent the explosion. Literature indicates that TFE decomposition reactions fed by liquid TFE create pressure-piling effects that cannot be vented reliably via a rupture disk arrangement.

5. References

[1] Couture, M.J., "TFE Hazards," Plastics Department, E. I. Du Pont De Nemours & Co., January 27, 1969.

[2] Urban, P.G., ed., *Bretherick's Handbook of Reactive Chemical Hazards,* 5th ed., Vol. 1, Butterworth Heineman, Oxford, 1995.

[3] Kelly, Richard R., "Review of Safety Guidelines for Peroxidizable Organic Chemicals," Chemical Health and Safety, September/October, 1996.

[4] Gozzo, F. and Camaggi, G., "Oxidation Reactions of Tetrafluoroethylene and their Products – I: Auto-oxidation," Tetrahedron, Vol. 22, 1966.

[5] "Specification for Alloy Steel Bolting Materials for Low-Temperature Service, SA-320/SA-320M," ASTM Specification A 320/A 320M-99, 1999.

[6] Epstein, Michael, Hauser, George M., Tilley, J. Brian, Barron, C. H., Wise, James L., Thistleton, Paul, Harper, Robert L., Couture, Maury J., and Blair, B. Robert, "A computer model for the estimation of peak pressure for sonic-vented tetrafluoroethylene

decompositions," Journal of Loss Prevention in the Process Industries, Vol. 3, October 1990.

[7] Lisochkin, Ya A., Ozol, S. I., Poznyak, V. I., and Rykunov, V. A., "Explosion-Proof Method of Storage of Liquid Tetrafluoroethylene," Russian Journal of Applied Chemistry, Vol. 68, No. 6, Part 2, 1995.

Additional references not cited:

[8] *Kirk-Othmer Encyclopedia of Chemical Technology,* 2nd ed., Vol. 9: Ferroelectrics to Foams, John Wiley & Sons, Inc., 1966.

[9] Yaws, Carl, *Matheson Gas Data Book,* 7th ed., Matheson Tri-Gas, Parsippany, NJ, 2001.

Flammable Liquid Process Tank Fire Investigation

James R. Reppermund
James Reppermund Consulting
15 Livingston Drive
Howell, NJ 07731
Phone: 732-367-1269
Fax: 732-367-8674
jreppermund@msn.com

ABSTRACT

In June of 1998 there was a fire incident at a coatings facility in the UK. It was reported that two contractors were injured, both had been transported to the hospital, one was not expected to live.

The facility was shut down pending the outcome of the investigation. Local police and Fire Service personnel were in charge of the official investigation. In addition, the government safety regulatory agency, The Health and Safety Executive (HSE), was involved. In the UK, industrial accidents involving fatalities are considered to be criminal investigations with the police in charge.

Although the fuel source and the ignition source appeared to be somewhat obvious at first blush, the final results were anything but obvious. Open cooperation with local fire and police resulted in the company's incident report being accepted by the authorities.

1. INCIDENT DESCRIPTION

1.1 Background

At approximately 1438 local time, there was a fire incident at a coatings facility in the UK during a tank cleaning operation. At this facility, tank cleaning was done by a contractor. It was reported that two of the contractor's employees were injured, both had been transported to the hospital, one was not expected to live.

The investigation began at 0700 local time the next day. The facility was shut down pending the outcome of the investigation. Local police and Fire Service personnel were in charge of the official investigation. In addition, the government safety regulatory agency, The Health and Safety Executive (HSE), was involved.

The Fire Service informed the company that, since one victim was not expected to live, the incident would be handled as a potential criminal investigation. Local management asked for assistance with their investigation and for a liaison to interface with the various government agencies doing the official investigation. Based on my experience with flammable liquid fire investigations and my knowledge of coatings manufacturing operations, I was asked to fill this role.

This facility had recently been acquired (within the previous 12 months) and was in the process of being integrated into the parent company. This integration process was proceeding at a pace that was intended to be invisible to the plant's customers and have minimal impact on the plant operation. Many procedures and practices had yet to be examined in detail and updated to comply with the parent company's procedures and practices. Tank cleaning fell into this category.

1.2 Tank Cleaning Procedure

This incident occurred on a Monday afternoon as two contractor's employees were working at cleaning a tank that was used to manufacture coatings (paint). This system of tank cleaning had been used at this site for several years and was done by two people who entered the tank wearing appropriate personal protective equipment, a supplied breathing air respirator and a rescue harness. The procedure was to scour the coating residue from the tank interior surfaces using a flammable solvent, in this case xylene, and a section of a non-woven scrubber pad. After this residue was removed, the tank surface was wiped clean using paper towel-type wipers. This system resulted in a very clean tank interior with no residual material.

The tank was approximately 3 meters (9.6 ft.) high and slightly over 2 meters (7 ft.) in diameter. To reach the upper portions of the tank, a wooden ladder was placed in the tank. The solvent, xylene, was available to the cleaner in a 4-liter metal can that had a plastic strap for a handle and was suspended from a rung on the ladder by means of a homemade "S" hook. This 4-liter can was refilled as necessary from a 20-liter metal pail of xylene that was located outside of the tank, adjacent to the tank access port. The contractors took turns in the tank, working 20-30 minutes at a time. Figure 1 shows the approximate configuration of the contractor inside the tank

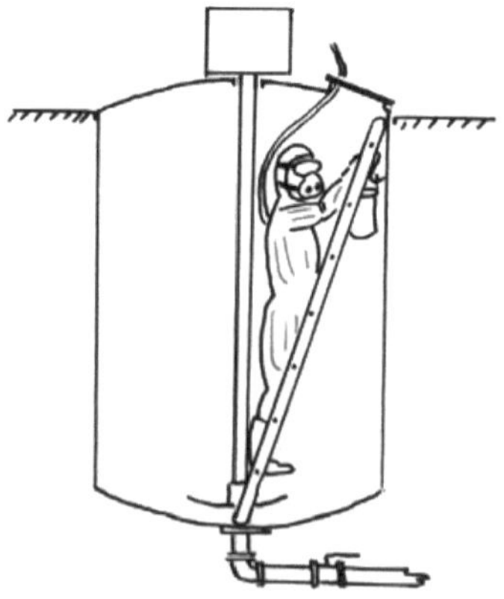

Figure 1

2. THE FIRE

The first contractor worked his 30-minutes in the tank, cleaning the dome. Shortly after the second contractor entered the tank, there was an ignition and the tank interior deflagrated. Fire damage to the facility and equipment was minimal, however, the contractor who was outside of the tank received some second degree burns on his hands and face while the contractor that was inside the tank at the time of ignition suffered third degree burns to approximately 80% of his body surface. This man succumbed to complications from his injuries after about 2 weeks in the hospital. Upon his death, the incident became a criminal investigation under the jurisdiction of the local police.

As most fire investigators know, fires in closed tanks typically self extinguish. In a closed tank, since the combustion process consumes oxygen more rapidly than the fuel, the tank interior becomes oxygen deficient very quickly, about a second or so for a tank of these dimensions, and the fire extinguishes itself. However, the atmosphere in the tank is now quite hot and expanding. This hot, fuel rich and oxygen deficient gas will exit the tank via the access port. As it exits, it mixes with ambient air, becomes a flammable mixture once again, and is re-ignited by the hot gas. The resulting event would most likely be described by an observer as "the tank exploded" (Figure 2).

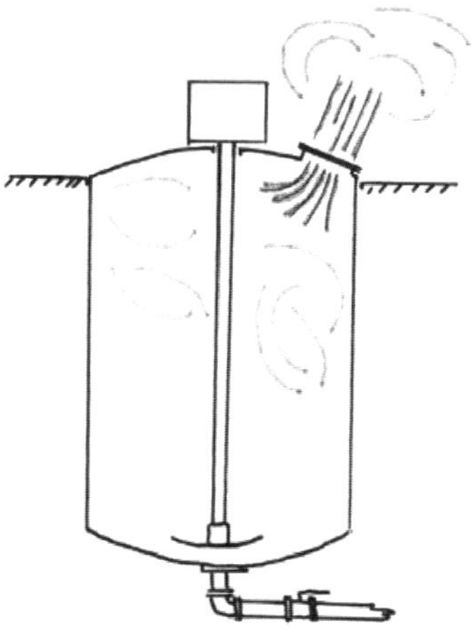

Figure 2

The self-extinguishment of the fire was confirmed the day after the incident when an internal inspection of the tank revealed unburned solvent in the bottom of the tank along with several unburned paper towel-type wipers. Tragically, if the contractor who was working in the tank had stayed in the tank during the event, he would most likely have been unharmed. But, of course, the survival instinct is screaming, "get out of here", and he did.

The investigation resulted in the following ignition scenario.

Outside the tank, the second contractor was looking into the tank when the ignition took place. He (probably) reflexively threw himself backwards, away from the access port. In the process of doing this, he (probably) knocked over the 20-liter pail of xylene sitting beside the access port. This spilled xylene was almost immediately ignited. This pool of burning xylene was the situation that contractor inside the tank exited into. At this point, it is my opinion that this worker was covered with burning xylene either by a spill from the 4-liter pail in the tank or by slipping in the pool fire outside of the tank. He walked out of the fire to the nearest safety shower, which was some distance away. The whole time he was walking to the shower, his clothing was burning and smoldering. It was this extended period of time with the burning/smoldering clothes on that no doubt caused the extensive, and ultimately fatal, burns.

This scenario was found to be consistent with the physical evidence and with most of the eyewitness statements.

3. THE INVESTIGATION

The investigation consisted of determining –

a. What was the fuel source?
b. What was the ignition source?

3.1 Determination of Fuel Source

The answer to the first question seems to be very obvious, until one looks at the fire properties of xylene and the ambient temperature at the time of the incident. Ambient temperature in the vicinity of the tank at the time of ignition was approximately 16.5° C (55° F). When you look at the Flash Point (FP) of xylene, 32° C (90° F), it becomes clear that something else ignited first. Gas Chromatograph (GC) analysis of liquid drained from the tank discharge piping, showed the presence of a significant amount of Methyl Ethyl Ketone (MEK). Subsequent investigation revealed that MEK was, on occasion, added to a dirty tank and left for a period of time to soften the paint residue and thus speed up the cleaning process. The fire properties of these 2 flammable liquids are:

	Xylene	MEK
Flash Point	32° C (90° F)	-9° C (16° F)
Lower Flammability Limit	1.1% @ 32° C (90° F)	1.7% @ -9° C (16° F)
Upper Flammability Limit	7.0% @ 62° C (144° F)	11.4% @ 23° C (73° F)
Minimum Ignition Energy	0.20 x 10^{-3} Joules	0.27 x 10^{-3} Joules

At this point, it was clear that MEK vapor was the fuel ignited inside of the tank.

3.2 Determination of Ignition Source

The ignition source was somewhat less obvious. An extensive list of possible ignition sources was developed. Included on this list were friction, electrical sparking, chemical reaction, metal-to-metal impact sparking, and others. A discharge of static electricity was high on this list. Possible ignition sources were eliminated when they were found to be inconsistent with the facts. A static electricity discharge scenario appeared to be the most likely ignition source. The investigation showed that the 4-liter metal pail, suspended by its plastic handle from the ladder, was an ideal isolated conductor. This conductor could be electrostatically charged by the action of "sloshing" or rinsing the section of non-woven scrubber pad around in the non-polar flammable liquid – xylene. The charged xylene would induce a charge on the metal can. If the can was jostled and contacted the grounded tank wall, there would be a discharge; and if the spark discharge had sufficient energy the MEK vapor would be ignited.

3.3 Verification

To prove this theory, a local (UK) expert in static electricity was retained to re-create the charging scenario and take measurements. The testing proved conclusively that this was a viable ignition scenario. Laboratory and field-testing showed a substantial quantity of charge was generated in the xylene when a section of the scrubber pad was rinsed in it. It was found that as much as 2.5 x 10^{-8} Coulombs of charge was generated each time the pad was rinsed and squeezed to remove excess liquid. The capacitance of the 4-liter pail was measured and found to be between 12 and 64 x 10^{-12} Farads depending on pail proximity to the grounded tank wall (see Figure 3).

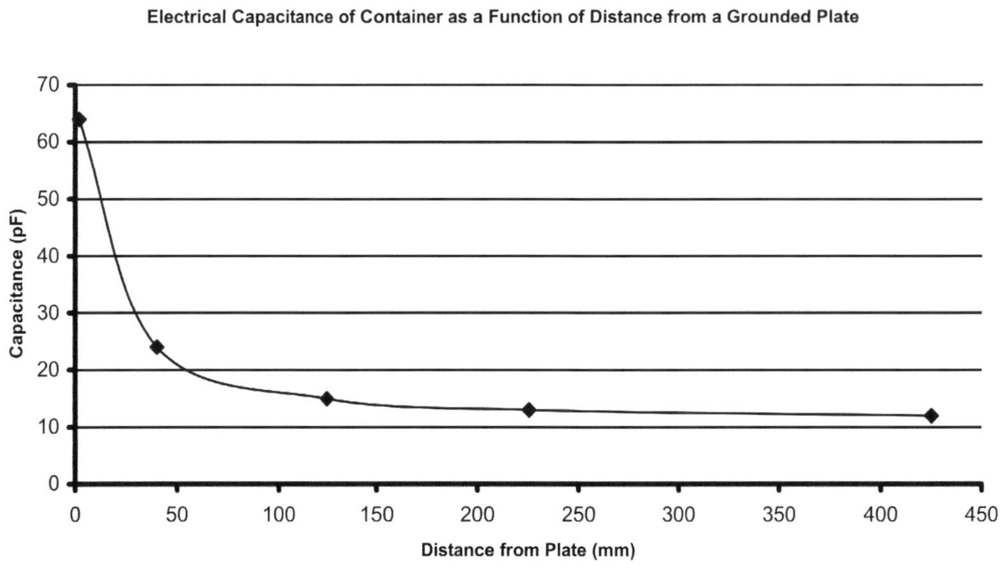

Figure 3

Knowing the capacitance of the pail, the minimum ignition energy of MEK, and the amount of charge generated by each rinse cycle, it is possible to calculate the number of rinses it would take to raise the electrical potential of the pail sufficiently to produce a spark with enough energy to ignite a flammable MEK vapor/air mixture.

The formula for calculating the energy stored on an ungrounded conductor is,

$$\mathbf{W} = \tfrac{1}{2}\,\mathbf{C}\mathbf{V}^2 \quad (1)$$

Where **W** is stored energy in Joules
C is capacitance in Farads
V is potential in Volts

Electrical charge (**Q**) on a conductor, in coulombs, can be expressed as the product of it's voltage and capacitance,

$$\mathbf{Q} = \mathbf{V}\,\mathbf{C} \quad (2)$$

Solving formula (1) for the voltage necessary on the pail to produce an ignition capable spark and plugging that value for voltage into formula (2) and solving for Q, the amount of charge on the pail necessary to raise it to this voltage, it was determined that between 3 and 7 rinse cycles would result in the pail being charged to a level that could result in a spark discharge energy sufficient to ignite a flammable mixture of MEK vapor and air

4. KEY LEARNINGS

This investigation stretched out over approximately 6 months and involved dozens of individuals. There were many different views and versions of what happened and many ideas of how it happened. The seriousness of the potential fallout from this incident required that no loose ends or unanswered questions remained after the report was issued.

The local authorities, the HSE, the Chief Investigator of the regional Fire Service and the local officials accepted the company's report and conclusions as accurate, impartial and the most likely explanation of the incident. Fines were imposed but there were no criminal charges.

Some of the actions that were taken early in the investigation that contributed to the report being accepted by the various government agencies and officials and by the employees include the following key activities:

1. Preserve the scene of the incident as much as possible until everything is documented - use barricades, guards, barbed wire, anything to isolate it.

2. Locate a secure area to store evidence, debris, etc. so it will not disappear

3. Document the scene of the incident as soon as it is possible to re-enter the area using photographs, videos, handmade drawings, etc. Document where people, equipment, and debris was located using a plan of the area.

4. Get statements from everyone in the vicinity of the incident. Document what everyone saw, heard, smelled, felt, etc. This is must be done as soon as possible before people start to compare stories and/or forget critical information.

5. Get local governmental and/or regulatory officials on board as soon as possible if they are going to be involved.

6. Be as open and cooperative as possible with the authorities, involve them as much as possible in the investigation.

7. Involve workers in the investigation if possible. This will help to mitigate rumors that will no doubt start.

8. Get the best, most knowledgeable people you know involved in the investigation. If you are over your head, get an expert. Your investigation and report must have credibility.

Fired Heater Damage Due to Inadequate Management of Change

Lee N. Vanden Heuvel, Randal L. Montgomery, and Donald K. Lorenzo
ABS Consulting
10301 Technology Drive
Knoxville, TN 37932 U.S.A.
Phone: 865-966-5232
LNV@absconsulting.com

Abstract

During the restart of a fired heater following a unit outage, the heater experienced extensive tube damage from overheating. Gas flow through the tubes was lost when the compressor feeding the heater tripped on low oil pressure. During the outage, the facility installed a distributed control system (DCS). But the commissioning of the DCS was incomplete. Numerous indications and alarms were inoperable during the unit startup, including those for the heater compressors. As a result of the inoperable alarms and incomplete training, about 1½ hours elapsed before the operators realized that the compressor had tripped.

As a result of the investigation [1], the facility discovered a number of underlying process safety management (PSM) element issues that contributed to the incident. Recommendations generated from this investigation should enhance the efficiency and effectiveness of the facility's PSM processes.

1. Overview of the Incident

During the restart of a fired heater following a unit outage, the heater experienced extensive tube damage from overheating. The heater tubes were damaged because the heater was fired without gas flow through the tubes during the reduction step of the platformer catalyst regeneration process just prior to unit startup. Gas flow to the heater had stopped because the gas compressor had shut down as a result of low lube oil pressure. During the unit outage, the platformer compressor instrumentation was switched to the new distributed control system (DCS), which was being implemented in phases. The operators were unaware that the compressor had shut down because the alarms for the newly installed compressor instrumentation had been suppressed during a previous phase of the switchover to the DCS.

After noticing that the heater firing rate had increased without a corresponding increase in the process temperature, the operators began investigating the reason for the heater malfunction and noticed that the firebox and tubes were overheated. A short time later, the operators discovered that the compressor had shut down. Limited instrumentation prevented a more rapid diagnosis of the problem. This incident resulted in significant equipment damage to the heater and an extended unit shutdown.

The underlying causes of the incident are related to management of change, pre-startup safety review, procedures, computer displays and alarms, and communications.

2. Investigation Methodology

2.1 Data Collection

A root cause analysis (RCA) team comprised of both company personnel and ABS Consulting personnel was formed shortly after the incident. The RCA team collected data about the incident from several sources, including (1) interviews with company personnel, (2) records generated around and at the time of the incident (e.g., production logs, operator notes, DCS data historian), (3) engineering information (e.g., piping and instrumentation diagrams [P&IDs], alarm review reports), and (4) written procedures.

2.2 Data Analysis

The data were analyzed using a number of methodologies. The data were initially summarized in a timeline. This was then converted into a causal factor chart, which provided a chronology of the events leading up to, and immediately after, the incident, along with the identification of the direct causes (causal factors).

Why Trees and the Root Cause Map™ [2, 3] were both used to identify underlying management system weaknesses (root causes). Recommendations were developed for each causal factor and root cause.

3. Process Overview

At the time of the incident, the operators had just finished the reduction step of the platformer catalyst regeneration process and were preparing to bring the platformer unit online. Figure 1 is a simplified process flow diagram of the platformer. During the reduction step, the three platformer reactors (R-1, R-2, and R-3) are heated to an outlet temperature of 800 °F with pure hydrogen flowing through the reactors. The hydrogen flow is provided to the reactors via the recycle line from one or both of the platformer compressors (i.e., C-1 or C-2). During catalyst regeneration, only one platformer compressor is required, while both compressors are needed during normal operation.

Once the reactor outlet temperature of 800 °F is achieved, the reactors are held at that temperature for an hour and then cooled to 700 °F. While the reactors are cooling, the operators prepare to bring the platformer unit online by introducing feed into the two platformer towers. Once the reactors are cooled to 700 °F, feed is introduced into them. Initially, the hydrocarbon feed is charged to the platformer at half the design rate. Once level is established in the stabilizer, the platformer is then brought online by raising the inlet temperatures to 880 °F.

The only automatic compressor shutdown is on low-low lube oil pressure. In addition, the platformer heaters are equipped with low flow interlocks that shut them down when low process flow to the heater is detected. However, these low flow interlocks can be bypassed for catalyst regeneration and unit startup.

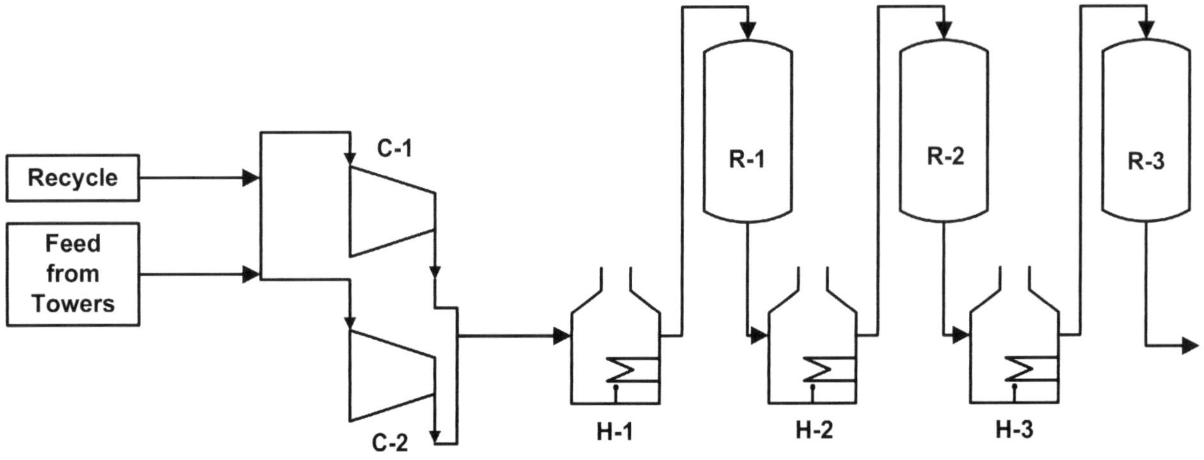

Figure 1. Overview of the platformer process

The DCS instrumentation for each platformer compressor includes (1) a handswitch to load and unload the compressor valves, (2) a compressor start and stop handswitch, (3) a high lube oil temperature switch with alarm, (4) a high motor temperature switch with alarm, (5) a low lube oil pressure switch with alarm, (6) a low-low lube oil pressure switch with alarm and shutdown, and (7) a run/stop indication.

4. Detailed Sequence of Events

4.1 Initial Conditions

The facility was at the end of an outage to regenerate the catalyst, repair equipment, and implement Phase 3 of the DCS conversion. The facility was in the process of starting up the platformer unit.

4.2 DCS Installation

4.2.1 New DCS Capabilities. During the unit outage, another phase of the DCS project was implemented. The new DCS provided the operators with a variety of new information, including tube skin temperature alarms and indications for the fired heaters and low pressure alarms for the compressors. The operating limits table and procedures had not been updated to include information on the tube skin temperature devices and alarms. As a result, the operators did not know what a normal tube skin temperature should be. Also, information about the compressors was not displayed on the same screen that showed an overview of the platformer operation. As a result, the operators did not normally monitor the screen with the new compressor data. Finally, the units for gas flow through the heaters were changed. The flow rate, which had previously been in arbitrary flow units, was changed to thousands of standard cubic feet per hour (mscfh). Even after the compressors tripped (and there was no gas flow), the DCS displayed a non-zero value for gas flow. The operators, having no "feel" for the magnitude of the flow represented by the value in mscfh, did not interpret the value as essentially zero.

The DCS provided operators with information that would have helped them diagnose the condition of the platformer. However, the operators either did not understand the information presented to them or did not monitor the data because of its location.

4.2.2 Testing of the DCS. In preparation for the upcoming outage, instrumentation and control (I&C) personnel and the Process Control Engineer worked together to install and check out the platformer unit compressor instrumentation and input/output (I/O) points. The I&C personnel performed the field activities (e.g., wiring, activation of switches), and the Process Control Engineer verified switch and alarm operation on a similar DCS console for the crude unit. The crude unit's DCS console was not configured to respond to platformer alarms; therefore, the Process Control Engineer was not able to completely verify alarm operation (e.g., audible alarm, alarm banner operation).

Prior to the outage, the plant personnel had suppressed the compressor indications and alarms on the platformer console. Plant personnel suppressed the alarms so they would not be a nuisance until the instruments were switched to the DCS. Personnel performing the subsequent platformer compressor DCS changeover did not have procedures to ensure that the instrumentation and I/O points were properly implemented (e.g., all alarms worked and were operational). The tests did not specifically verify that the alarms were operational on the platformer console, and a pre-startup safety review (PSSR) was not performed to verify that the DCS instrumentation and I/O points had been installed and were functioning as designed. As a result, the compressor alarms remained suppressed during the incident.

4.2.3 Operator Training Related to the New DCS. The formal training that the operators received on the new DCS was very limited and consisted of classroom training only. A training terminal was installed in the control room, but it was not used during the training. The training provided a basic overview of the operation. However, it did not address issues such as alarm management and suppression and restoration of alarms. Also, there was no training on a few modifications related to the way information was presented to the operators. A key issue related to this incident was the change in the process gas flow units. The unfamiliar units made it difficult for the operators to "get a feel" for the flow rates they were observing.

4.2.4 Status at the Time of Startup. At the time of the startup, the exact status of the DCS points, alarms, and interlocks was not known to the operators. They knew that all of the points had been tested in some manner. However, the results of this testing (showing which points had been successfully tested and were operational and which points had unresolved issues) were not communicated to the operators.

4.3 Startup

Most operators used an overview screen to monitor the performance of the platformer. They relied upon the DCS alarms to direct them to other screens as needed.

During startup, the compressor tripped on low oil pressure. The compressor has an oil filter that requires periodic cleaning. This is especially important during startup and following oil changes. Shortly before the incident, the operators did not clean the filters as recommended (every 8 hours) because they were occupied with other duties.

As the oil filter clogged, the compressor oil pressure decreased. The low oil pressure setpoint was reached at about 2322 hours the evening before the compressor tripped (about 5½ hours prior to the compressor trip). However, the low oil pressure alarm was suppressed, so the only indication the operator had of this condition was a change of the oil pressure indication from green to blue on a screen that the operators did not normally monitor. The operators were also having problems with other equipment. They used two of the three DCS screens to monitor the other equipment. This left one screen available to monitor the platformer. This screen did not display the compressor status.

Operators also performed routine rounds to monitor the operation of equipment, including the compressors and the fired heaters. The operators typically used log sheets to guide their rounds. One shift before the incident, the operators used the last of the log sheets in the control room. The operators requested additional blank copies from their supervisor (they were not available on the company network and no master blank forms were maintained). However, they were not supplied with additional log sheets over the next two shifts. Given the additional monitoring, trending, and alarm capabilities of the new DCS, the operators were unsure if the use of the log sheets on rounds was still required. As a result, the operators did not use the log sheets to guide their equipment monitoring and did not record parameter values (such as gas flows). Thus, the normal monitoring of equipment was not effective in rapidly identifying the loss of gas flow and increasing heater firing.

Another contributor to the lack of operator rounds was insufficient plant staffing. During most startups, additional plant staff was available to assist with some of the startup activities. However, due to some delays in completing some of the turnaround activities, additional personnel were not available or scheduled. The operators at the platformer console were on their twelfth 12-hour shift in a row.

At 0500 hours, the compressor low-low lube oil pressure switch activated and tripped the compressor. Again, the operators did not detect that the compressor had shut down because they were not monitoring the DCS screen that displays compressor status, the compressor alarms were not functional, and they were not performing routine rounds that included the compressors.

Under normal operating conditions, the fired heaters would have tripped on low gas flow. However, during startup, operators bypass the trip because low gas flow is an anticipated condition. As a result, the loss of the compressor did not result in the tripping of the unit.

4.4 Overheating of the Heater Tubes

Heater H-3 operated for about 1½ hours without any flow through it. During this period, the tubes were heated until they were glowing red. The tubes were damaged beyond repair and required replacement.

4.5 Detection of the Overheated Tubes and Shutdown

About an hour after the compressor tripped (at about 0600 hours), the board operator noticed that the reactor associated with heater H-3 was cooling faster than the other two reactors and that the heater was firing harder than the other two heaters. The operator began making adjustments to the heater in response to the indications he was seeing.

The board operator then went outside to check the status of the heater damper. The operator was aware that the heater damper had been out of position during a recent incident and it caused similar high skin temperature alarms. He verified that the damper was in the correct position and continued to try to diagnose the situation.

At this point (at about 0630 hours), the next shift arrived in the control room. The operators discussed the indications they were observing. Some of the operators went out to the heater and discovered that the tubes were overheated. At about the same time (between 0630 and 0700 hours), the high and high-high skin temperature alarms activated on the other two heaters.

The oncoming shift operator believed that the running compressor had shut down. By 0715 hours, the operators had verified that both compressors were shut down, and they attempted to restart C-1. However, the low-low oil pressure interlock prevented it from restarting. The operators then attempted to start C-2, but it would not start either. The operators continued to work to start a compressor and eventually got C-2 running. The operators then unsuppressed the alarms associated with both of the compressors (C-1 and C-2). About 30 minutes later, C-2 shut down on low-low lube oil pressure.

About 5 minutes later, the low lube oil pressure interlock was cleared on C-1 and the operators were able to start it. By feeding nitrogen into the heaters, the operators were able to safely cool the unit down.

5. Recommendations

The RCA team analyzed the data to identify causal factors and root causes. For each causal factor and root cause, the team considered the development of four levels of recommendations. The four levels considered were as follows:

1. Address the causal factor (correct the human error or repair the damaged equipment)
2. Address the underlying cause of the causal factor (typically a system fix)
3. Address generic implications of the cause (this typically looks for potential application to other processes, equipment, or procedures in the facility or organization)
4. Address the root cause of the issue (typically this is a management system fix)

A total of 34 recommendations were identified by the team. They ranged from specific recommendations to general management system recommendations. Example recommendations included the following:

1. Develop a DCS instrument and (I/O) checkout procedure and check sheets for the future phases of the DCS project.
2. Develop and implement a testing schedule for "critical" discrete alarms for all the units recently converted to the DCS in order to verify alarm operation.
3. Conduct a PSSR of the platformer unit and other units that were recently switched to the DCS.
4. Develop and implement a procedure for approving, identifying, and communicating suppressed DCS alarms. Integrate this procedure with the current safety device bypass procedure.
5. Add compressor lube oil pressure readings and cleaning the lube oil filter to the log sheets for operator rounds.
6. Review other unit compressors and other similar rotating equipment with lube oil systems and filters to determine the need to monitor lube oil pressure.
7. Develop guidelines for operator rounds and log sheets, including the activities required and the specific content of the activities (e.g., readings needed to ensure proper equipment operation).
8. Conduct refresher training on DCS alarm suppression and management, and all other DCS changes, once the DCS training materials are updated. Also, define the frequency with which DCS refresher training should be held in the future.
9. Develop testing materials to qualify personnel on DCS operation and test personnel operating units recently changed to the DCS.
10. Develop alarm settings for the different platformer operating modes: regeneration, normal, and other operating modes.
11. Review other units to determine whether different alarm settings need to be developed for different unit operating modes. If the need for different alarm settings is identified, develop alarm settings for the affected operating modes.
12. Review and revise the platformer regeneration graphics to include the status of critical equipment and critical process parameters.
13. Review operating procedures for changes that may have resulted from the DCS conversion.
14. Place an electronic copy of the regeneration log sheets on the plant intranet for operator access, and communicate this to the operators.
15. Develop a policy/procedure for communicating, evaluating, and managing deviations in unit staffing levels.
16. Review the heater shutdown interlocks on the platformer and other units to determine whether other process parameters (e.g., compressor shutdown status, high-high skin tube temperature) should cause a heater to shut down.

6. Conclusion

This incident resulted in equipment damage only. There were no personnel injuries or environmental insult. The equipment damage did have a significant economic impact on the facility because an extended shutdown was required to replace the heater tubes.

Investigation of equipment damage incidents often reveals breakdowns in the functionality of PSM systems. By performing a detailed root cause analysis, many underlying management system issues can be identified and addressed. Specifically, the investigation identified issues related to the following PSM elements:

- Process safety information
- Operating procedures
- Training
- Pre-startup safety review
- Management of change

Effective resolution of the investigation's findings significantly strengthened the facility's PSM program.

7. References

[1] *Investigation of Fired Heater Tube Damage* (incident investigation report); confidential client.

[2] *Root Cause Analysis Handbook*; ABS Consulting; Rothstein Associates, Inc.; 2005.

[3] *Root Cause Analysis/Incident Investigation* course; ABS Consulting, Inc.

215a Apriori Alarm Management during Project Feed Stage Using the Results from the Hazop

E. Michael Stafford and **Harry West**. Shawnee Engineers, London, United Kingdom

1 ABSTRACT A recent Front End Engineering Design (FEED) on a major plant expansion included a significant effort in apriori alarm management which represents innovative engineering. This consisted in implementing alarm classifications directly after the HAZOP, using the results of HAZOP prioritization and grouping to suggest Alarm priorities

User Approval of Safety Instrumented System Devices

Angela E. Summers, Ph.D., P.E
Susan Wiley
SIS-TECH Solutions, LP
Houston, Texas 77034
Phone: 281-922-8324

Prepared for Presentation at
American Institute of Chemical Engineers
2006 Spring National Meeting
8th Process Plant Safety Symposium
Orlando, Florida
April 24-26, 2006

February 2006

UNPUBLISHED

User Approval of Safety Instrumented System Devices

Angela E. Summers, Ph.D., P.E
Susan Wiley
SIS-TECH Solutions, LP
Houston, Texas 77034
Phone: 281-922-8324

Abstract

User approval is a simple process that supports design, engineering, configuration management, and management of change. User approval assesses manufacturer information and installed performance data to justify that a device is suitable for use in a safety instrumented system (SIS) application. The process supports configuration management and management of change activities by providing a means to determine whether a new or revised device is acceptable. This paper explains the concept of user approval as documented in ANSI/ISA 84.00.01-2004, ANSI/ISA TR84.00.04, and the Center for Chemical Process Safety book, Guidelines for Safe and Reliable Instrumented Protective Systems.

1.0 Introduction

This paper is distributed "as is." All warranties, either expressed or implied, are disclaimed as to quality, performance, or merchantability, whether expressed, implied or statutory, hidden defects, or fitness for any particular purpose. The author makes no representations about the suitability of the information in this paper for any purpose. Reader bears the entire risk relating to the use of this paper.

A paragraph similar to the above is contained in many end-user license agreements accepted during the installation of user software and with the hardware manufacturer's terms and conditions. The underlying concept is that the "user" of the product bears the risk associated with determining whether a product is fit for a particular purpose.

The manufacturer limits its liability associated with effects beyond its control, such as the operating environment. Other limitations are generally addressed in the installation and maintenance manual (or in the safety manual for safety instrumented system devices). Product manuals provide warnings concerning the installation, commissioning, maintenance, and testing requirements. These agreements, terms, and warnings establish the boundary for the manufacturer's claims. Deviation from the manufacturer's recommended practices generally invalidates all warranties, either expressed or implied.

No manufacturer's claim, third-party analysis, or certification report reduces the user's responsibility for determining the product is fit for purpose. This user responsibility was acknowledged in ISA 84.01-1996, which used the term "user approved" as "hardware, software, procedures, etc., that the user has evaluated and determined to be acceptable for the application."

2.0 The User Approval Process

A user approval process should be established to examine evidence of the suitability of devices for the application and operating environment. A device can be user approved and considered suitable for use when it can be demonstrated that the device meets the core attributes associated with its protection layer. The device boundary includes the hardware and software elements necessary for the device to perform its design intent.

The user approval process examines the device using analysis and testing and demonstrated performance in the operating environment. The outputs of the process should be an approved manufacturer's list and an installation, commissioning, and maintenance plan (e.g., the safety manual for the SIS) that outlines how the approved devices should be implemented.

User approval requires a close relationship between the manufacturer and the owner/operator as a product moves from early development through obsolescence. Technology evolution results in some devices moving rapidly toward obsolescence. For example, the lifespan of some programmable devices is less than 10 years due to the rapid development of computer technology, the increasing demand for interconnectivity, and a desire for uniform configuration tools. Such rapid evolution results in the manufacturer selling products that have some of the desired hardware and software features with a migration path for future upgrades. In some cases, significant changes in the product may be required to correct problems or to provide requested functionality.

The owner/operator should have a process to control the replacement of components in any way other than "like" for "like." When the manufacturer offers a new version of an approved component or recommends a replacement, these new components should go through the user approval process. For new versions of approved products, this process can be simplified when the manufacturer provides documentation explaining changes to the products and highlights any configuration changes.

Configuration management can be challenging, but is essential to achieve the core attributes. Success requires that SIS product specialists serve as an information resource for the owner/operator. These SIS specialists should be highly familiar with SIS requirements, receive and review advisory notices and product information from the manufacturer, and make decisions regarding the addition or removal of devices from the approved manufacturer's list. It is recommended that the SIS product specialist be independent of the project team seeking approval of the device.

Field performance and lessons learned in applying a device should be shared with the manufacturer to assist in future product development and in the dissemination of information regarding critical failures. Monitoring field performance and sharing failure reports with project and plant personnel is important for performance improvement.

3.0 Operating Environment

The operating environment can be identified by drawing an imaginary bubble around the device as installed in the process. The bubble establishes the boundary of the device. The operating environment may include a variety of items that affect device operation:

- External environmental conditions,
- Process operational conditions,
- Communications and interconnectivity,
- Human interfaces,
- Access security,
- Support systems, e.g. instrument air and electricity.

For Programmable Electronic (PE) logic solvers, the operating environment also covers the embedded software, the hardware architecture, application software, I/O configuration, communication to other systems, operator and engineering interfaces, and access security.

4.0 Analysis and Testing

Devices should be subjected to analysis and testing to evaluate the device's design, manufacture, and validation procedures, as well as the manufacturer's quality and change management systems. The rigor of the analysis and testing is related to the complexity of the device and the maturity of its technology. When completed, there should be a better understanding of how the device functions and fails. This analysis should consider the device boundary and should result in the following:

- Description of the analysis boundary.
- Identification of potential dangerous failures.
- Identification of means to detect dangerous failures by diagnostics or testing.
- Description of assumed diagnostic coverage factor and any user requirements necessary to achieve the diagnostic coverage, e.g., configuration requirements and the need for external diagnostics.
- Statement of the probability of safe and dangerous failures with any assumptions regarding maintenance activities and testing intervals.

Analysis and testing should also include evaluation of the device's conformance to applicable codes, standards, and practices. Meeting the area classification (e.g., hazardous environments) and following applicable electrical codes is important for all instrumented system applications. Devices that do not meet the area classification can become ignition sources should a loss of containment occur. For SIS, the requirements of IEC 61511 Clause 11.5 should be met. Guidance on complying with this clause is also provided in ANSI/ISA TR84.00.04 Annex L.

Manufacturers, who wish to advertise their products as safety products, analyze and test their devices for compliance with the intent of IEC 61508. An IEC 61508 evaluation may yield a "certification" if the product is evaluated by a recognized certification body. A certified device can be labeled with an "SIL (Safety Integrity Level) Claim Limit." In the United States, certification to IEC 61508 is issued by Nationally Recognized Testing Laboratories (NRTL), such as those listed on the US-OSHA website (www.osha.gov). The OSHA website has a complete listing of NRTLs. Other countries establish certification requirements by policy or regulation acceptable certification bodies.

When a device is certified according to IEC 61508, the manufacturer supplies a safety manual providing the criteria for its implementation. Successful implementation of SIS devices often requires a specific configuration, the addition of external diagnostics, the provision for inspection and maintenance, mandated proof test intervals, and specific installation details. Deviation from the safety manual may invalidate the SIL Claim Limit.

It is important to note that IEC 61508 is a generic functional safety standard that applies to safety-related systems used in many different industry sectors, including process, rail, machinery, and medical. Since this standard applies to many sectors, there are devices that meet the requirements of IEC 61508, but are not sufficiently robust for the operating environment in many process industry facilities. A device developed for use on a manufacturing floor may not be appropriate for use in a refinery SIS.

Manufacturers sometimes provide field performance reports based on an installed base at other owner/operator sites. More often, manufacturers supply analysis and testing results and documented predictive calculations of their product's probability to fail on demand and spurious trip rate based on the product's "shelf-state" design and manufacture. For any new technology, the analysis and testing is generally based on very limited field operating experience. Thus, the manufacturer evaluation represents the highest performance that can be expected from the device when it is implemented in accordance with its safety manual.

Field device and non-PE logic solver performance is generally dictated by the operating environment and the owner/operator's inspection and maintenance practices. For example, a sampling of data for pressure transmitters from various manufacturers show 600 to 800 year mean time to failure dangerous (MTTFD), however, owner/operator prior use data show a range of 75 to 200 year MTTFD.

Many important failure modes are excluded from the manufacturer's boundary. This includes the process connections, manner of installation, power supplies, and communication interfaces. Failures due to the operating environment can be significantly greater than failures due to device manufacture. Failures in the interfaces between the device, the process, and other protective systems should be considered when determining its suitability for addition to the approved manufacturer's list.

Some owner/operators develop installation details for each type of technology. These owner/operators then analyze the installation itself to ensure that the core attributes of a protective system are achieved, e.g., independence, functionality, integrity, reliability, auditability, access security, and management of change. This provides a consistent operator and maintenance interface and allows the analysis and testing for device approvals to be limited to the device and its ability to work within the existing approved installations.

For PE logic solvers, performance is generally dictated by how well the safety manual is followed, how recommended upgrades are executed, and whether the manufacturer's recommended operating environment is maintained. When implemented according to manufacturer's recommendations, PE logic solvers tend to achieve the reported level of safety. This does not necessarily mean that the PE logic solver has the robustness to provide the trouble free service that may be desired. In general, PE logic solvers tend to fail-safe more frequently than reported by analysis. This is largely due to violation of the operating environment requirements or human error.

5.0 Prior Use History

In general, manufacturer failure rate data is 3 to 10 times better for programmable devices than the actual performance observed in the process industry. For mechanical devices, the ratio is 30 to 100 times better. Programmable electronic systems are not immune, with some certifications assuming diagnostics coverage factors that are simply not achievable using current practices. Because of the discrepancy between manufacturer and owner/operator environments, devices should be selected based on demonstrated history in a similar operating environment.

The intent of prior use is to gain an understanding of the device's failure modes so that the design and operating basis can take them into account. Previous operating experience provides valuable information for the selection of field devices, because it identifies how the operating environment degrades the theoretical performance claimed by the manufacturer. For some devices, especially field devices, evidence of successful operating experience in similar process applications is very important. In IEC 61508, this is called "proven-in-use;" while in IEC 61511, it is called "prior use."

There are specific requirements and limitations regarding the prior use evaluation in IEC 61511 Clauses 11.4 and 11.53 through 11.5.6. The requirements vary depending on the device type (sensor, logic solver, final element) and whether the device uses programmable elements. If the device is programmable, the requirements also vary dependent on whether fixed programmable language or limited variability language is used to configure the device. These requirements are further discussed in ISA TR84.00.04 Annex L.

For SIS, IEC 61511 requires devices be selected based on their expected performance in the operating environment. Maintenance records are a valuable information source. The plant maintenance tracking system can be used to flag devices that have repeatedly failed so that more detailed analysis can be performed. In general, it takes approximately three years of operating time to gain sufficient understanding of the failure modes of a field device. This operational time can be obtained in the actual or similar operating environment in process control, other non-SIS applications, or at other owner/operator sites.

Operational time can be gathered during alpha and beta testing where owner/operators work in close association with the manufacturer. Alpha testing is conducted to demonstrate the basic functionality of the device and general compatibility with the service. A successful alpha test leads to beta testing, which is usually conducted using multiple installations to gather information on the device failure modes in a variety of operating environments.

However, alpha and beta testing results are not sufficient for approving devices for SIS service.

Manufactures often provide references to other owner/operators who use their product. These references can often provide valuable insight into product application and use. In addition, the manufacturer user's group can provide valuable networking opportunities to gather information concerning product performance. The owner/operator may choose to rely on the manufacturer and other owner/operator's field experience or may choose to supplement this evidence with more direct experience. This may include bench testing and field trials in process control or low hazard services. Generally, the more unfamiliar or complex the technology is, the more time should be spent understanding how it works and how it fails.

Proven performance takes on special significance for SIS, because most operate as dormant systems that only take action when a process demand occurs. In contrast, control system devices are expected to operate frequently, if not continuously, so failures in control system are rapidly detected. If these failures cannot be managed by the control system or operator, an independent SIS is often implemented to mitigate any unacceptable process risk.

One of the strongest arguments for separation of the SIS and the control system is the common practice of using the latest technology in control applications. Many owner/operators who strive to use the newest control technology often find that the products are not as well developed as originally thought. New control system implementation can turn into an unplanned product development project. While necessary for production or quality reasons, this can be a hazardous practice for SIS. Proven operating history cannot be underestimated.

6.0 Summary

User approval is an important aspect of complying with ANSI/ISA 84.00.01-2004. User approval relies of predictive evidence, such as analysis and testing reports, balanced with field operating history. While some manufacturers are now providing devices that are "certified for use" in safety instrumented systems (SIS), these devices are relatively limited in their range of application and technology. Many are not field proven and some are not demonstrating the robustness necessary to survive a chemical plant environment. Within an IEC 61508 analysis, an unreal operating environment is often assumed: one that does not include the process impact, ambient conditions, stress, manufacturing defects, software errors, installation issues, electrical disturbances, instrument air or other support system quality problems, etc.

Consequently, owner/operators in the process sector should rely heavily upon operating experience for selection of devices for SIS, especially for those devices installed in the process environment. User approval seeks to get the proper balance of analysis/testing information and field experience. The amount of information or experience required and the degree of rigor associated with their evaluation may vary, depending on the nature of the device and the required safety integrity level. The key point to the user approval process is to establish sufficient evidence to justify that the device can and does achieve the required performance in the intended operating environment.

7.0 References

[1] Instrumentation, Systems, and Automation Society (ISA), ANSI/ISA S84.01-1996, "Application of Safety Instrumented Systems (SIS) for the Process Industry," Research Triangle Park, NC (1996)

[2] Occupational, Safety and Health Administration (OSHA), "Process Safety Management of Highly Hazardous Chemicals; Explosives and Blasting Agents, 29 CFR Part 1910." Federal Register 57,36, Washington, DC (1992).

[3] International Electrotechnical Commission (IEC), IEC 61508, Functional Safety of Electrical/Electronic/Programmable Electronic Safety Related Systems, Parts 1-7, Geneva, Switzerland (1999-2001).

[4] International Electrotechnical Commission (IEC), IEC 61511, Functional Safety: Safety Instrumented Systems for the Process Sector, Geneva, Switzerland (2003).

Advances in Abnormal Situation Prevention in Refineries and Petrochemical Plants

Ahmad Hamad and Ravi Kant
Emerson Process Management
Eden Prairie, MN 55344

Abstract

Abnormal situation prevention is of high importance in the operation of refineries and petrochemical plants. Abnormal situation may result from equipment malfunction, process upsets, operator errors, etc. There have been several approaches to deal with those situations. The proposed abnormal situation prevention approach in this work targets the timely and early detection of abnormal situations to prevent undesired upsets, shutdowns and catastrophic consequences. The approach uses advanced statistical control techniques, smart field devices, and multivariate analysis to analyze trends of process variable and detect faults or deviations. The approach has three phases: learning, model development, and monitoring. All these phases can be implemented online or by using off line data. The capabilities of the proposed abnormal situation prevention approach will enable refineries and petrochemical plants to run safer and cleaner while achieving higher economic benefits.

Introduction

Abnormal situations can be defined as process disturbances, upsets, and events that can cause the process to deviate from its normal operating state and may lead to damaging consequences. Abnormal situations can vary in severity and consequences from off-spec product to catastrophic failure of equipment. Cost of abnormal situation consequences increases as the severity increases. Marsh Risk Consulting (2001) identified 379 abnormal events in the Hydrocarbon industries with individual losses exceeding $10 billion in the period from 1970 to 1999. Average annual losses were approximately $700 million with twenty-five of these events having individual losses of $150 million or more. Unfortunately, these economical losses are sometimes also accompanied by loss of lives and environment incidents. Till now, the main approach for preventing damaging consequences of abnormal situations is preventive maintenance. However, this approach has major disadvantages. First, this approach is costly since maintenance is done on regular basis regardless of the status of the process or equipment. This approach may include unnecessary plant or unit shut down.

Second, this approach does not detect abnormal situations. This is a major disadvantage because abnormal situation can occur between scheduled maintenance periods. Another disadvantage includes the direct and complete dependence on plant operators. To overcome these disadvantages, several efforts have been done by various industries and researchers to come up with diagnostic techniques to detect and prevent abnormal situations and reduce the cost of maintenance. Roemer and Kacprzynski (2000) presented a real-time, integrated health monitoring of gas turbine engines that can detect, classify, and predict developing engine faults. Kubiak et al. (2004) introduced a hybrid fault pattern approach for diagnosing gas turbine component degradation. Reising et al. (2005) discuss how abnormal situation management (ASM) is used by plant operators to recognize the presence of failures before alarms are issued. Eryurek et al. (2005) discusses how smart field devices are used for abnormal situation prevention. This paper presents a novel on-line methodology for abnormal situations prevention in the process industries. The proposed methodology builds-on several tools to come up with a novel abnormal situation prevention structure that is robust and operator-friendly. The methodology is embedded in Emerson's Plantweb technology and relies on using smart devices, high frequency transmitters, statistical-based analysis methods, on-line data gathering and analysis, and on-line process automation using operating software such as DeltaV. Examples of applying this approach to refineries and petrochemical plants will be presented.

Process Control and Abnormal Situation Prevention

Before talking about the proposed abnormal situation prevention methodology, it is important to differentiate between process control and abnormal situation prevention. Abnormal situation prevention technology is intended to provide early detection and timely corrective action. Consider Figure (1) where abnormal event severity is presented as a function of time. The severity of the abnormal situation increases as time progresses. It is traditional ASM approach that when the value of the process variable reaches unacceptable region an alarm is issued to alert the operator. The operator takes an action in an attempt to return the process into the normal operation region. The operator is doing process control in this case based on prior knowledge or rules. In many instances this approach is ineffective because the detection is too late and may lead to catastrophic situations or unplanned plant/equipment shutdown. In addition, this approach fails in most cases to identify the source of faults. The operator gets trapped in fixing the symptoms of the fault by manipulating process variables. The main goal of the operator is to bring the process back to operate within the acceptable region. The operator may fix the fault symptoms but fails to identify and eliminate the fault that caused these symptoms.

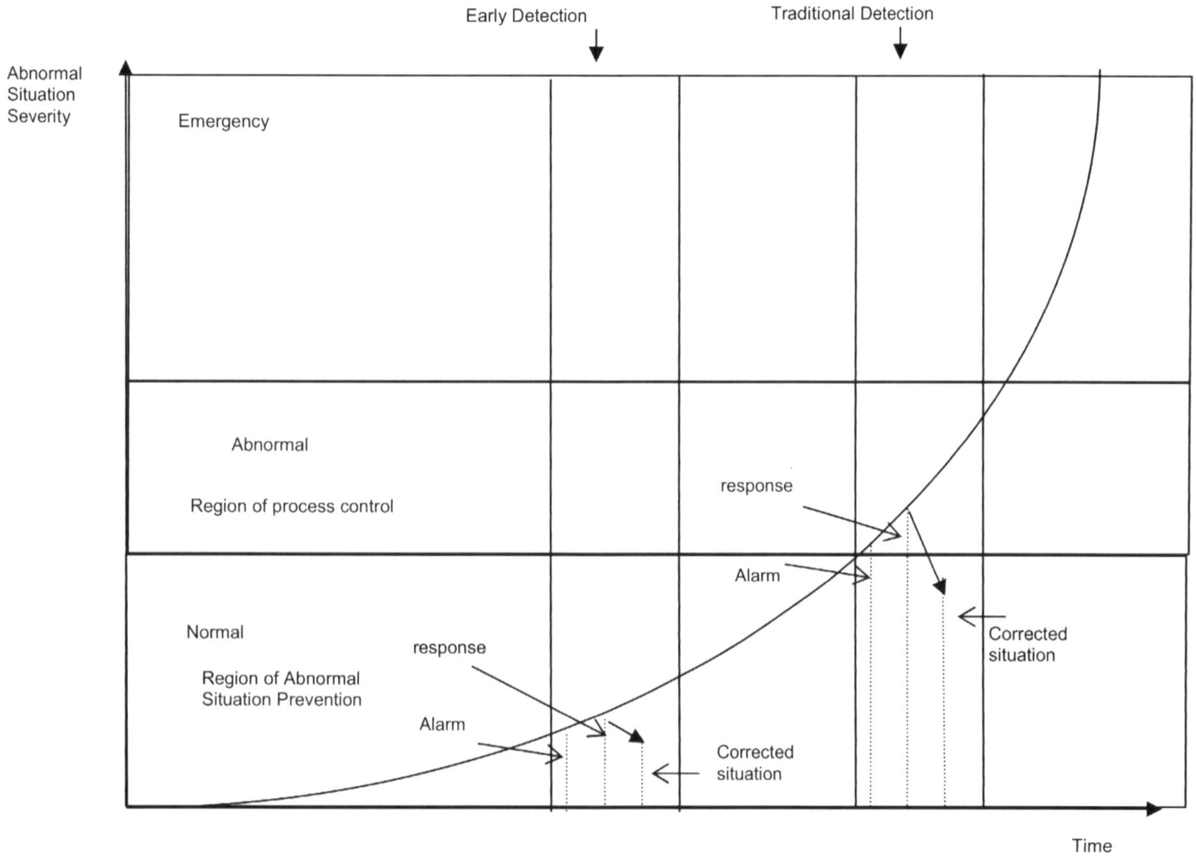

Figure 1: Typical Response Diagram for Abnormal Situations

Process control in this situation may take the process variable back to a point in the normal region that is not far from the abnormal situation. The proposed abnormal situation prevention approach, on the other hand, identifies the abnormal situation at an earlier time (Figure 1) and identifies the source of the fault. The abnormal situation is identified as it starts to develop. The whole diagnostic, detection, and corrective actions take place in the normal operating region. How is this done? This is done by comparing the actual real time operating conditions to the "ideal base." Ideal base here refers to the normal process operating conditions without any deviation or abnormal situation. The proposed approach uses advanced pattern recognition technologies, signal analysis, statistical process control, as well as expert system based solution with the smart devices for improved process operation, and early abnormal situation detection and prevention.

Technology Advances In Abnormal Situation Prevention

The Fluid Catalytic Cracker (FCC) unit is used to illustrate advances in abnormal situation prevention technology development. Figure 2 presents the cost of abnormal situation consequences resulting from slide valve failure in the catalyst transfer lines of the FCC unit. Valve failure is due to irregularity in catalyst circulation (abnormal situation) in the transfer pipes between reactor and regenerator. The abnormal situation consequences in this case progress can be as follows:

1. Stage 1: represents pressure irregularity due to catalyst erratic circulation in transfer pipes
2. Stage 2: represents loss of production due to regenerator fouling
3. Stage 3: Valve malfunction because of catalyst build up in transfer pipes
4. Stage 4: FCC unit shut down due to pressure loss across transfer pipes
5. Stage 5: Plant shut down

As the stage order increases, the cost severity associated with the abnormal situation increases and may result in major damaging consequences. Circle (A) in Figure (2) represents two corrective approaches, preventive maintenance and knowledge-based diagnostic and fault detection. The preventive maintenance takes a corrective action with respect to physical entity such as cleaning the valve or the transfer pipe. The knowledge-based diagnostic uses historian data to develop a multivariate model to detect the behavior of the process. This approach can get very complex. It uses low frequency filtered data and relies on historian data. It does not have the capability to gather data on-line to update the monitoring models. This approach fails to detect abnormal situations if the abnormal situation is very fast or with a high frequency unless a damaging consequence is evolved. Furthermore, this approach fails to predict health of machinery, devices, and transmitters used in the process. It can, though, give a good indication of process variable deviation from normal operation condition. The response time domain for circle A is slow and can be too late.

Circle B represents the use of smart field devices and field instruments. Eryurek et al. (2005) illustrates that improved process measurements and real time analysis/detection could have prevented or at least substantially reduced the damage of approximately 25-40% of reported abnormal situations. With advances in computing and measurement device technologies and advances in communication such as the digital Fieldbus technologies, today's smart field devices and transmitters are capable of providing much more information regarding the processes and its conditions. Smart field devices and instruments manufactured by Emerson Process Management have built-in ASP™ blocks (Figure 3) which captures high frequency process data (22 Hz) and have data analysis capabilities. Abnormal situation prevention algorithm now can have access to data that was lost with a low frequency analysis (Figure 4).

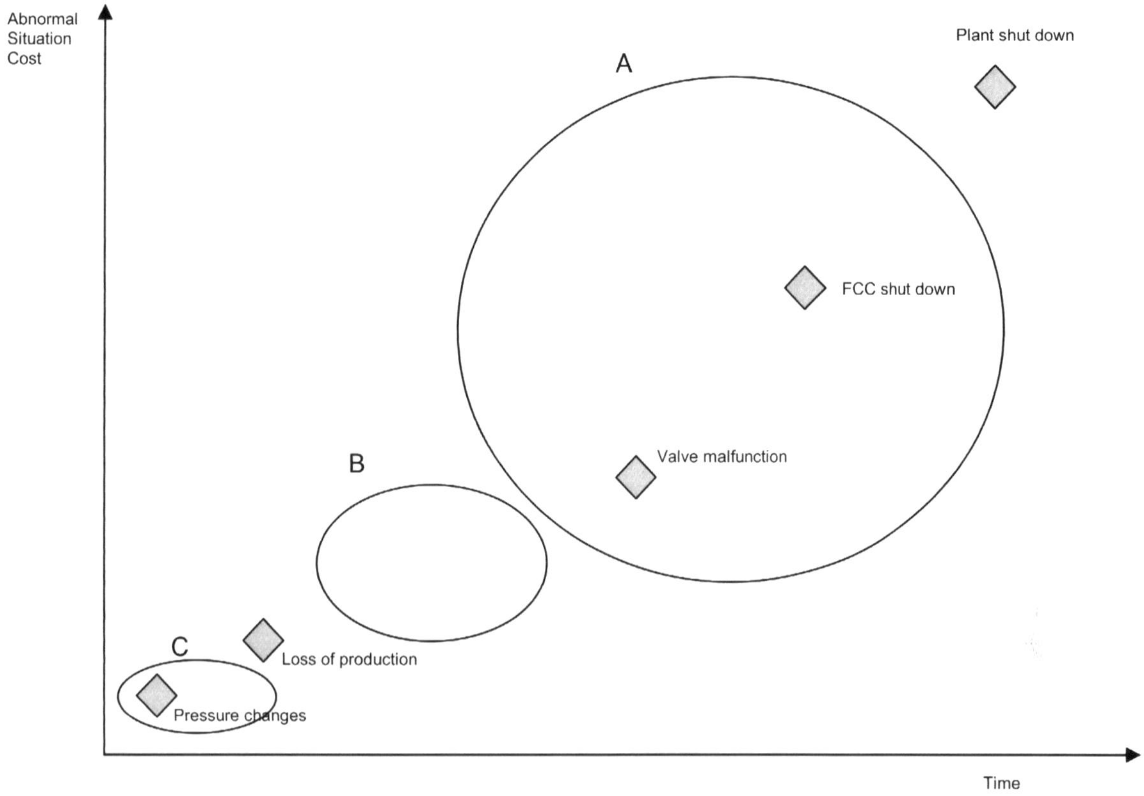

Figure 2: Abnormal Situation Cost For Slide Valve in FCC Unit

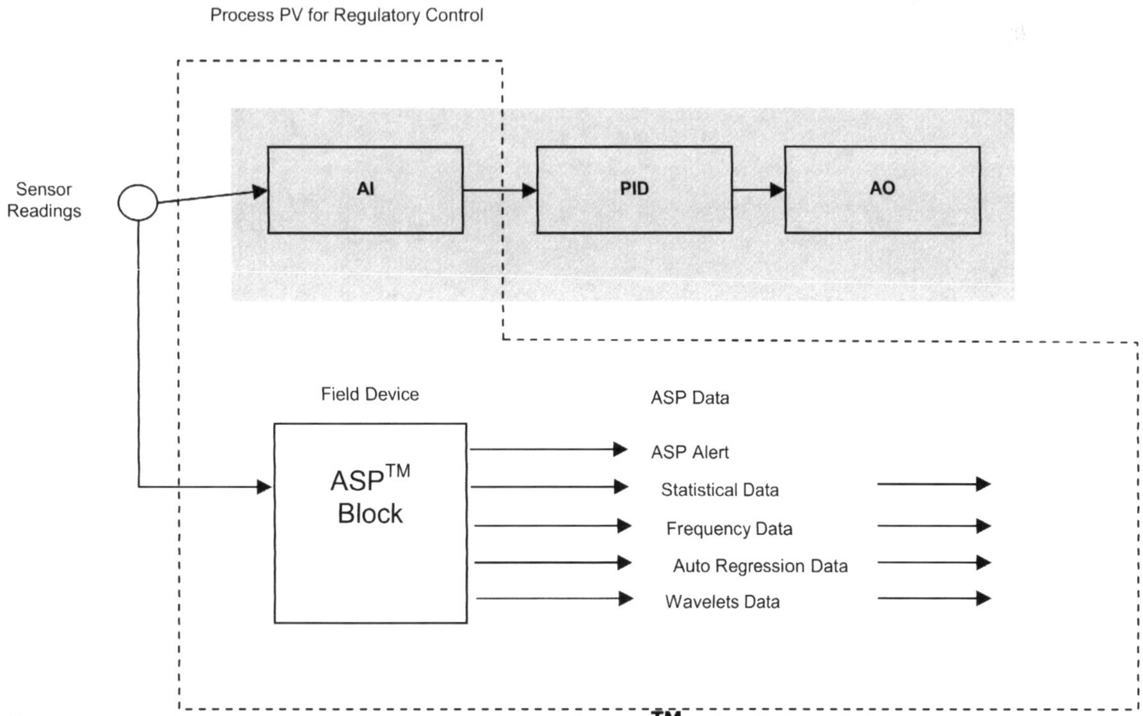

Figure 3: Schematic Representation of ASP™ Block in Field Devices and Instruments

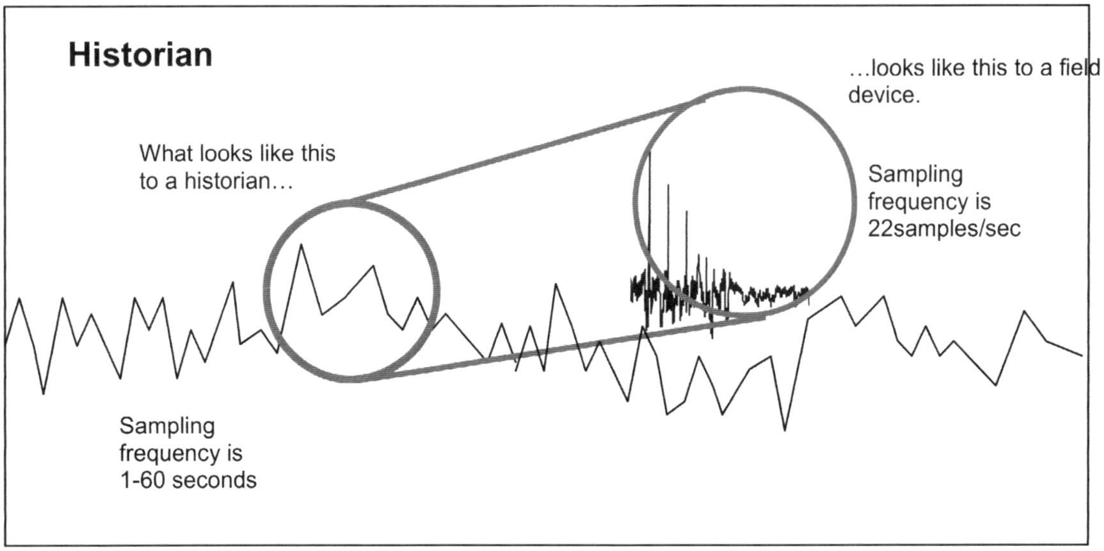

Figure 4: Data (Pressure, Temperature, etc.) Frequency Sampling in Smart Field Transmitters

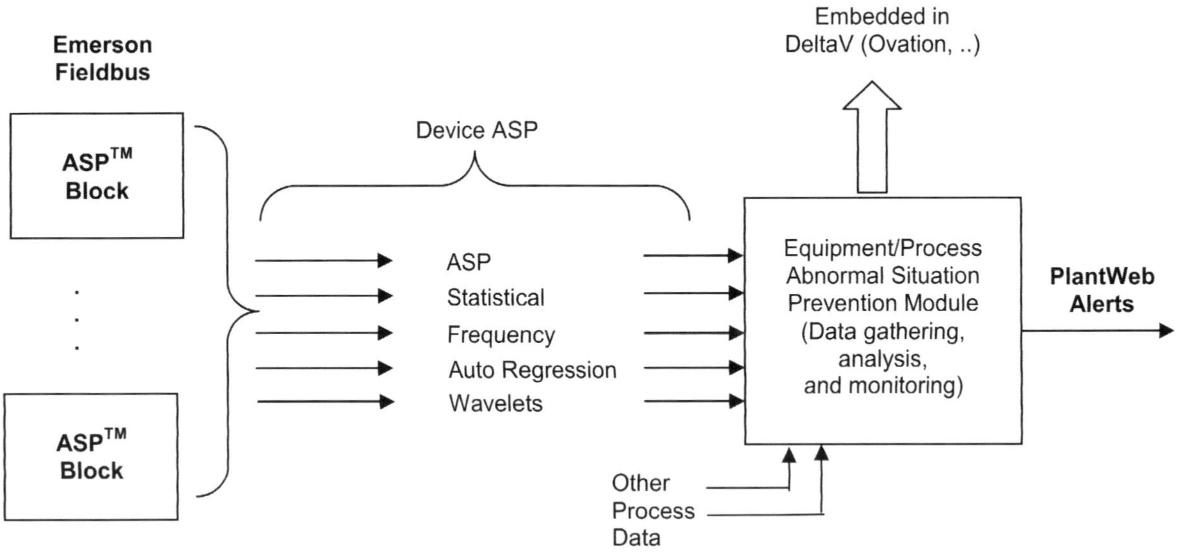

Figure 5: Equipment and Process Abnormal Situation Prevention Approach Using Plantweb Technology

This data could provide information about process performance and abnormal situations. Smart devices and instruments can provide the operator with an assessment of their own health by providing high frequency statistical data such as the variance and mean. This methodology still cannot go to detect abnormalities due to process equipment fouling or process-related abnormalities.

Circle (C) in Figure 2 represents the proposed equipment and process abnormal situation prevention approach in this work (Figure 5). This approach utilizes the benefits of knowledge-based, smart field devices and instruments, advanced process control techniques, and other diagnostic techniques to detect and prevent abnormal situation in early stages while the process is running in the normal operation region (see Figure 1). The proposed abnormal situation prevention approach extracts high frequency real time plant data using Plantweb components and applies on-line diagnostic techniques to analyze this data and identify any abnormal situation or behavior in the plant. Plantweb is a digital plant architecture that delivers proven results, and the DeltaV system makes it easy. Designed with this digital architecture in mind, the DeltaV system fully integrates "smart plant" capabilities including HART®, FOUNDATION™ fieldbus, high-speed discrete buses, embedded advanced control, and advanced unit and batch management. Such tools can be used in the proposed approach to make online analysis and data extraction easier and faster. The following section illustrates the phases of the proposed ASP approach.

Phases of Abnormal Situation Prevention Approach

The proposed abnormal situation prevention approach has three phases to detect faults and abnormal situations in the process industry. The phases include learning, model development, and monitoring. In the learning phase, the module collects data (input and output) as needed and applies various statistical techniques to relate input (load variable) and output (monitored variable) data. The models relating outputs and inputs can be as simple as variance with a threshold, or linear, or polynomial, or any other form of interest. Multivariate analysis techniques can also be employed to relate inputs and outputs. Selected model type is user defined. During the learning phase, data are collected for both input and output variables. Two time domains are important here and both are user specified. The first domain is time interval between data collected. It could range from fractions of seconds to minutes or hours. The shorter the interval the more data is collected. The second domain is the length of the learning phase. It could be hours, days, weeks, or months. It is recommended to have a learning period long enough to capture all anticipated variances in process operation. It is important to realize here, that the models developed during a learning phase are good enough for the range of data used in the regression process. If the process shifts to a new region of operation, then new models are needed. This means that a new learning phase is needed. The models are developed during the learning phase and then used in the monitor phase to detect and prevent abnormal situations. Operator may monitor point or pattern deviation of the process variable. Smart field devices and instruments are necessary to have robust data gathering and analysis of process variables.

Implementation of Abnormal Situation Prevention Approach

One major advantage of this approach is that it is easy to implement. The developed abnormal situation prevention algorithms or modules can be implemented in any of the following platforms:

1. Field Device Interface Module (e.g. Rosemount 3420)
2. Control System (e.g. DeltaV, Ovation)
3. Standalone software application

Various abnormal situation prevention modules can be developed which may be process equipment based or process unit based. Operators or plant engineers will have the flexibility to tailor ASP modules to fit the needs of their process unit application.

Case Studies

Various case studies are presented below to illustrate the effectiveness of the proposed abnormal situation prevention methodology.

1. Coke Detection in Refineries

The Delayed Coker is used in refineries to process the heavy residue (Coke) from the vacuum distillation unit. Although there are many components to the Coker unit, a component with great value to Abnormal Situation Prevention is the Coker Heater. This fired heater takes the feedstock from the vacuum bottoms, and heats it prior to sending it to the Coke Drums. The feed enters the unit at the left, and is divided into multiple passes (number of passes is user defined). A load-balancing control regulates the flow though each of the passes. One problem that can occur in the Coker Heater is the build-up of coke on the inside of one or more of the passes. When this happens, the performance of the furnace begins to degrade, and the other passes need to take more loads, in order to compensate. If coking continues to build up even more, eventually the affected passes, or even the entire unit, will need to be shut down. Thus, it is desirable to know when coke buildup begins to happen, in order that a corrective action can be taken.

Module Summary

The Coker abnormal situation prevention algorithm monitors each pass in the furnace of the Delayed Coker to check for high coking. Generally, a decrease in either the flow or heat transfer rate for a given pass indicates that there is high coking in that pass. During the learning phase, data is collected for each pass.

At the end of the learning phase, a regression model is developed which can predict the value of each monitored variable as a function of the load variable. The value predicted by the model is considered the normal operating condition for that monitored variable. When the actual value differs from the predicted value by more than a certain threshold, an abnormal situation (high coking) is detected. This algorithm is implemented in DeltaV (Figure 6) and is used in industrial trials.

2. Catalyst Circulation In FCC Units

Catalyst circulation in the FCC unit is extremely important. Without "proper" catalyst circulation the FCC unit deteriorates significantly. Detecting circulation problems requires good understanding of pressure balance around the reactor-regenerator system. It is important to detect erratic circulation; otherwise FCC unit may experience unscheduled and costly shutdowns.

<u>*Module Summary*</u>

A lower or higher differential pressure across the slide valve than "normal" slide valve opening is a good indication of abnormal situation development in the catalyst transfer pipe. Catalyst circulation can also be limited by mechanical problems with the slide/plug valve. The abnormal situation prevention module consists of two steps:

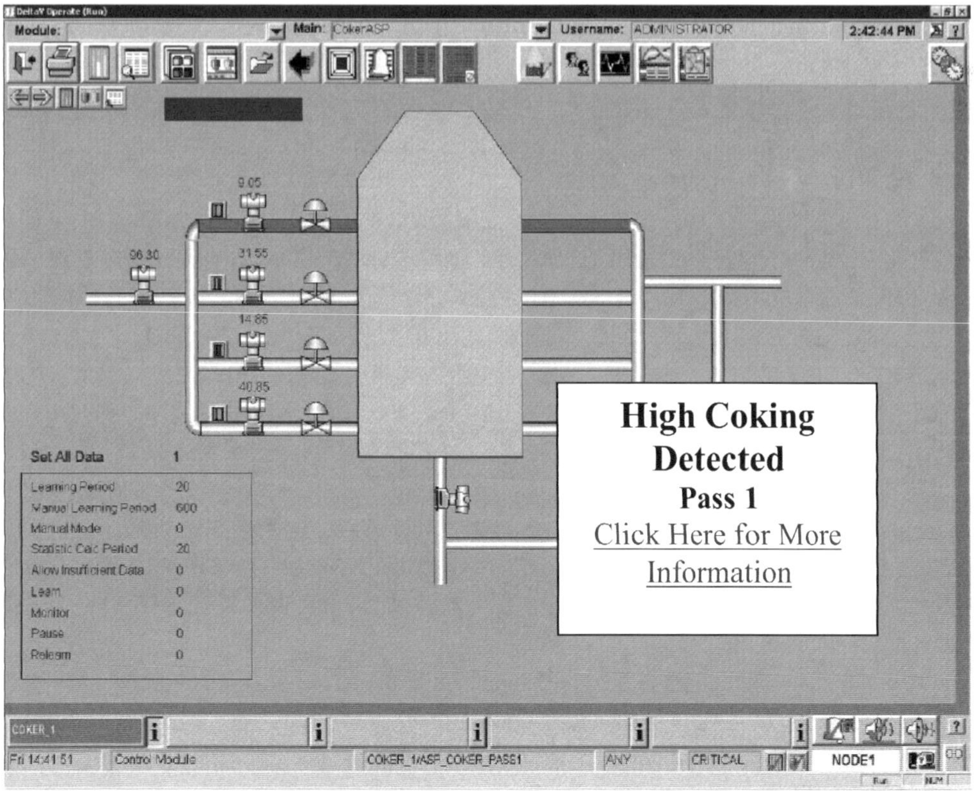

Figure 6: High Coking Detection as Implemented in DeltaV

a. Using smart field valves such as DVC6000f. This valve will keep operators informed of the health of the valve to avoid any unplanned shut down.
b. Pressure balance (drop) across the valve will be monitored using smart pressure transmitters to detect abnormality. With this information, operators will be able to take corrective action early in the problem development before reaching the abnormal situation failure (Figure 1).

3. Maltrays Detection in Crude Columns

One of the major problems that face crude columns and distillation columns in general is maltrays. Maltray could result from flooding, foaming, lost trays, etc. The major technology that is used currently for column diagnostics is Gamma ray. This technology is not handy and is expensive to use. As an example of maltrays, let's discuss column flooding. Flooding is excessive accumulation of liquid inside the column. This accumulation is generally caused by various mechanisms such as downcomer backup flooding, spray entrainment flooding, etc. Flooding can be recognized by some symptoms including excessive column differential pressure, loss of bottoms, loss of separation, etc.

Module Summary

The objective of this algorithm is to detect conditions in column operation that lead to flooding or to predict flooding conditions as early as possible to avoid entire column or tray flooding. Different algorithms can be developed to identify flooding conditions depending on existing instrumentations and potential future instrumentations.

4. Gas Turbines Abnormalities

One of the major problems in gas turbines (GT) is fuel supply and efficient usage of fuel. Uncontrolled and/or unburned fuel in the combustor could lead to various abnormal situations in gas turbines and could cause sudden shut downs. Abnormal situations include loss of fuel efficiency, damaged blades (due to high temperatures, localized explosions), and damaged nozzles (due to high temperatures). The objectives of the proposed abnormal situation prevention method in this work are:

1. Control and monitor fuel supply to prevent abnormal situations in GT
2. Detect abnormalities in fuel system to prevent abnormal situations in GT
3. Detect abnormalities in expander to prevent abnormal situations in GT and prevent unscheduled shutdowns

Module Summary

In the learning phase, the method collects data and applies various modeling techniques to relate input and output data. The models developed in the learning phase are used in the monitor phase to identify abnormal situations. Each model is assigned a region of confidence for operation. The region of confidence is user specified.

Conclusion

Preventing abnormal situations in the manufacturing processes (refineries, petrochemical plants, etc.) can lead to substantial operational, economic, safety and environmental benefits. The proposed abnormal situation prevention approach diagnoses and analyzes process variables to detect abnormal situations as they start to evolve. This advantage gives plant operators/engineers the opportunity to eliminate the source of the abnormal situation before it causes process variables to deviate from the normal operating region. The proposed abnormal situation prevention approach has three phases: learning, model development, and monitoring. The developed modules can be implemented in field devices, DeltaV, or stand alone software.

References

1. Eryurek, E. P. Sharpe, and D. White (2005), "Abnormal Situation Prevention through Smart Field Devices," presented at the NPRA 2005 Annual Meeting (AM-05-49), San Fransisco, CA
2. Kubiak, J. S., A. G. Garcia, G. B. Urquiza, and G. R. Gonzalez (April 2004), "Hybrid fault patterns for diagnosing gas turbine component degradation," Hydrocarbon Processing, p. 71 – 74
3. Marsh Risk Consulting (Feb. 2001), "Large Property Damage Losses in the Hydrocarbon Industries, A Third Year Review
4. Reising, D., T. DeMaere, J. Errington, K. Harris, and P. Bullemer (October 2005), "Establishing Operator Performance Improvements and Economic Benefit for an ASM Operator Interface, presented NPRA Plant Automation & Decision Support Conference, Grapevine, TX
5. Roemer, M. J., and G. J. Kacprzynski (May 2000), "Advanced Diagnostics and Prognostics for Gas Turbine Engine Risk Assessment," presented at IGTI/ASME Turbo expo, Munich, Germany

"Strategic Proof Testing"

Dr. William M. Goble
Exida
Sellersville, PA 18960
wgoble@exida.com
+1-215-453-1720

Introduction

Safety Instrumented Systems (SIS) are designed to provide automatic protection functions. A typical design includes a sensor to detect the potentially dangerous condition, a safety PLC and a remote actuated valve capable of taking action to prevent an accident. In a well designed chemical plant the potentially dangerous conditions occur very infrequently. Hence these equipment sets may be idle for long periods of time and operate only once every few years. In such a situation failures of the equipment that prevent a safety response may go undetected for years without a periodic "proof test."

Proof testing is an important part of the maintenance activities needed to assure safety. Proof testing is done on a periodic schedule using procedures specifically developed during the design. If the designer is following a safety lifecycle (SLC) per IEC 61511, an international functional safety standard, the periodic proof test interval is calculated based on a safety integrity level (SIL) design target. A SIS design process is shown in Figure 1.

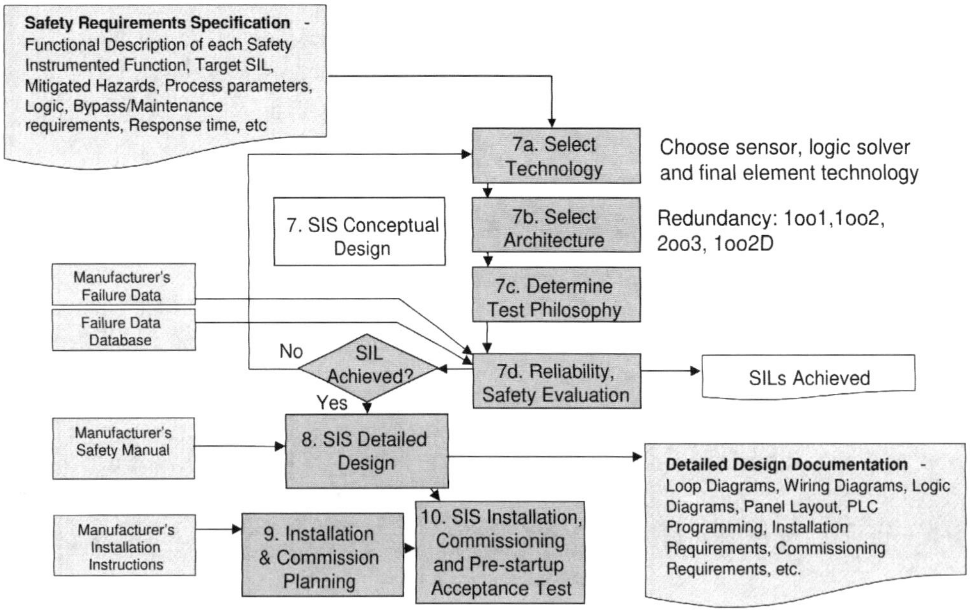

Figure 1: SIS Design Phase Activities (copyright exida 2001, from Safety I course notes)

During the design verification analysis (step 7d) the probabilistic calculation includes a number of variables including the proof test interval and the proof test effectiveness. The traditional approach to proof testing is to perform some action which simulates the dangerous condition and observe to the extent possible that the safety function performs its duty. A typical test (simplified) for a single channel design might be:

Procedure:
1. Procure all management of change permission forms
2. Block valve from closing.
3. Move input signal above trip point.
4. Verify that valve attempted to close.
5. Move input signal back to normal below trip point.
6. Remove valve block.
7. Document test results and restoration to normal.

An assumption is often made, especially when using simplified equations for the analysis, that the proof test is 100% effective. But how effective is that test?

FMEDA Analysis Techniques

In the late 1990's techniques were developed to determine the diagnostic coverage capability of smart instruments with automatic diagnostics (Ref. 1). These techniques were created from the need to more precisely determine the quality of proposed diagnostic tests in PLC product (Ref. 2). The objective of the analysis is to measure how many of the potential failures are detected by a diagnostic. This is measured by a metric called the "coverage factor." A number of 100% means all failures are detected and that is the goal.

The FMEDA technique involves studying every failure mode of every part and recording if a particular test can detect that part failure mode. The method is very systematic and complete. The same methods can be used to determine the effectiveness of any proposed manual proof test. Given a list of all components and failure modes one can reasonably determine which items are detectable. Using a weighted average based on failure rates the effectiveness is calculated.

The techniques have evolved and are now being used on instruments (pressure transmitters, temperature transmitters, etc.), PLC products and even mechanical products (solenoids, actuators and valves). An example from a ball valve is shown in Table 1.

Table 1: Partial FMEDA of ball valve partial stroke test (PST).

Valve component	Failure mode and effect	Lambda (FITS)	PST Credit	Lambda (detected
Ball	Damage Pin slot resulting in failure to close	31.50	0	0
Ball	Surface Damage resulting in failure to close	126.00	0.8	100.8
Ball	Surface Damage resulting in Internal leak, major	189.00	0	0
Drive dowel pins	Break/damage resulting in failure to close	25.00	0	0
Stem	Break/deformation resulting in failure to close	18.75	0	0
Stem	Damage/Deflection resulting in failure to close	112.50	0.8	90
Trunnion	Damage/Deflection resulting in failure to close	210.00	0.8	168
Seat Seals	Damage resulting in Internal leak, major	20.00	0	0
Stem Bearings	Deformation resulting in failure to close	80.00	0.8	64
Thrust washers	Deformation resulting in failure to close	20.00	0.8	16
Adapter Bolting	Break resulting in failure to close	1.00	0	0
Adapter Plate Pins	Break/deformation resulting in failure to close	2.00	0	0
Actuator Adapter Plate	Break/deformation resulting in failure to close	1.50	0	0
Seat Ring	Damage resulting in Internal leak, major	180.00	0	0
Seat Ring	Damage resulting in failure to close	60.00	0.8	48
	TOTAL DANGEROUS FAILURE RATE FOR VALVE	**1077.25**		
	DANGEROUS DETECTED FAILURES			**486.8**
	TEST EFFICIENCY			**45.19%**

Note that in this case the result shows an effectiveness of 45%, far from the 100% assumption often made.

Instrument manufacturers are proposing a range of proof test techniques as part of their IEC 61508 safety certification. For example, a page from the safety manual of a pressure transmitter is shown in Figure 2. This shows a relatively simple test that can be remotely with a handheld communicator. It has a proof test effectiveness of 65%. Another more extensive test is specified with an effectiveness estimated to be greater than 989%. One may wonder why the first test is even mentioned. Remember that the first test is far simpler and less costly to run. It is likely that a combination of the two tests, each run on a different schedule can provide good coverage with optimal cost.

SAFETY INSTRUMENTED SYSTEMS

Additional Safety Instrumented Systems information is available in the Rosemount 3051S Series Pressure Transmitter reference manual (document number 00809-0100-4801). The manual is also available electronically on www.rosemount.com or by contacting an Emerson Process Management representative.

Installation

The 3051S SIS can be used in safety instrumented functions up to SIL2 with a single instrument and SIL3 with hardware fault tolerance of 1. No special installation is required. Follow standard installation practices as outlined in this document.

Set-up

The security switch must be in the **ON** position during normal operation.

Configuration

No special configuration is required. Verify configuration as outlined in this document. Use any HART-compliant master to communicate with and configure the 3051S SIS. DD revision 3051S SIS Dev. 1 Rev 1 is required.

Operation and Maintenance

Proof Test and Inspection:

The following proof tests are recommended. Proof test results must be documented with corrective actions taken in the remote possibility that safety functionality is not correct.

To perform a HART Loop Test follow the "Loop Test" instructions available in the 3015S reference manual (document number 00809-0100-4801).

Five-Year(1) Proof-Test

This proof test, when combined with a HART Loop Test, will detect more than 65% of DU failures not detected by the automatic diagnostics programmed into the 3051S SIS.

1. Enter the milliampere value representing a high alarm state
2. Check the reference meter to verify the mA output corresponds to the entered value.
3. Enter the milliampere value representing a low alarm state
4. Check the reference meter to verify that the mA output corresponds to the entered value.
5. Execute the Master Reset command to initiate start-up diagnostics.

Ten-Year(1) Proof-Test

This proof test, when combined with a HART Loop Test, will detect over 98% of DU failures not detected by the automatic diagnostics programmed into the 3051S SIS.

1. Perform a minimum two point calibration check using the 4-20mA range points as the calibration points.
2. Check the reference mA meter to verify the mA output corresponds to the pressure input value.
3. If necessary, use one of the "Trim" procedures available in the 3051S reference manual to calibrate.
4. Execute the Master Reset command to initiate start-up diagnostics.

(1) May be a longer proof test interval as justified by PFDavg calculation.

Figure 2: Safety Manual page showing recommended proof tests along with effectiveness (used with permission of Rosemount)

Using FMEDA results to pinpoint effective proof tests

The FMEDA results can identify the component failure modes that cause the failure of the safety function. Given that these are now clearly identified why not design a proof test focused on discovering that failure?

This is effectively a reversal of thinking on the objective of a proof test. Conventionally a test is devised to show that the safety function works properly. It is assumed that any equipment failure that prevents the safety function will be discovered during the test. This is generally true but what about transient failures? What about marginal failures? What about the failure of an automatic diagnostic? Rarely are these revealed during a conventional proof test.

The reverse method of thinking says that a test should be focused on finding a functional failure. The assumption then is that if there are no functional failures then the safety function works. There are some advantages to this approach. One can be more precise in the testing. The tests can be designed to detect failures of any automatic diagnostics. And tests can be designed to reveal any weak or marginal parametric failures. One significant advantage that seems to be appearing is that the testing can be done with less time and effort if designed correctly.

Conclusions

If proof testing is designed NOT to see if a function is working but instead to see any component of a function has failed then it appears possible to create tests that are more effective and cost less to perform. In addition a side benefit is that more precise estimates of proof test effectiveness can be done allowing more conservative SIL verification calculations.

The FMEDA technique can be used for this purpose and several instrument manufacturers are beginning to provide more effective proof test recommendations.

References

1. Cheddie, H.L. and Goble, W.M., *Safety Instrumented System Verification – Practical Probabilistic Analysis*, Chapter 7, ISA, 2005.
2. Goble, W. M., Bukowski, J. V. and Brombacher. A. C., "How diagnostic coverage improves safety in programmable electronic systems," *ISA Transactions*, Vol. 36, No. 4, The Netherlands: Amsterdam, Elsevier Science B. V., 1998.

FIRED HEATER SAFEGUARDING SURVEY

Edward M. Marszal, Kenexis, Columbus, OH

Abstract

Refiners use a wide array of process heaters. The operation of these heaters poses a number of hazards that must be controlled, and are typically controlled using safety instrumented functions. The number and type of safety instrumented functions that are employed vary to a great degree from refiner to refiner and from site to site. This paper presents a summary of some of the results of a benchmarking study that was performed to determine the most prevalent practices in industry and establish the degree of adoption for some of the more common safety instrumented functions.

1.0 Introduction

Most refining organizations are very concerned with meeting the requirements of safety related standards and best practices in addition to their overall commitment to continuous improvement of safety and reliability. One common area for standardization among refining companies is the engineered safeguards that are implemented on fired heaters. There is a great benefit in standardizing the design of these systems because there are so prevalent. Like many refinery standard practices, standards developed to address fired heater safeguarding are often extensions of a related industry consensus standard – in this case American Petroleum Institute Recommended Practice 556 (API RP-556). API standards are extremely well adopted in the refining industry, and respected by both practitioners and regulators. Although the API RP-556 document is an excellent foundation, like most industry consensus standards it does not provide a completely definitive basis for design. API RP-556 contains many clauses where multiple options are available for meeting requirements and other clauses where the recommendations are not as strict as others (i.e., should instead of shall). Refiners typically address this limitation of consensus standards by issuing corporate standards that supplement other industry practices instead of developing a completely separate practice.

The author's organization performed a benchmarking study of the standard practices of refining organizations with respect to fired devices. The study was developed so that it would not only address common and well understood failure modes, but also address potential new approaches required as a result of new conditions and equipment, such as low NOx burners, and recent incidents and accidents related to fired heaters. Over the past several years, a number of accidents have occurred in the refining industry that were in whole or in part related to the following conditions.

- Improper use of bypass around heater fuel shut down valves
- Individual burner valves being left in the open position after occurrence of a shutdown resulting in inadvertent gas admission during subsequent light off
- Introduction of excess air into a fuel rich firebox subsequent to an operating anomaly

1.1 Benchmarking Survey Questions

The benchmarking was performed in accordance with a survey form that contained the information presented in *Table 1*.

Table 1 – Survey Questions

Which Automatic Trips are Implemented?
Fuel Gas High / Low Pressure Shutdown
Pilot Gas High / Low Pressure Shutdown
Loss of Flame on Main Burner - If yes, what method? (Flame rod, scanner, other)
Loss of Flame on Pilot Burner - If yes, what method? (Flame rod, scanner, other)
Low Process Flow (Pass Flow)
Low Excess Air / High Combustibles / Improper Fuel/Air Ratio - method for action?/detection?
Combustion Air Low Flow / Fan Not Running
Alarms
High Draft Pressure
High Bridgewall Temperature / Stack Temperature
High Process Outlet Temperature
Low Oxygen / High CO or High Combustibles
Permissives
Purge flow permissive – Automtic purge? – Material Used for Purging?
Fuel Gas Shutoff valve position permissive
Individual burner valves proved closed prior to light off (how?)
Pilot flame proven
Shutoff Valves
Double block and bleed on main / pilots
If a single valve, control valve or special purpose valve, what type?
Are bypasses used around shutoff/control valves for light-off or minimum firing
Pilots
Continuous pilots - Pilots sourced from a separate supply (i.e., natural gas)
Post Purge
Automatic Steam addition for purge upon initiation of a heater trip

The survey also developed information with regards to both new equipment projects and existing equipment installations, and the content of standards versus application of those standards in the field. Information regarding the content of corporate standards was requested in a yes or no question. Summary results for standards questions were then presented in terms of percentage of yes responses to the question. Information regarding the degree to which a practice has been implemented (both in new construction and existing equipment) was ranked from 1 to 3 where 1 corresponds to 'almost never' implemented, 2 corresponds to 'sometimes' implemented, and 3 corresponds to 'almost always' implemented. Summary results for implementation questions were then presented as the arithmetic mean (average) of all responses (one response per operating company).

While the form was oriented toward natural draft heaters, several of the questions were related to forced draft furnaces, specifically the questions related to combustion air flow shutdowns. The response from the survey indicated that there are some significant differences in how forced draft heaters are safeguarded as opposed to natural draft heaters. In some cases the author made revisions to survey forms returned by respondents, based on their comments, to ensure the focus of the study was natural draft heaters.

It is important to note that the commentary to the surveys is as important as the statistical data that was obtained. The rationale for why some companies chose certain practices over others gives a deeper understanding of how to obtain overall functional safety. Some statistical figures may be deceiving at first glance and should not used outside of the context of a full survey including all participant commentary. For instance, one respondent organization does not have fired heater standards at the corporate level, leaving all individual sites to set their own practices. As a result, a reported figure of less than 100% of companies require a certain safeguard as a standard practice, may actually mean that 100% of the companies that <u>have</u> standard require the safeguards – the remainder do not have standards, and thus would not require the safeguard as a standard practice. Furthermore, some hazards can be addressed in multiple ways. For instance, loss of combustion air flow can be detected using an air flow measurement, fan motor status, or fan shaft speed. If the survey reports that only fraction of respondents have a low combustion air flow shutdown that does not mean that the balance of respondents do not care about loss of combustion air, it means that other ways of detecting loss of combustion air are utilized at their sites.

1.2 Survey Participants

In order to understand the survey and put it in proper context, it is important to understand the demographics of the surveyed companies. Several operating companies responded to the survey (in some cases, multiple responses per company were received – in this case the responses were averaged for that company). All of these companies operate at least one refinery in the United States and are in the business as either pure-play refiners, or integrated oil and gas production, refining, and petrochemical companies. The survey team selected companies that have large market capitalizations, operate multiple refineries and have a large crude oil processing capacity. Significantly smaller companies with very low oil processing capabilities were not included in the study due to the potential that their practices may not reflect the best practices in industry.

It should be noted that the results presented in this survey are based on a somewhat limited data set. Although the survey includes companies that represent the preponderance of crude oil processing in the United States, typically only one or two respondents per operating company completed survey forms where the organization might have more than a dozen refineries. Furthermore, many refining organizations have recently been built up through a series of mergers and acquisitions, leading to a wide variety of safeguarding practices within a single operating company. While most refiners are working diligently to standardize their operations, that task is not complete yet. While the author has made great efforts to provide a comprehensive analysis of operating company practices, data from every site in every organization could not be obtained. For operating companies where the author knew that practices varied to a great degree from site to site, we made an extra effort to survey multiple sites to get an average response.

3.0 Survey Results Summary

A summary of some of the results of the survey is shown below. The information presented below is simply a statistical compilation of the responses to the survey forms and notes containing general commentary on the results and the comments that the respondents provided in their survey forms. While the original survey analyzed a number of design scenarios, the summary results presented here only represent the application of these design features on new equipment.

3.1 Shutdowns

Fuel Gas Low Pressure Shutdown

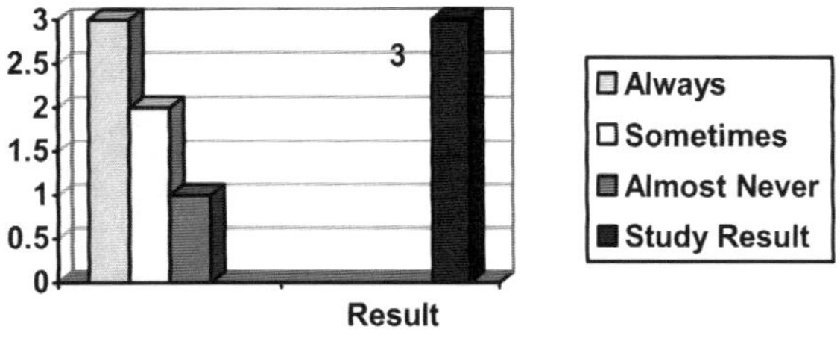

The low fuel gas pressure is the most common shutdown employed on fired heaters. All companies 'almost always' implement this shutdown on their new and existing equipment. Of the surveyed companies that have corporate standards, only one does not require that low pressure trip be implemented, but even that standard requires that the shutdown must be 'considered' based on a risk assessment and further requires that the low pressure trip shall be implemented unless continuously operated separately fuel sourced pilots are utilized on the heater.

Fuel Gas High Pressure Shutdown

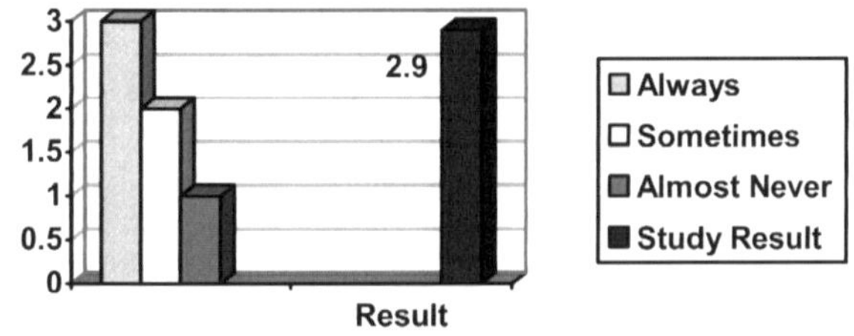

Of the operating companies that have fired heater safeguarding standards, all require implementation of a high pressure trip in their most current form. The high level of adoption of the high pressure trip is based on the knowledge that hazards resulting from high pressure fuel gas situations can not (in all cases) be safeguarded by other means such as flame scanners and continuously operating pilots. Even though standard practice dictates that this shutdown be implemented, existing equipment and even in some cases new equipment may not employ this shutdown. One respondent noted that even though their standard requires the high pressure shutdown, some projects may choose not to implement the trip. If this decision is made it is based on knowledge (and testing) of the burners to determine their stable operating range and verification that the fuel supplied to the burner can not reasonably exceed that pressure based on the design of the fuel supply system.

Pilot Gas High/Low Pressure Shutdown

High Pressure

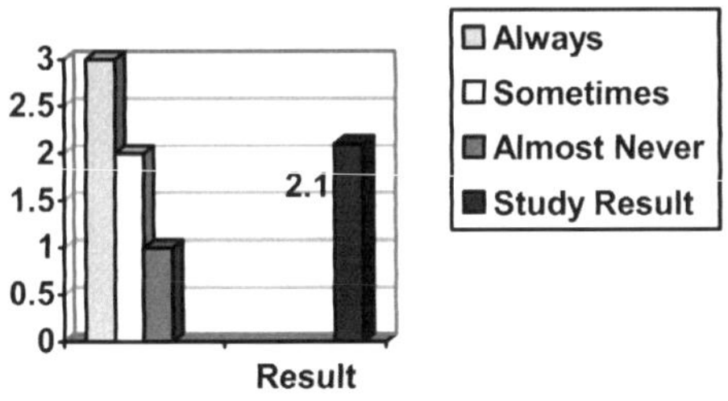

Low Pressure

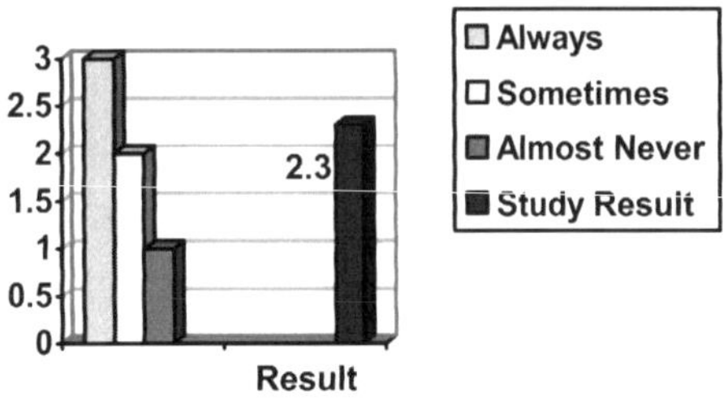

The figures reported for adoption of high pilot gas pressure shutdowns are difficult to interpret without considering the context in which they were generated and the respondent notes regarding how pilots are utilized at the various operating company sites. While the figures show that the preponderance of operating companies do not require shutdown due to high pilot gas pressure, that should not be taken to imply that high pilot gas pressure has been deemed to be an insignificant risk. Not all companies employ continuously operating pilots, and as such their standards would not require a high pressure pilot gas shutdown, not because the risk is insignificant, but because the pilots are not typically in operation. Of the operating companies that elect not to employ a high pilot gas pressure shutdown, their rationale is that the risk is tolerably low. A high pressure event in the pilot during normal operation would typically be mitigated by action of the main flame, leaving only startup and pilot-only operational modes as situations where the plant is at risk. In addition, many deem the likelihood of generating a pressure high enough to extinguish the pilot flame as very unlikely.

Many of the same considerations in analyzing the response to the question of high pilot gas pressure shutdowns apply to low pilot gas pressure shutdowns. Low pilot gas pressure shutdowns are more common for users of continuous pilots than high pressure shutdowns due to the increased likelihood that a low pressure pilot gas condition can occur and generate a hazard.

Loss of Flame on Main/Pilot Burner Shutdown

Main

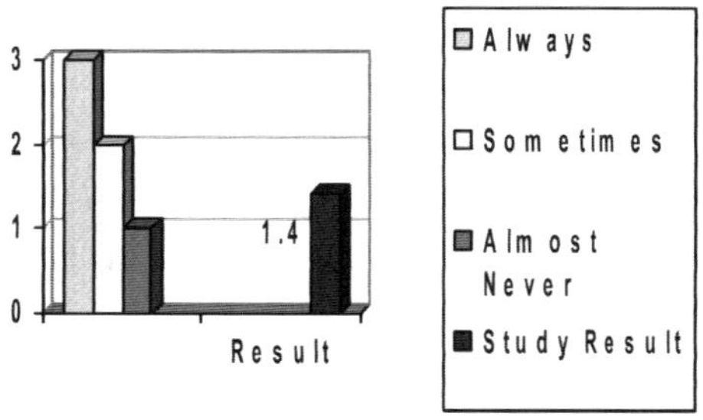

Pilot

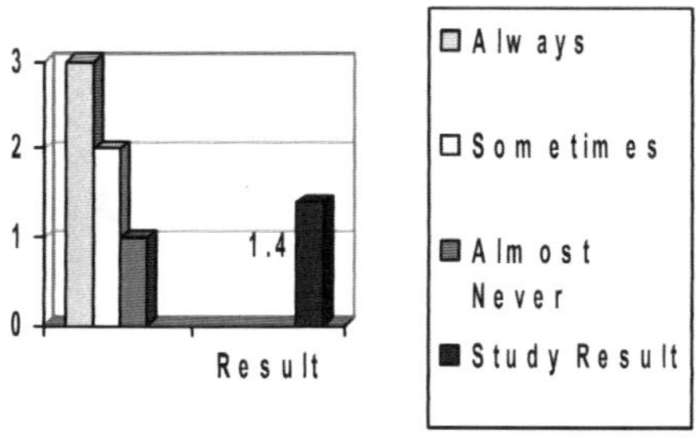

Shutdown on loss of flame at the main burner (as detected by flame scanners, flame rods, pattern recognition cameras, or other device) is not a very common shutdown, and many refiners feel that the hazards that would be prevented utilizing this function are adequately safeguarded using other means. It was noted by several participants that a loss of flame on the main burners' shutdown is not required to be measured or cause a shutdown if continuously operated pilots that are sourced with a supply of fuel that is independent of the main burners.

It should be noted that the figures presented above represent natural draft process heaters. Use of flame scanners is much more prevalent on induced and forced draft process heaters, and almost always done on boilers (true purpose-built boilers, as opposed to process heaters where steam generation is incident to the process purpose of the heater).

Shutdowns based on flame detection on pilots are also not extremely common. Although not common, one operating company employs pilot burner loss of flame shutdowns as a standard practice and almost always implements this shutdown in practice. The rationale of this operating company is that if the pilot burner is in operation, then a lot of the hazards that might occur as a result of main burner fuel system anomalies will be effectively mitigated by the pilot. As a result, they place a high degree of importance on ensuring that the pilot flame is always available to perform its safety purpose.

Low Process Flow (Pass Flow) Shutdown

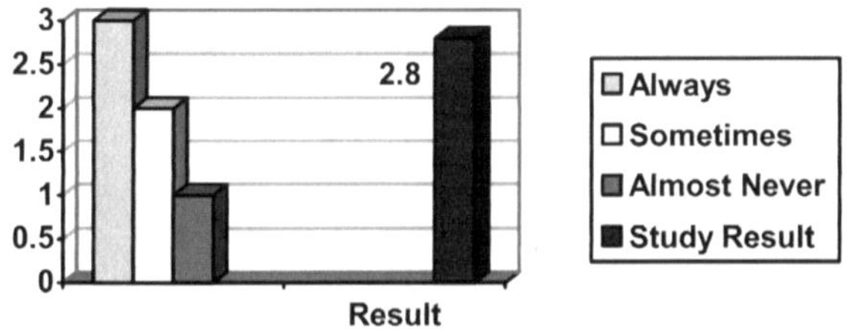

Heater shutdown based on low flow of process material through heater passes is very common. For those respondents whose standard practice does not require this trip, analysis is required to ensure that loss of flow through the heater tubes does not pose a significant hazard and are still recommended in coking services. Even for those whose standards require the shutdown, in some cases exceptions are allowed if an analysis of the type of heater and the process fluid is performed to determine the detrimental effects of loss of flow is performed that shows the risk posed by loss of flow is tolerable.

Low Excess Air / High Combustibles / Improper Fuel/Air Ratio Shutdown

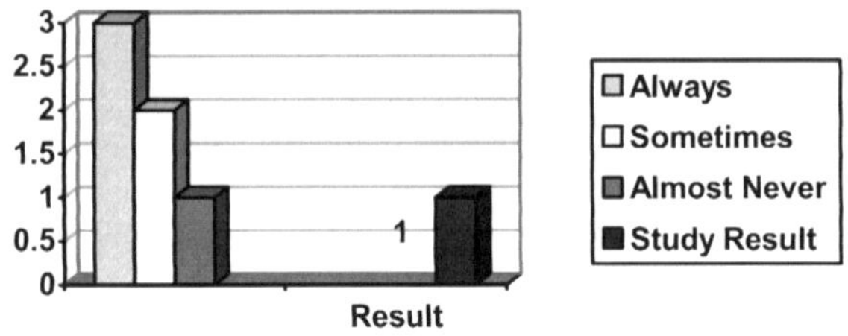

None of the operating company respondents employ any automatic actions to mitigate the hazards posed misoperation of a fired heater resulting in an oxygen deficient fuel rich atmosphere in the heater firebox. The basis of safety for preventing realization of fuel rich firebox hazards is through good control of furnace draft, in some cases employing closed loop control based on oxygen analyzer measurement, and thorough operating training to ensure that if an oxygen deficient firebox condition were to occur that the proper control action of cutting fuel gas, as opposed to adding air, is used.

Combustion Air Low Flow / Fan Not Running Shutdown

Low Air Flow

Fan Not Running

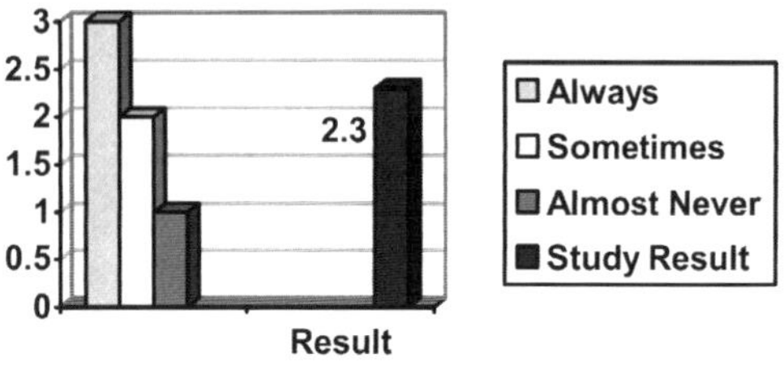

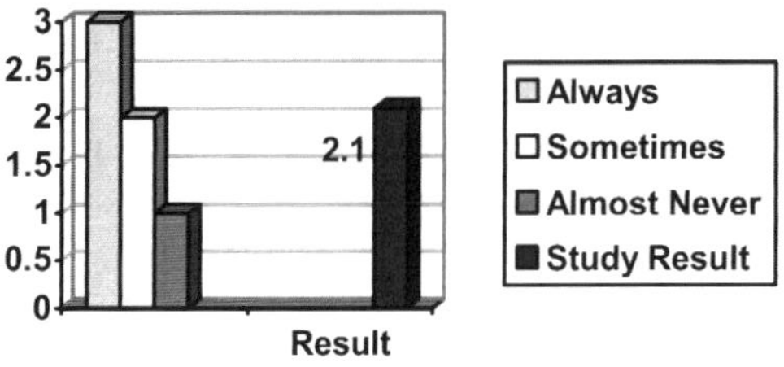

It was noted that in almost all cases, if a heater is equipped with a forced or induced draft air supply system, some sort of shutdown was employed to detect a loss of air condition. While the reported numbers for standard practice and implementation of either low combustion air flow trip or air fan not running trip were near 50%, these figures are not indicative of their importance to process safety, but to the preference of what type of measurement is used for causing the shutdown. The two primary means of measurement were direct measurement of combustion air flow or an auxiliary contact on the combustion air fan motor starter. For new installations there is a strong preference to move toward direct measurement of combustion air flow due to the more comprehensive and direct indication of a hazardous condition. In some cases, both fan not running and low air flow were used as measurements that can result in a heater shutdown. In the case of one of the respondent operating companies, their preference for detecting insufficient combustion air was the measurement of the speed to the fan motors.

3.2 Alarms

Alarms are very consistently applied across all operating company respondents, and very frequently applied. The only alarm where application was more represented by 'sometimes' than 'almost always' is the High CO or High Combustibles alarm. It was noted that application of this alarm is becoming more prevalent as regulatory requirements increase.

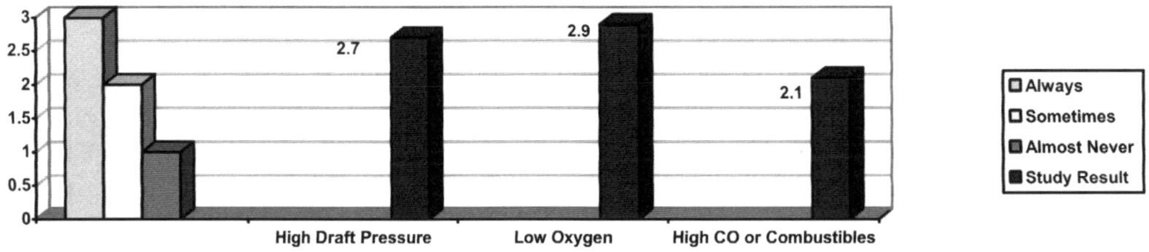

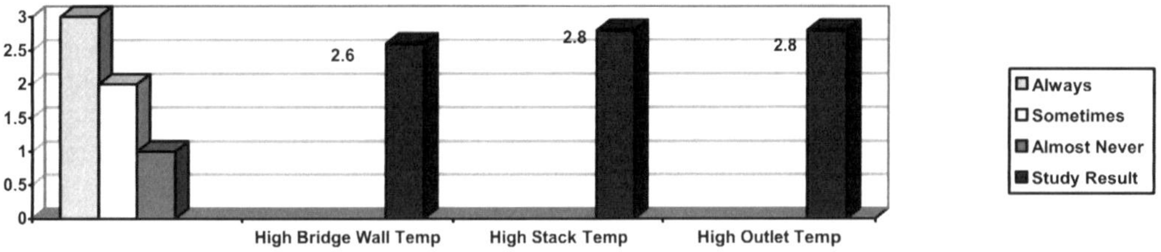

3.3 Permissives

When reviewing the permissives section it is important to note that the intention of the survey was to primarily review the operation of natural draft heaters, as opposed to forced or induced draft heaters. It was noted that the degree to which permissive functions are implemented is much greater for heaters with forced or induced draft fans and also for boilers.

Purge Flow Permissive

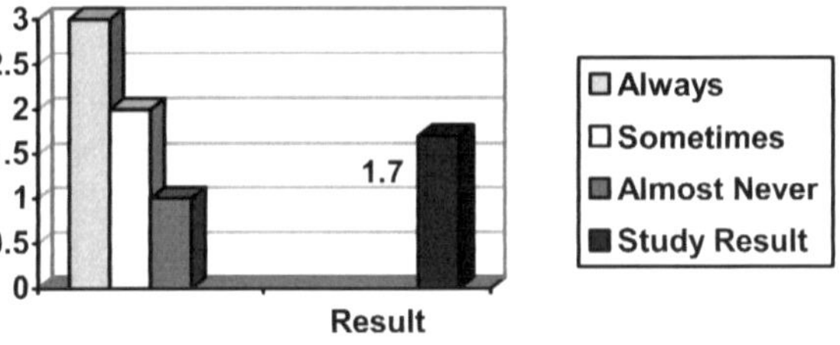

Operating company respondents that employ purge flow permissives were in the minority, but application was significant enough to warrant consideration. It is important to draw a distinction between this function and a subsequent question regarding automatic firebox purge. If automatic firebox purge is employed, then a purge flow timing permissive is virtually always used, but automatic purging is not required for a purge flow timer permissive. Some operating companies employ this permissive by measuring the heater draft or purge steam flow and ensuring that the measured variable is of a sufficient quality for the preset time interval.

Fuel Gas Shutoff Valve Position Permissive

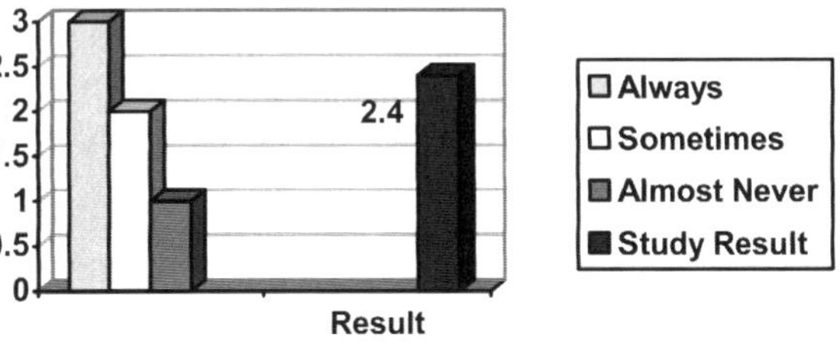

The fuel gas shutoff valve permissive is quite common in practice while not as often directly required in corporate standards. Where this shutoff is not employed in an automatic fashion it is always performed through operator procedure. When the permissive is employed in an automatic fashion it is typically performed using limit switches. No use of valve positioner signals for this purpose was noted.

Individual Burner Valves Closed (Seal Test) Permissive

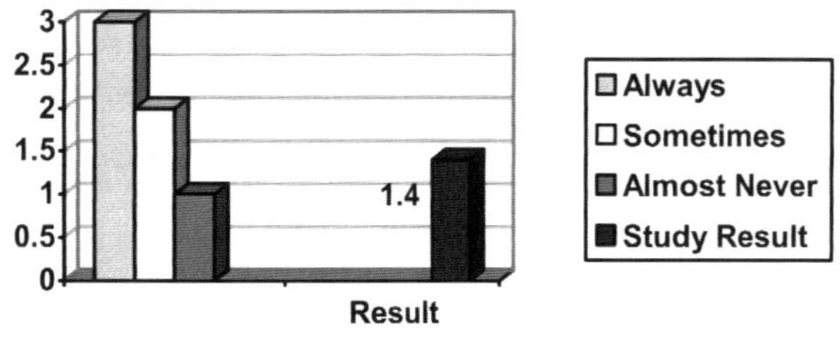

Closure of individual burner valves is always ensured prior to light off of the heater, but it is rarely performed automatically, unless the heater is of a certain type that has a very large number of burners, such as a steam reformer or an ethylene cracker. Only one respondent has historically been doing this type of permissive on a regular basis and only two respondents employ this type of permissive in the standards for new equipment (on all heaters). Of these companies, neither requires a 'seal test' type permissive on all heaters – and typically does not use the seal test methodology unless there is an extremely large number of burners (e.g., more than 100).

Automatic Firebox Purge

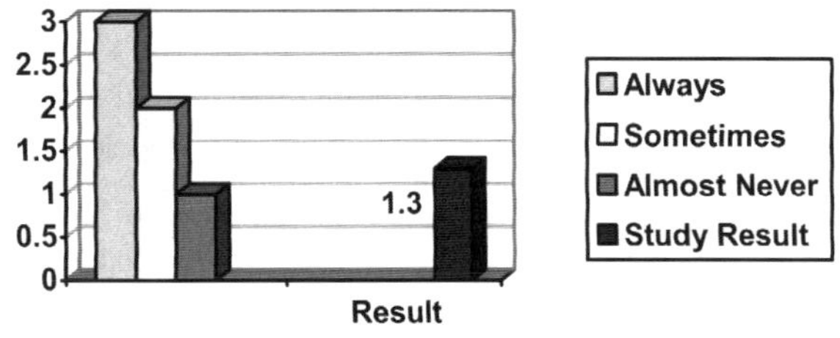

The results presented above show that implementation of automatic purges on natural draft heaters is very rare; in fact, only one site of one respondent organization is known to regularly utilize this technique.. Only one respondent incorporates this as standard practice, and most of their existing heaters do not employ the permissive. Where forced draft heaters are used, the automation of purging is more frequent, but still would only be characterized as "sometimes" for existing equipment. More new installations of forced draft heaters implement automatic purging of the firebox utilizing the forced/induced draft fans, but not enough to categorize the implementation as "almost always".

The medium used for purging is extremely consistent with the type of heater being used. For natural draft heaters, steam was virtually always used. For forced/induced draft heaters, combustion air was almost always used. No significant usage of any other purging medium (i.e., nitrogen) was identified.

Flame Proven Permissive

Pilot

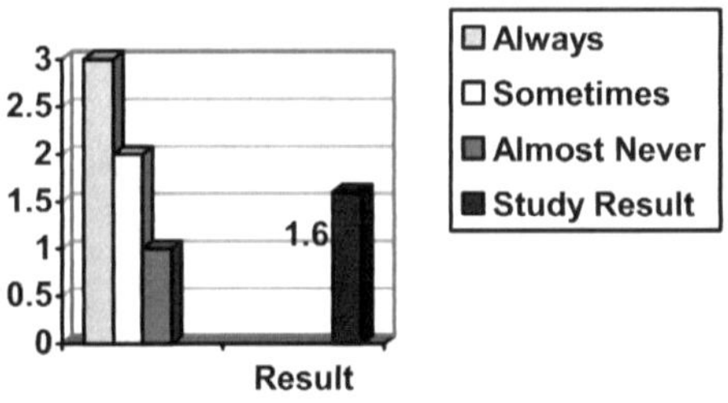

Main

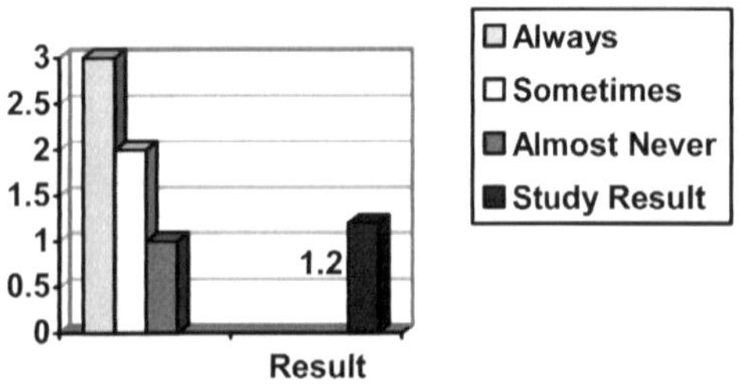

The survey indicated that flame proven permissives are not common. Flame proven permissives are typically performed automatically when there is a loss of pilot flame trip, but not performed if there is no associated trip.

3.4 Shutoff Valves

Double Block and Bleed Shutoff Valves

Main

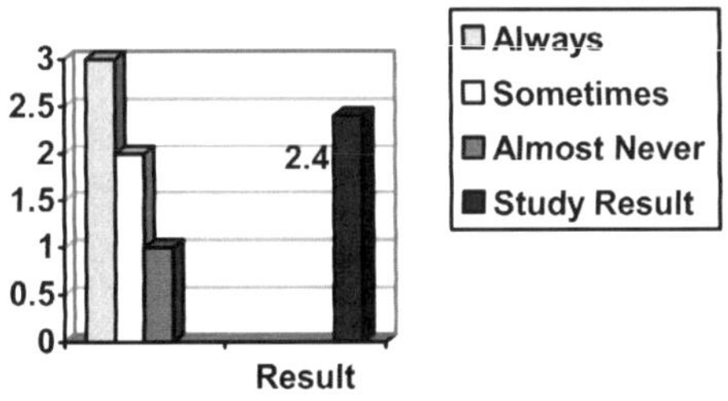

Pilot

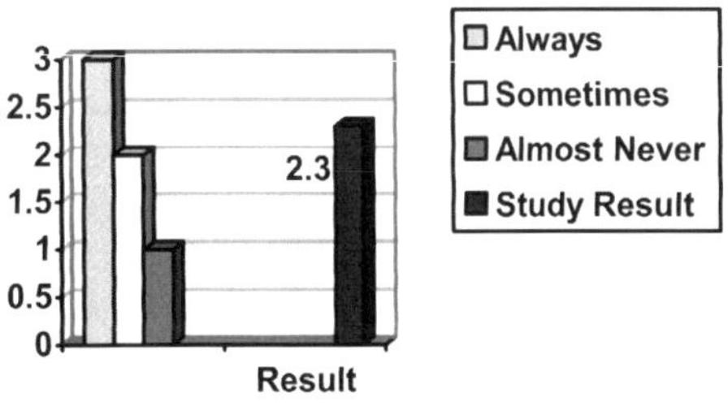

Use of double-block-and-bleed arrangements on heater fuel gas supplies is not uncommon now and is becoming even more common. Most of the respondents are requiring double block and bleed as standard practice, and many require in their standard that double block and bleed be considered during the SIL analysis process. In actual practice, use of double-block-and-bleeds is also prevalent.

Use of a double-block-and-bleed arrangement provides two benefits over a single valve. First, a positive isolation is ensured, preventing hazards resulting from fuel gas leaking across a single valve in to the firebox when the valve is closed. Second, the use of two devices to perform the shutdown (in a 1oo2 vote) decreases the probability of failure on demand and increased the safety integrity level (SIL) that can be achieved by the functions. These two goals can and are being achieved by measures other than use of double-block-and-bleeds by a number of respondent companies.

For positive isolation, some refiners isolate the fuel when not in use as a standard practice by blinding and rolling a spool piece in the gas supply. The main fuel source is typically not reconnected to the main burners until after the pilot burners have been lit, virtually eliminating the risk of leaking main shutdown valves and open individual burner valves for the case of a cold start. One refiner noted a move to quick disconnects to facilitate the practice of isolation.

To decrease the probability of failure on demand of gas shutoff some refiners are electing to use increased diagnostics and more frequent testing in lieu of increasing the number of valves. Where required to meet elevated SIL targets, partial stroke testing is sometimes utilized. Partial stroke testing is performed commonly either using manual jamming devices or with purpose built automatic testing devices – no strong preference between the testing types emerged in the survey.

In some cases, the use of double block and bleed arrangements of pilot gas systems was noted to be a matter of consistency between shutoff practices between the main burners and the pilot burners, and not for risk reduction in and of itself.

Bypasses Valves around Shutoff Valves (for Light-Off or Minimum Firing)

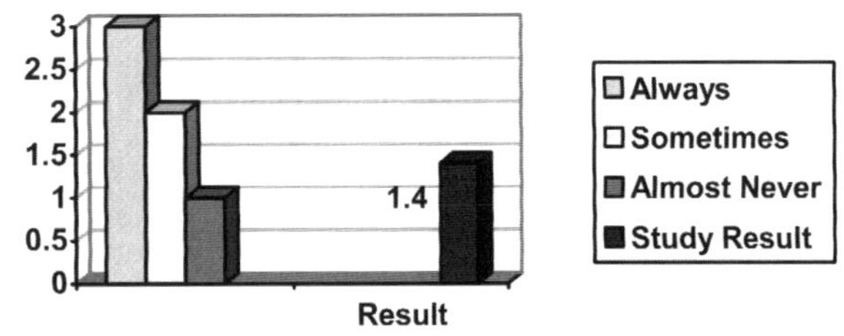

Bypasses around shutoff valves are not very frequently used and are falling out of favor in situations where they were traditional used. One respondent noted that the only instance where bypasses are allowed are for 'seal testing' and in this case the bypass was a manual 'dead man' valve and the bypass itself was only 1/4" tubing. The preferred design is moving toward either ensuring that SIL targets can be met with equipment that only needs to be tested at a turnaround interval (i.e., double block and bleed) or use of partial stroke testing.

3.5 Pilots

Continuously Operated Pilots

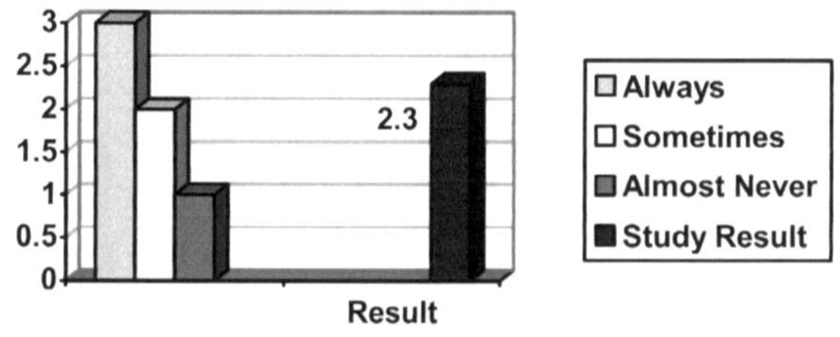

While the use of continuously operated pilots is not required by the preponderance of operating companies, actual use of continuously operating pilots is very common. Many of the standards that do not strictly require that this practice be implemented do require that it be considered. These standards require that if continuous pilots are not used that other means of enhancing safeguarding be employed, such as loss of flame detection.

Some operating companies tend to have very strong opinions about the use of pilots that permeate through their organizations. One operating company relies on continuous pilots as a primary means of safeguarding, including heater shutdown if loss of pilot flame is detected. Two operating companies that responded to the survey do not employ continuous pilots as a standard practice. One of these companies uses pilot burners to light the main burners, but then takes them out of service. Another company does not install pilot burners at all; instead, they light the main burners directly. This company stated that they find pilot burners to be very unreliable and hard to maintain due to the increased propensity toward clogging due to the smaller orifice size and lower flow rates. Their preferred means of ensuring safe operation is through strict control of burner pressure and well established testing and knowledge of the safe operating limits of the burner system.

Pilots are Source from a Separate Supply (i.e., Natural Gas)

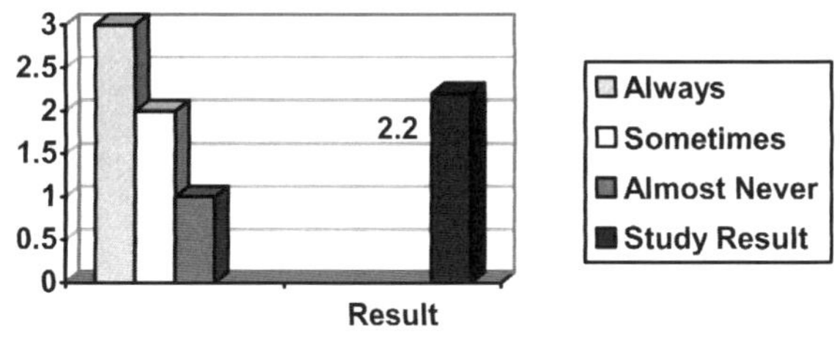

Use of a separate source of pilot gas is somewhat common where practicable, but not to the degree where it can be considered to be 'almost always'

employed. It should be noted that the figures shown here are somewhat skewed because those respondents who do not use pilots at all are also indicated as not using a separate supply. This means that use of a separate supply of pilot gas for those respondents who use pilots is actually somewhat higher than what is shown.

3.6 Post Purge

Automatic Steam Addition upon Natural Draft Heater Shutdown

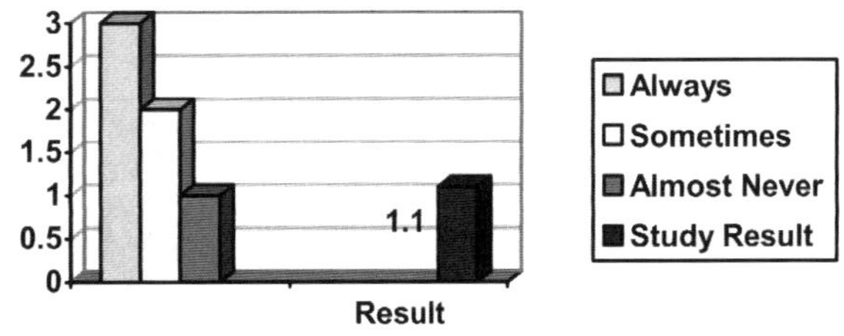

The surveys indicate that this action is virtually never done automatically. Only one respondent noted that this is done automatically in some situations, and another respondent noted that this practice is employed regularly at one of their sites, but never at the others. In fact, some respondents noted that this practice is intentionally avoided due to the potential that spurious operation of the function or use of the function in non-optimal conditions may actually cause damage to the heater.

4.0 Conclusions

Benchmarking is an effective tool in assisting refiners to determine how to implement portions of standard practices where the requirements are ambiguous, open to interpretation, or multiple options are available. In many cases the practices that are implemented are remarkably consistent and form the basis for "recognized and generally accepted good engineering practice".

CSB Investigation of the Explosions and Fire at the BP Texas City Refinery on March 23, 2005

Don Holmstrom, Francisco Altamirano, Johnnie Banks, Giby Joseph, Mark Kaszniak, Cheryl MacKenzie, Reepa Shroff, Hillary Cohen, and Stephen Wallace

U. S. Chemical Safety and Hazard Investigation Board
2175 K Street, NW, Suite 400
Washington DC 20037
Telephone 202-261-7633
Fax 202-974-7633

Prepared for Presentation at
American Institute of Chemical Engineers

2006 Spring National Meeting
40th Annual Loss Prevention Symposium
Orlando, Florida
April 24-26, 2006

UNPUBLISHED

CSB Investigation of the Explosions and Fire at the BP Texas City Refinery on March 23, 2005

Don Holmstrom, Francisco Altamirano, Johnnie Banks, Giby Joseph, Mark Kaszniak, Cheryl MacKenzie, Reepa Shroff, Hillary Cohen, and Stephen Wallace

U. S. Chemical Safety and Hazard Investigation Board
2175 K Street, NW, Suite 400
Washington DC 20037
Telephone 202-261-7633
Fax 202-974-7633

Abstract

At approximately 1:20 p.m. on Wednesday, March 23, 2005 a series of explosions and fire occurred at the BP Texas City oil refinery during the startup of an isomerization (ISOM) process unit. Fifteen workers were killed and about 170 others were injured. Many of those killed were working in or around office trailers located near a blowdown drum and stack open to the atmosphere. CSB found that because of problems initiated during the ISOM process unit start-up, the blowdown atmospheric vent stack experienced a sudden, geyser-like release of flammable hydrocarbon liquid and vapor. This release created a flammable vapor cloud, which ignited, causing as many as five explosions and fire in and around the ISOM unit. CSB identified multiple possible ignition sources in and around the blowdown drum area, including idling vehicles; however, the exact ignition source remains unknown.

The U.S. Chemical Safety and Hazard Investigation Board (CSB) deployed a team of investigators to BP Texas City to conduct a root-cause investigation immediately after the incident. This team examined blast patterns to determine the explosion origin; reviewed the design of the isomerization unit equipment; and examined plant safety and operating procedures, past accidents, maintenance procedures, and oversight and inspection. Investigators also reviewed the adequacy of applicable regulations and industry standards for the placement of temporary structures such as trailers in refineries. This paper presents the preliminary findings from the CSB investigative team regarding the technical and underlying causes of the incident and discusses three urgent recommendations issued by the CSB stemming from the BP incident.

This paper has not been approved by the Board and is published for general informational purposes only. Every effort has been made to accurately present the contents of any Board-approved report mentioned in this paper. Any material in the paper that did not originate in a Board-approved report is solely the responsibility of the authors and does not represent an official finding, conclusion, or position of the Board.

1.0 Introduction

The March 23rd incident involved a sudden release of flammable hydrocarbon liquid and vapor from an atmospheric vent stack in the refinery's ISOM unit. Temporary work trailers were positioned 100 to 150 feet away from the vent stack. Their purpose was to provide office and storage space for contractors performing maintenance (turnaround) work at an adjacent unit The fifteen fatalities were contractors who were in and around these trailers at the time of the explosions and fire.

2.0 Texas City Refinery Background

The BP Texas City refinery is the third largest in the United States. Amoco Corporation built and operated the refinery until January 1999 when Amoco Corporation merged with BP. The Texas City refinery can produce about 11 million gallons of gasoline a day. The refinery has 30 process units spread over a 1200 acre site. In addition to gasoline, the refinery produces jet fuels, diesel fuels, and chemical feed stocks. The Texas City refinery also employs about 1800 staff and hundreds of contractors.

3.0 Process Description

The incident occurred in the raffinate section of the ISOM unit. The ISOM unit converts the hydrocarbons pentane and hexane into isopentane and isohexane in order to boost the octane rating of gasoline. The unit works by heating the pentane and hexane in the presence of a catalyst. The raffinate section is comprised of a splitter tower and associated process equipment, which are used to prepare the hydrocarbon feed for the isomerization reactor.

4.0 Incident Description

On the morning of March 23, the raffinate splitter was being restarted after a maintenance outage. During the startup, liquid hydrocarbons were pumped into the splitter tower for 3 hours without any of this liquid being taken out and sent to storage. As a result, the 164-foot tall distillation column was overfilled and eventually overflowed hydrocarbons into the vertical piping off the top of the tower. In addition, unknown to operators controlling the process, critical alarms and control instrumentation were malfunctioning, providing the operators with erroneous process control information.

The hydrocarbon level in the vertical piping built-up and exerted pressure on the emergency relief valves 150 feet below. Over a period of several minutes, the pressure, as measured at the bottom of the overhead piping, rose from about 20 pounds per square inch to about 60 pounds per square inch.. This pressure activated three pressure relief valves used to vent high pressure in the column; the relief valves were set to open at 40, 41, and 42 pounds per square inch. The relief valves remained open for six minutes, according to computerized records. During this period, hydrocarbon liquid and vapor was discharged to the blowdown drum. The blowdown drum overfilled with hydrocarbon liquid, leading to a geyser-like release out the stack. The flammable hydrocarbons pooled on the ground releasing vapors which ignited, resulting in multiple explosions and fires. This blowdown system was of an antiquated design; it was originally installed in the 1950s, and it had never been connected to a flare system to safely combust flammable vapors released from the process.

5.0 Key Issues

At this point in its investigation, CSB has identified six key issues:

1. Occupied trailers were sited in an unsafe location, too close to a process unit handling highly hazardous materials. All 15 fatalities occurred in or around the two trailers positioned closest to the blowdown drum and atmospheric vent stack.
2. The ISOM unit should not have been started up with existing malfunctions of the tower level indicator, level alarm, and control valve. Known problems with instrumentation and equipment were not repaired prior to start up.
3. The raffinate splitter tower had a history of abnormal start ups. It became the norm to fill the splitter tower with liquid above the range of the indicator for long periods of time throughout the startup process; therefore, it was impossible for operators to discern the actual level of liquid in the tower.
4. On the day of the incident, an unsafe blowdown drum and stack vented highly flammable material to the atmosphere. The drum was never connected to a flare since its construction in the 1950s. The first rule of oil refinery safety is to keep the flammable hazardous materials inside the piping and equipment. A properly designed flare system would safely contain the liquids and combust flammable vapors, preventing an unsafe release to atmosphere.
5. Between 1995 and March 23, 2005, there were at least four other serious releases of flammable material from the ISOM blowdown drum and stack. Even though these serious near misses revealed the hazard of the blowdown design, no effective investigations were conducted or changes made.
6. The refinery missed a number of opportunities to replace the hazardous blowdown drums. In 1992, OSHA cited a similar blowdown drum and stack at the refinery as unsafe, but the citation was dropped and the drum was not connected to a flare. In 1997, the ISOM blowdown drum and stack were replaced with like equipment due to corrosion even though corporate safety standards required replacing blowdown drums with flares when major modifications were made to existing equipment. In 2002, the refinery evaluated connecting the ISOM blowdown system to a flare as part of an environmental initiative, but did not pursue this option.

6.0 Preliminary Findings

The following are some important facts that relate to the causes of the hydrocarbon release, the subsequent fire and explosions, and the reasons for the injuries and the fatalities. Since the investigation is not complete, CSB will refer to these items as preliminary findings.

6.1 Trailer siting

The first preliminary finding concerns the siting of the occupied trailers too close to hazardous process areas. All of the fatalities, and many of the critically injured, were in or around the nine contractor trailers that were positioned near process units and as close as 121 feet from the ISOM blowdown drum and stack. The ISOM unit contained large quantities of flammable hydrocarbons, and had a history of releases, fires, and other safety incidents over the previous two decades.

At the Texas City refinery, trailers have been periodically sited in and around process units handling highly hazardous materials for reasons of convenience, such as ready access to work areas. BP had sited trailers for contractors in the same location near the ISOM unit for a number of years. Trailers such as these did not need to be sited as close as they were to process areas in order for the contractors to perform their job duties.

Under BP's siting policy, trailers used for short periods of time, such as turnaround trailers, were considered as posing little or no danger to occupants. This approach conforms to the safety guidance published in Recommended Practice 752, by the American Petroleum Institute (API).

The American Petroleum Institute is the primary safety standard setting trade association for the oil industry. API 752 states that each company may define its own risk and occupancy criteria. There are no defined minimum protections. API 752 provides no safe minimum distances from process areas for the location of trailers used in refineries and other chemical facilities.

Over 40 trailers were damaged in this incident. Workers in adjacent units were injured in trailers 480 feet from the ISOM blowdown drum. A number of trailers, some as far as 600 feet from the blowdown drum, were heavily damaged. Trailers can be easily relocated to less hazardous areas.

Trailers are not generally designed to protect occupants from fire and explosion hazards. In contrast, occupied buildings such as control rooms and operator shelters located within a process unit are typically permanent and constructed to be blast and fire resistant. For these reasons, CSB issued an urgent recommendation to API to advise their members to remove trailers away from hazardous areas in process plants.

6.2 Unit start up and mechanical integrity issues

The next preliminary finding addresses unit start up and mechanical integrity issues. The raffinate splitter tower was started up despite malfunctioning key process instrumentation and equipment on the day of the incident, including the tower level indicator and sight glass, tower hard wire level alarm, blowdown drum high level alarm, and the tower three pound pressure valve. Both the tower level indicator and a hard wired high level alarm were malfunctioning during the events that led to the incident. The hard wired high level alarm and the level indicator were documented by work orders as malfunctioning, but were not repaired prior to start up. The sight glass on the raffinate splitter tower, which gives a visual check of tower liquid level and can help verify the accuracy of the level indicator, was reported prior to the startup to be dirty and non-functional. In addition, the high level alarm for the blowdown drum did not sound prior to the incident.

Prior to unit start up, operators informed management that the tower three pound pressure valve (e.g., a valve used to remove non-condensables and excess pressure) was not operating. Still, the valve was not repaired prior to start up. This led operators to use alternative methods to remove nitrogen and reduce tower pressure, such as opening an eight inch chain bypass valve to the blowdown drum and stack.

Post-incident testing confirmed that the three pound valve was not working properly. The proper working order of key process instrumentation was not checked as required by the start up procedure. This was an additional opportunity to verify that the instrumentation was working properly. However, operations personnel starting up the unit did not know of problems with the tower level indicator and high level alarm because checks to determine operability of these instruments were not performed. Verification of instrument operability is required in the procedures and a critical step in the start up process. However, unit operations management turned away instrument technicians and signed off on the checks as if they had been completed. The raffinate unit should not have been started up when important instrumentation and equipment was known to be malfunctioning.

Operations personnel did not open the tower level control valve at the time specified in the start up procedure. They did not balance the hydrocarbon flows in and out of the tower. The tower filled up because liquid was being added for nearly three hours without being removed. CSB is further investigating possible underlying causes such as workload, fatigue, communications and management oversight concerning this issue.

During the hours of 10:00 a.m. to 1:00 p.m., the liquid level was rising rapidly to the top of the 164 foot tower. However, the indicator that operators were using to determine the level in the tower showed the level to be significantly less, within the bottom 10 feet of the tower and declining further, during this same time period. The tower level indicator only indicates the bottom 10 feet. A false level indication showing the tower level declining was a factor in the delay in removing liquid from the tower. Operators relied on this level indication, and they took no action as they observed the level drifting back towards the normal operating range below the 10 foot indicator. Start up procedures did not address the importance of maintaining a balance of liquid flow in and out of the tower or the consequences of deviation. The tower was not equipped with additional instrumentation indicating tower level, such as a tower bottom pressure indicator, that could have provided the operators with additional data concerning tower level.

The tower had a history of abnormal unit start ups. In 16 start ups of the raffinate splitter tower from April 2000 to March 23, 2005, eight of these start ups of the raffinate splitter experienced at least two times the normal pressure, greater than 40 pounds per square inch versus 20 pounds, the normal operating pressure.

In 13 out of the 16 start ups, the pressure exceeded 30 pounds, the alarm set point for high tower pressure. In February 2003, the set points of the emergency relief valves were lowered from 70 to 40 pounds due to corrosion in the tower.

Since 2003, two splitter tower start-ups had pressure excursions of over 40 pounds that likely resulted in the emergency relief valves venting to the blowdown drum. Thirteen of the start ups of the raffinate splitter had liquid levels above the range of the level indicator, that is, greater than 10 feet, some lasting as long as four hours.

Also, in two thirds of the start ups, liquid feed circulation to the raffinate splitter was started when the tower was already above the range of the level indicator. Filling above the level indicator makes it difficult to know how much liquid is within the tower, thus making it much easier to over-fill. However, most of the previous start ups established liquid flow out of the tower much sooner, between three minutes and 45 minutes after flow was introduced, rather than the three hours it took in this incident. Moreover, preliminary analysis indicates that none of the previous start ups experienced a high level into the upper section of the tower, as occurred in the March 23 incident.

BP did not investigate these previous raffinate splitter tower start ups with high pressures and high levels as required by BP policy. The occurrence of high pressures and high liquid levels during the tower start ups had become part of the operating norm. It is important to investigate these near miss incidents, even those without serious consequences, because catastrophic incidents can follow if problems go uncorrected. Investigation of these incidents could have resulted in improvements to tower design, instrumentation, procedures, controls and training.

6.3 Management oversight and accountability

BP management did not assure that an experienced supervisor was in the unit during start up to provide oversight. BP policy requires experienced supervisory personnel to be assigned to process units during start up to assist in making important decisions. At 10:00 a.m., the supervisor in charge left the unit for a family emergency but no substitute with ISOM operating experience was assigned. The departing supervisor had many years of operating experience in the unit. The relief supervisor was not involved in the start up and had no ISOM operating experience. He could not provide the necessary level of guidance during the critical start up period. No other experienced supervisors were assigned to this start up.

6.4 Process design

The blowdown drum was not properly sized to handle the flammable material coming from the raffinate tower. The drum was attached to a 67 foot tall vent stack open to the atmosphere. The blowdown drum and stack design is outdated and unsafe because this system releases flammable hydrocarbons to the atmosphere. In fact, BP policies recommended phasing out such blowdown systems when major modifications were made. Modern emergency relief systems send hydrocarbons to a flare system that includes an adequately sized knock-out drum to capture liquids and a flare to safely combust the flammable vapors.

Amoco safety standards, last revised in 1994, state that blowdown drums should be connected to a flare when major modifications are made. In 1997, Amoco replaced the blowdown drum and stack with identical equipment rather than connecting the drum to a safer location such as a flare system. This replacement was due to corrosion, and was a major modification under Amoco's procedures. Amoco was the owner of the refinery in 1997. Consequently, at that time, the blowdown drum and stack should have been converted to a flare system. After the merger in 1999, BP adopted the Amoco safety standard for blowdown drums at the refinery. In 2002, BP evaluated connecting the ISOM blowdown system to a flare as part of an environmental initiative, but did not pursue this option.

Since the March 23, 2005 incident, BP has stated they plan to discontinue the use of blowdown drums open to the atmosphere at the facility. At the time of the incident, the raffinate splitter tower did not have an effective pressure control system to reduce high pressure and remove hydrocarbons to a closed system. This led, in part, to the dependence on a blowdown drum and stack to reduce high pressures that may build up within the tower. The tower should have had an additional layer of protection such as a pressure control valve to a closed system in order to remove hydrocarbons and reduce built-up excess pressures, thus lessening the dependence on the blowdown drum and stack relieving the atmosphere. The use of a closed system could have prevented or reduced the severity of the incident.

In 1992, OSHA cited and fined Amoco on the hazardous design of a similar blowdown drum and stack that handled flammable hydrocarbons at the Texas City refinery. The blowdown referred to was located in another unit, but was of similar design to the ISOM blowdown system. In the original citation and notification of penalty, OSHA suggested that the appropriate abatement method was to reconfigure the blowdown to a closed system with a flare. In a settlement agreement, OSHA withdrew the citation and fine and the refinery continued to use blowdown drums without flares. This early opportunity to remove hazardous blowdown drums was not acted upon.

6.5 Previous incidents

Since 1995, at least four releases from the blowdown drum sent hydrocarbons to the stack and sewer, generating flammable vapor clouds at ground level. During these releases, vapors escaped from the sewer and formed ground level flammable clouds. The drain valve off the bottom of the blowdown drum was chained open at the time of the incident and had been in this position for a number of years to allow for liquid hydrocarbons to flow into the process sewers. Since the hydrocarbon vapors coming out of the stack were heavier than air, material released descended towards the ground and also formed flammable clouds. Luckily, these previous releases did not find a source of ignition. CSB found that the four previous blowdown incidents were not properly investigated, nor were needed corrective actions implemented.

6.6 Vehicle traffic

The final preliminary finding we will discuss concerns BP's vehicle traffic policy. CSB found that vehicles played an important role in the incident. BP's vehicle policy allowed vehicles unrestricted access near process units. BP's policy does not establish safe distances from process unit boundaries for vehicles. Approximately 55 vehicles were located in the vicinity of the blowdown drum and stack at the time of the incident. Two running vehicles may have provided sources of ignition for the incident. One was within 25 feet of the blowdown drum and stack.

7.0 Urgent recommendations

In light of these findings, CSB issued three urgent recommendations which the agency believed needed immediate action such that it could not wait till completion of the full-root cause investigation. The first urgent recommendation called on the BP Global Executive Board of Directors to commission an independent panel that would review a range of safety management and culture issues. It is the first safety recommendation in CSB's eight-year history that has been designated as "urgent" and issued in advance of a completed investigation. In voting to adopt the recommendation, the Board said that identified safety management lapses pose an issue that "is considered to be an imminent hazard and has the potential to cause serious harm unless rectified in a short time frame."

Citing a series of serious safety incidents at the Texas City facility over the past two years, including two that occurred after the March 23rd incident, the Board recommended that BP establish and fund a diverse panel of experts, including employee representatives. The panel would review corporate safety oversight, safe management of refineries obtained through mergers and acquisitions, corporate safety culture, and management systems such as near-miss reporting and mechanical integrity programs.

The Board directed the other two urgent recommendations to two leading national trade organizations, the American Petroleum Institute (API) and the National Petrochemical & Refiners Association (NPRA), which represent most major domestic oil and petrochemical producers. API develops recommended safety practices that influence operations at thousands of petrochemical facilities around the country.

The first recommendation calls on API to develop new industry guidance "to ensure the safe placement of occupied trailers and similar temporary structures away from hazardous areas of process plants." The Board noted that the existing safety guidance, API Recommended Practice 752, does not prohibit the placement of trailers in close proximity to hazardous process units. The guidance, entitled "Management of Hazards Associated with Location of Process Plant Buildings," is widely used by U.S. oil and chemical companies to assess siting hazards, a regulatory requirement under OSHA's Process Safety Management standard.

As currently written, API 752 allows individual companies to define their own risk and occupancy criteria for trailers. Prior to March 23, BP had defined trailers used for short periods of time as posing little or no danger to occupants and allowed the placement of the trailers near the ISOM unit.

According to findings accompanying the Board's urgent recommendation, the explosions in Texas City injured workers in trailers as far as 480 feet from the source of the release, and trailers as far as 600 feet away were heavily damaged. Subsequent to the incident, BP announced it would relocate trailers at least 500 feet away from potential hazards and move nonessential workers into office space outside the refinery.

In many cases, trailers are positioned for convenience during maintenance and are not essential for facility operations. These trailers can be easily relocated to safe distances. Permanent buildings in refineries and chemical plants are often heavily reinforced to resist blast and fire damage, while most trailers and temporary structures provide little protection for occupants.

A separate urgent recommendation, directed jointly to API and NPRA, called on the organizations to immediately contact their members urging "prompt action to ensure the safe placement of occupied trailers away from hazardous areas of process plants," before the new API safety guidance is completed.

8.0 Ongoing Investigation Activities

CSB's investigation into the BP incident continues. The root and contributing causes to catastrophic events, such as what happened on March 23, 2005, are often complex and interrelated. For this reason, incident investigations require detailed and thorough analysis to uncover the underlying safety issues that may exist at the facility and within the industry in general. CSB is dedicated to that task.

The investigation team is aiming to publish its final report in September of 2006.

Lessons from Texas City
A Case History

Michael P. Broadribb, C.Eng.

Senior Consultant, Process Safety,
BP America Production Company,
Houston, Texas

Prepared for presentation at

American Institute of Chemical Engineers
40th Annual Loss Prevention Symposium

and

Center for Chemical Process Safety
21st Annual International Conference

Orlando, Florida

April 2006

Lessons from Texas City
A Case History

Michael P. Broadribb
Senior Consultant, Process Safety,
BP America Production Company,
Houston, Texas

A major incident occurred in March 2005 on the Isomerization Unit at the BP refinery in Texas City, Texas, one of the largest refineries in the USA.

An explosion occurred when heavier than air hydrocarbon vapors combusted after coming into contact with an ignition source, probably a running vehicle engine. The hydrocarbons originated from liquid overflow from a blowdown stack following the operation of the Raffinate Splitter overpressure protection system caused by overfilling and overheating of the tower contents. The severity of the incident was increased by the presence of many people congregated in and around temporary trailers, which were sited close to the process unit.

This paper will explain how and why the incident occurred, the general lessons learned, the actions taken to prevent recurrence, and highlight some wider messages for the industry.

1. INTRODUCTION

The Texas City Refinery is BP's largest and most complex oil refinery, with a rated capacity of 460,000 barrels per day (bpd) and production of up to 11 million gallons of gasoline a day. It also produces jet fuels, diesel fuels and chemical feed stocks. The refinery has 30 process units spread over a 1,200-acre site and employs about 1,800 permanent BP staff. It was owned and operated by Amoco prior to the merger of BP and Amoco in 1999 and largely uses Amoco safety management systems pre-dating the merger. At the time of the incident approximately 800 additional contractor staff were on site for significant turnaround work.

The incident occurred on the Isomerization Unit (ISOM) and involved the Raffinate Splitter (the Splitter), and Blowdown Drum & Stack. The ISOM converts low octane blending feeds into higher octane components for blending to unleaded regular gasoline. The unit has four sections, one of which is the Splitter, which takes a non-aromatics stream from the Aromatics Recovery Unit (ARU) and fractionates it into light and heavy components. The Splitter may be run in conjunction with the ISOM or independently to build inventory when the ISOM is shut down.

The Splitter was originally commissioned in 1976, but has been modified several times. It is a single fractionating column, 164 ft tall with 70 distillation trays, feed surge drum, fired heater reboiler, fin fan overhead condenser, and reflux drum. It has an approximate volume of 3,700 barrels, and processes up to 45,000 bpd of raffinate from the ARU. About 40% of the total raffinate fed to the unit is recovered overhead as C_5 / C_6 light raffinate. The remaining bottoms product consists of C_7 / C_8 heavy raffinate.

Figure 1

Raffinate Splitter

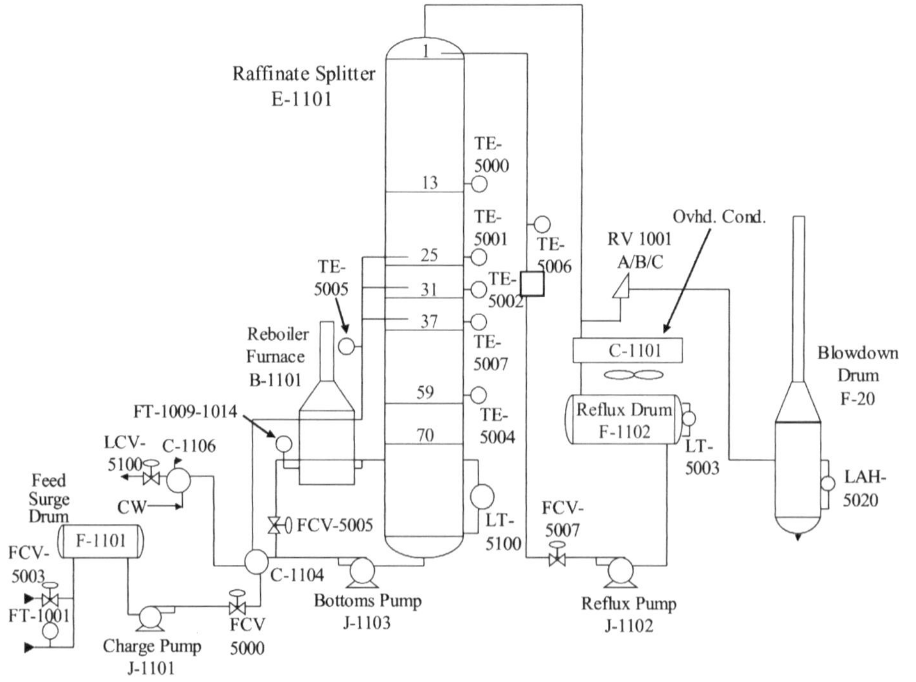

Figure 2 Flow Diagram of Raffinate Splitter

Hot hydrocarbon vapors and minor associated liquids from the ISOM relief, vent, and pump-out systems during upsets or shutdowns are disposed of to a blowdown system. The blowdown system consists of relief pipework headers (one of which is from the Splitter), the Blowdown Drum & Stack (F-20), and a pump-out pump. Vapors disperse from the top of the stack and liquids flow out of the drum through a gooseneck into the site's closed sewer system. F-20 was commissioned in the 1950s and has been modified several times over the years. It is a vertical drum of 10-ft diameter with a 113-ft-high stack.

Figure 3

Blowdown Drum and ISOM Unit, After the Incident

During startup of the ISOM on Wednesday, March 23, 2005, following a temporary outage, an explosion and fire occurred which killed fifteen and harmed over 170 persons. Many of those injured or killed were congregated in or around temporary trailers used for supporting contract workers involved in a turnaround taking place on the nearby Ultracracker unit. Several trailers were located between two operating units, the ISOM and the Naphtha Desulfurization Unit (NDU). The closest trailer, a double-wide trailer, was located within 150 ft of the base of F-20, and is where most of the fatalities occurred at the time of the explosion. Trailers had been sited in the same area on several occasions previously, as no concerns were raised regarding this location from prior siting studies.

The site was secured and an Investigation Team established the following day to investigate the circumstances surrounding the incident, determine the root causes, make recommendations to prevent a recurrence, and identify lessons learned.

The investigation was performed over 8 months, and included site inspections, witness interviews, document and record reviews, and chemical analysis of process samples. The computer hard drive from the process control system was secured and the data downloaded. Third party specialist companies were retained to document the explosion debris and effects, and to model the process and nature of the explosion. Process instrumentation and equipment, such as level indicators and relief valves, were tested, and an internal inspection of the Splitter was conducted.

2. DESCRIPTION OF THE INCIDENT

A double-wide trailer was installed west of the ISOM on September 1, 2004. The MOC for the siting of this trailer was approved to proceed (i.e., develop MOC for final approval) and a hazard review was conducted on October 6. The trailer was not approved for occupancy prior to the incident, but was occupied from late October/early November 2004. Subsequently, several other trailers were installed west of the ISOM for the Ultracracker turnaround. No MOC was initiated for these trailers.

On February 21, 2005 the Splitter was shut down for a planned temporary outage, because of work on another part of the ISOM and ARU turnaround. The Splitter was steamed out to remove hydrocarbons and some maintenance tasks were carried out during the outage. Condensate was drained from low point drains on March 14 in preparation for restarting the unit. Following pressuring with nitrogen at 22.5 psig for tightness testing, the Splitter was depressured on March 21.

The manpower on shift on the ISOM, NDU and AU2 units was doubled up for the period of the temporary outage from February 21, 2005, through March 23, 2005. On March 22, the Production Planning Department requested the Shift Supervisor to start up the Raffinate Splitter. Technicians were checking instrumentation on the Splitter when they were informed the unit would be starting up. It appears from interviews that they may not have completed all the checks prior to startup. The 3 psig vent system control valve (used to purge nitrogen from the Splitter) was stroked on March 22, although it is not clear from witness statements whether it functioned properly. There was no entry in the log book or work order raised for repair.

On night shift March 22/23, the Night Shift Supervisor told an operator to commence starting up the Splitter. The operator took control of the packing (establishing liquid levels) from the on-shift board operator and elected to pack the tower from the satellite control room. He brought in cold feed to the Splitter to establish levels in the Feed Drum and Tower, and to pack the Reboiler circulation loop. Prior to startup, the instrumentation on the tower was not checked as per the procedure. He commenced charging feed to the tower at approximately 15,000 bpd at 02:13 hrs March 23. By 02:38 hrs the Splitter base level sensor started to indicate a gradually increasing level. At 02:44 hrs he opened the Reboiler flow control valve to establish reboiler circulation and charge liquid to the reboiler circuit, causing the indicated level to fall back to 3% by 02:55 hrs (3% is equivalent to approximately 2 ft 9 in. height above tangent). Thereafter, the Splitter base level gradually rose again until the level indicator's high level alarm activated at 72% at 03:05 hrs (approximately 5 ft 5 in. height above tangent) as the tower was filled. The operator acknowledged the alarm, and at 03:08 hrs reduced the feed rate to the Splitter to approximately 10,000 bpd. The alarm remained on and acknowledged until after the incident, 11 hours later. The redundant hard-wired high level alarm, set at 78%, did not operate during this packing of the Splitter. The indicated Splitter base level continued to rise to 100% by 03:16 hrs, and at 03:20 hrs the ARU feed was routed to tankage. The operator closed the feed to the Splitter and the reboiler circulation to leave the remainder of the startup to the day shift.

This operator left the site at 04:59 hrs after making shift relief with the NDU/AU2 Shift Supervisor (not the Day Shift Supervisor). A handover did occur between the Night Shift Board Operator and the Day Shift Board Operator while the startup procedure remained in the satellite control room. At shift relief the tower had 4 psig pressure and a 100% base level indication (equivalent to 10 ft 3 in. height above tangent). The night shift did not report the faulty hard-wired high level alarm to the oncoming day shift either verbally or in the shift log. A work order was not initiated for repair of the alarm.

On arrival at around 06:00 hrs March 23, the day shift operators made their normal rounds and checked the unit lineup. The Day Shift Supervisor entered the site at 07:13 hrs. No pre-job safety review or procedure review was conducted. At 09:21 hrs the outside operators briefly opened the 8-inch chain-operated vent valve, around the tower overhead relief valves that vented residual nitrogen from the Splitter tower and dropped the pressure from 4 psig to nominally atmospheric pressure. This pressure gradually rose back to 0.5 psig by 10:08 hrs.

The Day Shift Board Operator started the reboiler circulation at 09:41 hrs, and at 09:52 hrs reintroduced the feed to the Splitter at a rate of 20,000 bpd. After stroking the Heavy Raffinate rundown control valve to verify the lineup to tankage, the Day Shift Board Operator closed this control valve in manual. (The startup procedure specifies 50% set point in automatic mode). The flow meter indicated a Heavy Raffinate flow of 3,000 to 4,700 bpd. This is believed to be a zero error on the meter, as the control valve was closed, and there was no heat exchange between the Heavy Raffinate and feed.

At approximately 10:00 hrs, two main burners were lit in the Reboiler fired heater. Shortly afterwards, the Day Shift Supervisor for the ISOM left the site due to a personal family matter. He later stated that he passed command to the NDU/AU2 Shift Supervisor, but this could not be confirmed. Two additional main burners were lit in the heater at 11:17 hrs, and the Splitter bottoms temperature continued to rise at approximately 75˚F per hour compared to the 50˚F per hour specified by the startup procedure. The exact number of main burners eventually lit is unknown as it is not recorded and operators interviewed variously described four, five, or six burners. Throughout this period, feed into the Splitter continued at about 20,000 bpd and the heavy raffinate rundown remained closed. No liquids were taken out of the Splitter despite the continuous feed input.

Several possible distractions occurred in the Main Control Room during the startup, when someone made several external telephone calls from the Control Board extension, and a safety meeting was held in the Control Room, around the control board for the ISOM.

Concerned at the continued absence of any indicated liquid level in the Reflux Drum by sometime late morning, the outside operators checked the bottom tap of the Reflux Drum level gauge and only vapor emerged. The Reflux Drum level transmitter continued to show 0% until 13:20 hrs.

The acting Superintendent, NDU/AU2 Shift Supervisor, an outside operator and a trainee operator left the site to get lunch and returned by 12:05 hrs. The acting Superintendent stated that he spent 75% of his workday on March 23 assisting the ARU turnaround.

By 12:20 hrs the Splitter base had reached the target temperature of 275˚F stipulated in the startup procedure, with the feed temperature to the Splitter still only 120˚F. (The normal feed temperature is 205˚F). The absence of heat exchange between the Heavy Raffinate rundown and feed at the Feed/Bottoms Exchangers at this time confirms the lack of any Heavy Raffinate rundown flow to tankage. The feed rate remained unchanged at 20,000 bpd.

By 12:40 hrs, the Splitter pressure had steadily climbed to 33 psig (normal operating pressure is about 20 psig) at the inlet to the overhead condenser, approximately 150 ft below the top of the tower. A trainee operator noticed the high pressure on the satellite building control board screen display, and brought it to the attention of the other operators. At this point the outside operators opened the 8-inch chain-operated vent valve for the second time in order to reduce the elevated pressure. The trainee reported seeing vapors that "looked like steam" venting from the top of the stack, but another operator told him not to worry as it was nothing unusual. After approximately 10 to 15 minutes the chain-operated valve was closed, and by 12:55 hrs the pressure had fallen to 22.6 psig.

By 12:40 hrs, the Splitter base temperature had reached 302°F (normal operating temperature is about 275°F). The feed to the Splitter continued at 20,000 bpd with no outflow, and the calculated level in the tower reached over 130 ft. At this temperature, modeling predicts some vaporization in the bottom of the Splitter, despite the head of colder liquid above. This would have lifted the liquid level higher than tray 13, but lower than the overhead line at the top of the tower. The cold liquid higher up the tower quenched these vapors and prevented them from distilling overhead.

After reviewing the unit status on the satellite control board, an operator telephoned the Day Shift Board Operator and told him that he needed to take a heavy raffinate rundown flow out of the tower. At 12:41 hrs the Day Shift Board Operator opened the Heavy Raffinate rundown control valve. The Heavy Raffinate rundown flow did not indicate a flow until approximately 13:00 hrs, and by 13:09 hrs had stabilized at 31,000 bpd. The feed to the Splitter continued at 20,000 bpd, and from 09:52 to 13:00 hrs approximately 2,500 barrels had been added to the tower. The calculated level peaked at 137 ft in the tower (based on a simplified calculation ignoring the effect of vaporization in the base of the tower). At this level, 57 of the 70 trays within the column were flooded, and the feed inlet at tray 31 was submerged.

During these operations, the level transmitter (designed to operate with a liquid gas interface) was fully submerged and displayed a signal on the DCS screen slowly drifting downwards to 80% before any liquid was removed from the Splitter. The DCS high level alarm remained in alarm mode throughout.

An Ultracracker turnaround meeting had been called in the doublewide trailer, and attendees had started arriving at around 13:00 hrs.

At 13:02 hrs the off-site Day Shift Supervisor telephoned the satellite control room from outside the Refinery, and spoke to an operator, who indicated he was busy and would call back. At 13:09 hrs the operator telephoned the Day Shift Supervisor at home, who, upon hearing of the pressure trend, suggested opening the 1½ inch vent valve around the Reflux Drum relief valve to vent nitrogen. This vent valve was opened and by 13:13 hrs the pressure at the inlet to the Overhead Condenser had fallen from 22.6 to 20.5 psig.

The Heavy Raffinate rundown stream exchanges heat with the incoming feed to the tower in the Feed/Bottoms Exchangers. At 13:01 hrs the feed preheat was 126°F, and rose to 260°F by 13:10 hrs. This abnormally rapid increase in temperature led to the rapid onset of vaporization of the feed at the inlet point (tray 31) to the tower. This would have lifted the liquid above the feed tray even higher, such that it quickly reached the top of the Splitter and flowed over into the 24-inch overhead line.

At 13:13 hrs the pressure at the inlet to the Overhead Condenser was 20.5 psig, but started to increase rapidly. This rapid pressure increase likely resulted from the rapid increase in feed preheat exchanged from the heavy raffinate rundown, which vaporized feed at the tower inlet, in conjunction with vaporization in the bottom of the tower, and lifted the excessively high liquid level over the top of the tower and into the overhead line. The liquid filled the 24-inch overhead line above the pressure transmitter and relief valves, located about 150 ft below the top of the tower.

Around this time an operator telephoned the Day Shift Board Operator and asked him to reduce firing on the Reboiler furnace due to the high temperature of 304°F on the Splitter base. The Board Operator trimmed the Fuel Gas control valve output at 13:14 hrs to reduce the Reboiler outlet temperature.

By 13:15 hrs the pressure at the inlet to the Overhead Condenser peaked at 63 psig, and two outside operators confirmed that the tower overhead relief valves (with set points of 40, 41 and 42 psig) had opened to relieve directly into the Blowdown Drum & Stack (F-20) through a 14-inch header.

At this point the fuel gas firing to the heater was stopped, and the outside operators blocked in the main burners. At 13:16 hrs the Reflux Drum low-low level alarm cleared, indicating liquid in the vessel for the first time, and the outside operators started a reflux pump at 13:17 hrs. The indicated reflux flow rate went off scale in excess of 35,700 bpd. The Reflux Drum low level alarm cleared at 13:19 hrs, indicating that the vessel was full of liquid (the drum normally runs in a flooded state). The DCS system shows that the second reflux pump was also started at 13:19 hrs, although none of the operators remembered doing so when interviewed after the incident.

At about this time there were radio messages from at least two witnesses, who saw vapors and liquid emerging approximately 20 ft above the top of the stack "like a geyser" and running down and pooling around the base of the Blowdown Drum & Stack. Vapors were seen evaporating from the liquid pool. The F-20 high level alarm alarmed for the first time at 13:20 hrs.

Alerted by the radio messages and the shouting of at least one eyewitness, several personnel in the area of the ISOM left the immediate vicinity before the vapors ignited. The evacuation alarm was not sounded. At least one witness saw a pickup truck parked just north of the Blowdown Drum & Stack with its engine racing and exhaust glowing, but it is not known if this was the source of ignition. Several witnesses described two or more explosions; the first minor explosion(s) followed rapidly by a louder, more powerful blast at approximately 13:20 hrs, although subsequent modeling suggests that there was only one explosion. The explosion severely damaged the doublewide and other trailers on the west side of the ISOM, and resulted in 15 fatalities and over 170 individuals harmed. The blast resulted in damage to the ISOM, causing a number of secondary hydrocarbon releases and fires.

The Site Emergency Response Team responded immediately and mounted a search and rescue operation. Mutual Aid and Lifeflight resources were requested and mobilized by 13:45 hrs. The feed to the Raffinate Splitter was not shut down, and stopped when electrical power went down at 14:45 hrs. The fires were brought under control after 2 hours, and injured personnel had been treated and/or transported to local hospitals, allowing ambulances and Lifeflight resources to stand down by 16:44 hrs. The final body was found at approximately 23:00 hrs, having been buried under debris.

3. CAUSAL ANALYSIS

Immediate and system causes were analysed using the evidence compiled. The evidence was broken down into discrete building blocks of events or conditions from which the Critical (Causal) Factors were identified. Critical factors are those events or conditions that, if removed, might eliminate or reduce the possibility of the event occurring, or reduce the severity of it. For each critical factor, possible Immediate Causes and possible Management System Causes (Root Causes) were identified using BP's Comprehensive List of Causes (CLC) methodology.

The CLC is a "pre-defined tree" technique for root cause analysis, which provides a systematic method of considering possible root causes associated with an incident. The investigator does not have to build the tree, but rather apply the critical causal factors to each branch in turn, and discard those branches that are not relevant to the specific incident. Further information on pre-defined trees is contained within the CCPS *Guidelines for Investigating Chemical Process Incidents*.

A number of underlying cultural issues were then distilled from the management system causes by using a "5 Whys" type of analysis.

3.1 Critical Factors

The following critical causal factors were identified without which the incident would not have happened or would have been of significantly lower impact:

Loss of Containment

Actions taken or not taken led to overfilling of the Raffinate Splitter and subsequent overpressurization and pressure relief. Hydrocarbon flow to the Blowdown Drum & Stack resulted in liquids overflowing the stack, causing a vapor cloud, which was ignited by an unknown source.

Raffinate Splitter Startup Procedures and Application of Knowledge and Skills

Failure to follow the startup procedure contributed to the loss of process control. Key individuals (management and operators) did not apply their level of skills and knowledge, and there was a lack of supervisory presence and oversight during this startup.

Control of Work and Trailer Siting

Numerous personnel working elsewhere in the refinery were too close to the hazard at the Blowdown Drum and Stack during the startup operation. They were congregated in and around temporary trailers and were neither evacuated nor alerted.

Design and Engineering of Blowdown Drum and Stack

The use of Blowdown Drum and Stack as part of the relief and venting system for the Raffinate Splitter, after several design and operational changes over time, close to uncontrolled areas.

3.2 Possible Management System Causes (Root Causes)

The following possible management system causes were determined for each critical causal factor. The bulleted items represent the title of the closest CLC branch:

Loss of Containment

- Poor judgment
- Inadequate training effort
- Inadequate leadership
- Inadequate adjustment/repair/maintenance
- Inadequate enforcement of policies/standards/procedures

Raffinate Splitter Startup Procedures and Application of Knowledge and Skills

- Fatigue
- Poor judgment
- Low mechanical aptitude
- Preoccupation with problems
- Improper supervisory example
- Inadequate training effort
- Conflicting roles/responsibilities
- Inadequate leadership
- Inadequate or lack of safety meetings
- Inadequate reference materials or publications
- Inadequate audit/inspection/monitoring
- Inadequate enforcement of policies/standards/procedures
- Inadequate vertical communication between supervisor and person
- Inadequate communication between work groups

Control of Work and Trailer Siting

- Inadequate leadership
- Inadequate correction of worksite/job hazards
- Inadequate identification of worksite/job hazards
- Inadequate evaluation and/or documentation of change
- Inadequate implementation of policies/standards/procedures due to deficiences
- Inadequate communication between different work groups

Design and Engineering of Blowdown Drum and Stack

- Inadequate leadership
- Inadequate correction of worksite/job hazards
- Inadequate performance measurement and assessment
- Inadequate design
- Inadequate evaluation and/or documentation of change
- Inadequate work planning
- Inadequate audit/inspection/monitoring
- Inadequate communication of safety and health data, regulations or guidelines

More information on the reasoning behind these possible management system causes is provided in the final investigation report.

4. LESSONS LEARNED

The significant number of possible management system causes identified above indicated many linked issues requiring further evaluation. In order to understand which recommendations would prevent reoccurrence it was decided to go deeper to the underlying cultural issues. The following issues were distilled from the system causes and illustrate a number of lessons learned, some of which will be applicable to a wider cross-section of the industry.

4.1 Business Context

There was a lack of clearly defined and broadly understood context and business priorities for the Texas City site. A clear view of the key process safety priorities for the site or a sense of a vision or future for the long term could not be identified.

- Focused on environment and personal safety, not process safety
- There was little ownership of PSM through the line organization
- The development of people was a low priority, with inadequate training
- In staff's minds this created no future

Over the years, the working environment had eroded to one characterized by resistance to change, and lacking of trust, motivation, and a sense of purpose. Coupled with unclear expectations around supervisory and management behaviors this meant that rules were not consistently followed, rigor was lacking and individuals felt disempowered from suggesting or initiating improvements.

- Created a poorly motivated workforce who behaved in a disempowered way
- Lack of enforcement of following procedures, etc.
- Lack of role models at supervisor and superintendent levels

- Little expectation of behaviors and performance
- No consequences of good or bad performance
- Reward structure leads to unintended circumstances
- Inward looking at plant and site level
- Lack of completion or follow-through
- No verification of actions
- Fear to challenge and say "no"
- Lack of teamwork evidenced by many behaviors and attitudes

Examples of this environment include:

- Failure to follow procedures was identified as a causal factor in several previous investigations, but there were not any consequences for this behavior at either supervisory/management or operator level.
- An explicit description of the desired behaviors for supervisory/management personnel was not readily available and evidenced by the absence of supervisors during key events such as critical shift handovers during a startup procedure.
- Supervisors did not reinforce the importance of following procedures.
- At its last inspection, appreciable corrosion and internal damage was discovered in F-20, an item of safety critical equipment, but it was neither repaired nor a work order submitted for later repair.
- A reward system that encouraged supervisors and operators to work for extended periods of time with no clear consideration of fatigue.
- Many examples were given where individuals felt that making suggestions for improvements had little value and, over the years, had moved into a mode where they would follow instructions in an incomplete and routine way without thinking.

The "check the box" approach to processes, such as MOC, was indicative of the poor motivation and reluctance to go beyond minimum compliance, and sometimes even the minimum was ignored. The inward-looking, closed environment resulted in corporate initiatives receiving limited attention as a "not invented here" culture was tolerated. The learnings from external incidents were largely ignored at the site.

4.2 "Safety" as a Priority

Process safety, operations performance and systematic risk reduction priorities had not been set and consistently reinforced by management.

Good safety is delivered through good line operations, underpinned by the right safety culture and values. The quality of basic operations had declined to the level where real safety interventions were necessary to ensure the right actions were being taken. Evidence of this was that shift changovers were inadequate, procedures were not followed, and line managers were unaware of operations that were underway.

There was no evidence of comprehensive and consistent business plans to reduce site risks. Existing plans focused on projects for reducing personal injuries and enhancing environmental compliance, but contained no plans for the systematic reduction of process risks or improving basic operations. For example, there were no plans to reduce or eliminate the use of blowdown stacks which vent to the atmosphere. A project considered an option to modify F-20 to tie into a flare system, but this was not progressed as the focus was on environmental compliance issues.

A number of interviewees noted that safety did not seem to be a priority, particularly as compared to cost management, for example. Although leadership stated "safety first", this was not evidenced or believed by many of the workforce. Lack of leadership visibility and poor communication through the complex, siloed organization did not assist in delivering the right messages. Examples to support this included the absence of reporting (in some cases) and investigation (in most cases) around loss of containment incidents and process upsets.

No clear plans were found for enhancing organizational capacity or capability for the site. The required training for compliance was generally being provided, but training and development for first-level supervisors and superintendents was incomplete. This was most evident in the case of step-up supervisors, where the training and development program for these individuals in their step-up roles was poor or nonexistent. The low investment in developing supervisory levels appears to have harmed the communication with, and behaviors of, the workforce.

In general, the further down the organizational structure, the less clear the picture was with regard to safety priorities and the future vision for the refinery. As a result, many employees felt as though it was not useful to raise safety concerns or think of future actions, ultimately reducing morale and pride in the site.

4.3 Organizational Complexity and Capability

Many changes in a complex organization had led to the lack of clear accountabilities and poor communication, which together resulted in confusion in the workforce over roles and responsibilities.

The Texas City facility is a large complex site, which had multiple levels within the organization, apparently to address the span of control across such a large site. This organization had many interfaces requiring clear accountabilities and good communication both horizontally and vertically throughout the organization. In reality, examples were found of a lack of accountability, unclear roles and responsibilities, and poor communication with employees tending to work within silos. This, in turn, created confusion around some of the many interfaces. As a result, the working environment was cluttered with many processes, committees, etc., such that it was relatively easy to lose sight of the basic fundamental requirements for safe efficient operation.

- Complex and unclear accountabilities
- Functional silos exist
- Communication within the plant is generally poor

There were numerous examples of unclear accountabilities between groups; for example, with regard to who was actually accountable for the area between the ISOM unit and the NDU unit, where the damaged trailers were located. Similarly, when asked who was accountable for siting the trailers in this area, a mixture of answers was received.

There was no evidence of communication of the Splitter startup to the adjacent work sites. One individual stated that contractors working inside the battery limits (ISBL) of the ISOM had been notified and warned to keep out of the area but, in the immediately adjacent catalyst warehouse area, no effort was made to notify or evacuate the people present in the trailers. The organization appeared to operate in walled silos.

The relationship between operations and engineering appeared inconsistent and fractured; evidence of this was the fact that engineering was not called during the startup problems; nor were engineering rounds being routinely conducted. In addition, the "independent" functions, such as training and PSM, had become underresourced and lacked the influence to ensure that standards were met.

The Shift Director's meeting should be a key cross-unit communication vehicle, but attendance was not mandatory and discussion for the entire site was compressed into about 15 minutes, which was not adequate for genuine communication. Communication worked better within immediate work groups but not well outward or upward.

The low investment in developing supervisory levels contributed to this lack of clarity and poor communication due to the low leadership skill levels of those involved. This was exacerbated by a lack of reward or recognition for those who stepped across boundaries and stood in personal leadership.

This complexity created an organization which required behaviors across boundaries, between different groups and companies, but the clear divisions seen ensured that people were unlikely to communicate with or influence across these boundaries. Because of this, natural teams were often incomplete or incoherent.

4.4 Inability to See Risk

A poor level of hazard awareness and understanding of process safety on the site resulted in people accepting levels of risk that were considerably higher than comparable installations.

Although some effort had been expended to raise awareness and understanding of process safety in the early 1990s, when OSHA promulgated the PSM rule, this basic training had not been effectively refreshed over the intervening years. There was no ongoing training program in process hazards risk awareness and identification for either operators or supervisors/managers.

- No risk reduction plan
- Passed down by experience, which was sometimes bad
- Gradually increasing risks and no process for systemic review
- Accepting incrementally lower standards throughout operations

There were not effective holistic plans to systematically reduce risks in the refinery. Examples of this included no plan in place regarding the ultimate replacement / reconfiguration of the blowdown stacks in the site. The most recent Major Accident Risk assessment for the site failed to address the risks associated with blowdown stacks. The ISOM unit and its associated blowdown stack did not feature in the list of the top 80 risks at the site. Similarly, the MOC risk review for the doublewide trailer placement did not include any consideration of the risks associated with the blowdown stack, located within 150 feet.

One consequence was that temporary office trailers were placed within 150 feet of a blowdown stack which vented heavier than air hydrocarbons to the atmosphere without questioning the normal industry practice.

Many examples of a high level of risk were observed within the site. At the most basic level was the significant number of vehicles allowed on the site, in close proximity to hydrocarbon processing units, and the extensive use of trailers for housing people in close proximity to hydrocarbon processing units. To visitors, this appeared unnecessary but had grown by custom and practice.

During the course of the investigation there were a number of (minor) fires within the site, in addition to a serious incident on the RHU. The general reaction of the workforce to these fires appeared to be not to worry, as fires were a fact of life in the refinery. Indeed, this was supported by the lack of investigation around previous fires. There were references to fires in Emergency Response Team logs without any documented investigation reports.

Other examples of high risk tolerance concerned the failure to conduct a review of the Splitter startup procedure with the crew, as required by the procedure; the absence of supervision during the startup, when it is common knowledge that startup is a higher risk activity; and the lack of reporting of significant process upsets, such as relief valves lifting during previous startups.

All of these observations point to both a high level of risk having become accepted and an inability to see key process risks.

4.5 Lack of Early Warning

Given the poor vertical communication and performance management process, there was neither adequate early warning system of problems, nor any independent means of understanding the deteriorating standards in the plant.

The fifth and final cultural issue was the lack of a holistic early warning system for process safety exposures. The site had numerous measures for tracking various types of operational, environmental and safety performance, but no clear focus on the leading indicators for potential catastrophic or major incidents. Numerous audits had been conducted at the site in line with regulatory and corporate requirements, but had generally failed to identify the systemic problems with work practices uncovered by the investigation.

- Vertical communication was poor.
- Many Key Performance Indicators, but not transparent or useful for loss of containment, showing Recordable Injury frequency improvement.
- Audit was process focused and did not gain verification of action.

The safety measures focused primarily on occupational safety measures, such as recordable and lost time injuries. This focus on personal safety had led to the sense that safety was improving at the site. There was no clear focus or visibility on measures around process safety, such as lagging indicators on loss of containment, hydrocarbon fires, and process upsets. Examples seen of this included that site leadership was not focused on trend analysis of measures that were likely to deliver an accurate sense of process safety at the site. Loss of containment incidents and process incidents did not get the attention they warranted. Many were not even formally reported or investigated, and thus corrective actions were not identified and addressed.

A large number of audit reports were reviewed by the Investigation Team, demonstrating that audits were being routinely conducted. The reports indicated that, for the most part, these audits were focused primarily upon review of processes and documentation. With the exception of the "Big 4" safe work practice audit, none of the audits reviewed focused on verification and assessed, for example, whether or not procedures were being followed, and whether work practices were actually consistent with the procedures.

5. FOLLOW-UP ACTIONS

The Investigation Report identified a large number of recommendations specific to the Texas City refinery, and the ISOM unit in particular. The site leadership team has developed prioritized plans to address all of the recommendations. However the following are some of the high priority actions already underway:

Developing People, Skills and Behavior

- Appointed a new refinery manager and created a new leadership position to ensure prioritization and tracking of key site initiatives

- Simplified the Texas City organization, so that everyone involved in maintenance and operations knows what they are accountable for and to whom

- Clarified and reinforced roles, responsibilities and expectations around startup, operating, maintenance and evacuation procedures

- Initiated specific education and communication actions to ensure personnel are following procedures across the site

- Started a program to design and deliver enhanced operator training and frontline leader education

- Retained a consulting firm to assist with leadership development and design of an improved workplace environment

- Co-located leadership in a common office location for improved communication

- Created the role of Group Vice President for Health, Safety, Environment and Technology in the Refining and Marketing segment

- Appointed a Group Vice President Safety and Operations to lead improvements in Safety, Health and Environmental Management

Ensure Safe, Reliable Equipment Operation

- Began a top-to-bottom, continuous inspection and evaluation of the refinery

- Removed the four high pressure units (above 1000 psi) from service and have underway a complete root and branch analysis from design through to current configuration. The units will not be run until that process is complete, and they are determined to be safe

- Heightened oversight by requiring supervisors be present for all start ups, shut downs and for other critical operations including requiring written shift handovers with supervisor present

- Strengthened control of work verification process to ensure work is stopped when deviations occur

- Leased a 100,000 square foot building in Texas City to provide offices for employees whose work does not require their presence on the refinery site

- Undertaking a complete relief valve revalidation study for all process units and a relief and flare system review

- Begun the process to replace blow down stacks with flares
- Initiated a process to identify and stop nonessential work
- Engaged an outside firm to help assess and improve current maintenance work processes
- Conducting a review of span of control and focus of operations leadership roles

Refine and Implement HSSE Policies

- Removed temporary office trailers from within the site.
- Drastically reduced the number of vehicles allowed within the refinery
- Instituted leadership audits to verify proper use of startup, shutdown and safe procedures
- Engaged a consultant to assist with management information around process safety
- Have underway training for employees regarding enhanced communications on non routine operations and expectations for sounding alarms

6. CONCLUSIONS

The incident was an explosion caused by heavier–than-air hydrocarbon vapors combusting after coming into contact with an ignition source, probably a running vehicle engine. The hydrocarbons originated from liquid overflow from the Blowdown Stack following the operation of the Raffinate Splitter overpressure protection system caused by overfilling and overheating of the tower contents.

The failure to institute liquid rundown from the tower, and the failure to take effective emergency action, resulted in the loss of containment that preceded the explosion. These were indicative of the failure to follow many established policies and procedures. Supervisors assigned to the unit were not present to ensure conformance with established procedures, which had become custom and practice on what was viewed as a routine operation.

The severity of the incident was increased by the presence of many people congregated in and around temporary trailers which were inappropriately sited too close to the source of relief. The likelihood of this incident could have been reduced by discontinuing the use of the Blowdown Stack for light end hydrocarbon service and installing inherently safer options when they were available.

While the site management had introduced improvement programs, had completed a site-wide Major Accident Risk assessment and, following previous incidents, had begun to introduce many improvements in the areas of training, audit, and culture, the investigation found many areas where procedures, policies, and expected behaviors were not met. The underlying reasons for the behaviors and actions displayed during the incident were complex. It is evident that they had been many years in the making and will require concerted and committed actions to address.

The author would like to thank BP for permission to publish this paper.

7. REFERENCES

1. *Fatal Accident Investigation Report, Isomerization Unit Explosion, Texas City, 23rd March, 2005*: BP America, Inc., December 9, 2005

2. *Guidelines for Investigating Chemical Process Incidents, 2nd Ed.*, New York: American Institute of Chemical Engineers, Center for Chemical Process Safety, 2003

Paper T5000-6f

The Accident in Bhopal: Observations 20 Years Later

Ronald J. Willey, Northeastern University, Boston, MA, r.willey@neu.edu
Dennis C. Hendershot, Chilworth Technology Inc., Plainsboro, NJ, dhendershot@chilworth.com
Scott Berger, CCPS Director, AIChE, New York, NY, scotb@aiche.org

Prepared for Presentation at
American Institute of Chemical Engineers
2006 Spring National Meeting
40th Annual Loss Prevention Symposium
Orlando, Florida
April 24-26, 2006

UNPUBLISHED

The Accident in Bhopal: Observations 20 Years Later

Ronald J. Willey, Northeastern University, Boston, MA, r.willey@neu.edu
Dennis C. Hendershot, Chilworth Technology Inc., Plainsboro, NJ, dhendershot@chilworth.com
Scott Berger, CCPS Director, AIChE, New York, NY, scotb@aiche.org

Abstract

The most influential process safety accident passed its 20th anniversary on Dec 3, 2004. At an international symposium to mark the event in Kanpur, India during the week of this anniversary, process safety practitioners from around the world assembled to discuss progress in resolving the Bhopal tragedy and in advancing the practice of process safety worldwide. This paper reports the main conclusions from the conference, and provides insight into the Bhopal site as attendees found it in December 2004. Since 1984, many positive steps worldwide have been made in regards to improvements in process safety and protection of personnel within chemical plants and of people in the surrounding communities. However, little progress has been made in decommissioning and decontaminating the Bhopal plant site, now under control of the Indian state of Madhya Pradesh. Many plant chemicals, abandoned there in 1985, were still at the site in 2004, mostly in sub-standard storage conditions. Mitigation recently commenced, but unconfirmed reports of the mitigation methods are concerning. The lesson learned: we all have a responsibility to insure that events which follow a chemical accident reach a proper conclusion; and that no further undue suffering results to the general public and our fellow employees.

1. Introduction

1.1 Background about the original accident

1.1.1 The pesticide plant and the MIC release

Much information about the original Bhopal accident is available through books [1, 2, 3], journal reports [4] case histories [5], documentaries [6, 7], proceedings from international conferences,[8, 9], and the Internet [10, 11, 12]. In essence, about 41 metric tons of methyl isocyanate was released from the Union Carbide India Limited (UCIL) pesticide plant in Bhopal just after midnight on December 3, 1984. This gas spread slowly southward from the plant site during the early morning hours with very stable weather conditions. Of the 900,000 population within the city, over 200,000 people were exposed to MIC tainted air. Documented death counts are listed at 3,787 [13]. The number of undocumented deaths will never be known, but estimates are over 10,000. Chaos surrounded the city afterwards. Thousands panicked. As the story of the disaster circled the globe, international aid began to flow into the city. Union Carbide, USA, was the majority owner of the plant (50.9%) with Indian investors owning the rest [14]. Warren Anderson, then chairman, made a personal trip to the India to reach out, only to be placed under arrest. Several years later Union Carbide settled with the Indian Government for over \$470 million dollars plus other considerations[15]. The event that caused the release was traced to a runaway reaction created by the contamination of a storage tank of methyl isocyanate with a substantial amount of water. Although the safety relief valve opened at design pressure, all downstream measures to mitigate an external release of MIC to the surroundings were on stand-by (a caustic scrubbing tower) or out-of-service (a flare tower). Furthermore, a water curtain spray designed to mitigate releases did not reach the elevation of the release plume. Thus, MIC exited at an elevation from ground level of approximately 35 meters for a period of 15 to 30 minutes. The cloud then descended to ground level (MIC gas is approximately 2 times as dense as air), infiltrating the surrounding residential areas (illegal shanty towns), and flowing slowly towards the center of Bhopal, located about 2 km to the south.

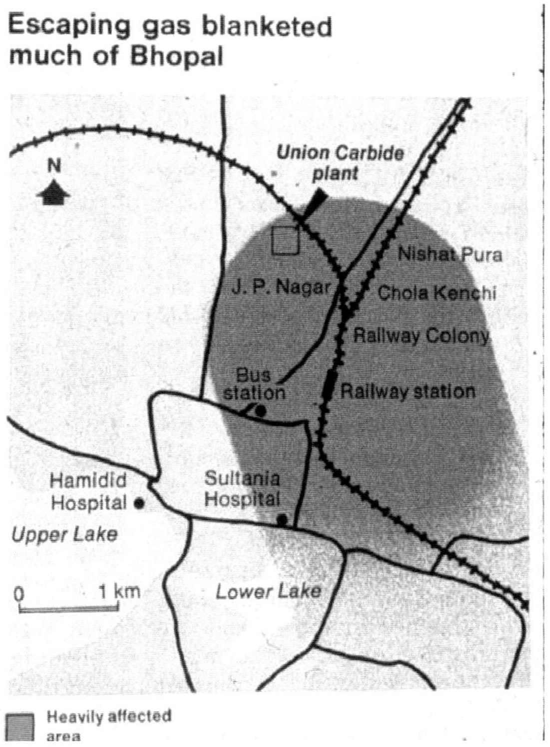

Fig. 1. Overview map of the Bhopal vicinity.

The UCIL plant manufactured Sevin®, a Union Carbide trade name for a pesticide, whose active ingredient is 1-napthyl-N-methylcarbamate, or the generic name carbaryl. The reaction involved two reactants, methyl isocyanate (MIC) and alpha naphthol. Methyl isocyanate is reactive, toxic, volatile, and flammable. The maximum exposure (TLV-TWA) during an 8-hour period is 0.02 ppm (20 parts per billion). By comparison, phosgene, another extremely toxic gas, has a TLV-TWA of 0.1 ppm (100 parts per billion). Individuals begin to experience severe irritation of the nose and throat at exposures to MIC above 21 ppm. The LC_{50} for rats exposed to MIC vapors in air for 4 hours is 5 ppm. In humans, exposure to high concentrations can cause enough fluid accumulation in the lungs to cause drowning. At lower levels of exposure, the gas affects the eyes and lungs. It acts as a corrosive agent, eating away at moist vulnerable tissue, such as mucous membranes and eye surfaces. Long-term effects also exist. MIC has a boiling point of 39.1°C and a vapor pressure of 348 mm Hg at 20°C. As such, it is quite volatile and it will easily enter into the surroundings at very high concentrations. With a molecular weight of 57, about 2 times that of air MIC has a higher vapor density compared to air.

1.1.2 Explanation of the Possible Causes

Several causes have been proposed, two of which have been subjected extensive examination. Cause 1 captioned in Fig. 2, is the admission of water via a water cleaning process where an isolation valve existed but the line had not been blanked off (the "Water Washing Theory). This isolation valve was located about 300 m from the storage tanks via a pipeline. Substantial water (500 kg plus line volume) with some head was required along with one other key valve leaking. Cause 2 was sabotage, deliberately connecting a water hose to piping that directly entered into the storage tank and deliberately admitted water. Cause 2 would have required intimate knowledge of piping around the tank, where to physically make the correct connection, and the removal of a pressure indicator and then the re-attachment of piping fittings.

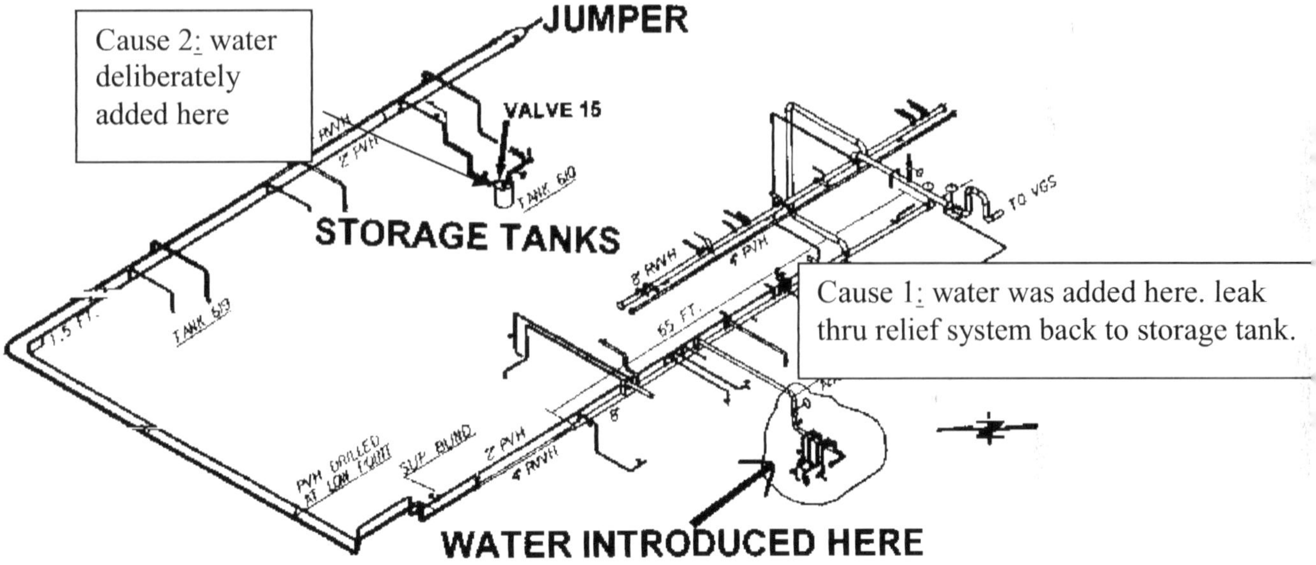

Fig. 2. Overview of the piping system.

However, neither is a root cause. The root causes were management decisions:

a decision to neglect a flare system in need of repair;

- a decision to place a scrubber system on stand-by to save on operating expenses;
- a decision to remove coolant from the refrigeration system used to cool the MIC storage tank.

And, there are additional root causes which made the incident more severe:

- inadequate emergency planning and community awareness;
- lack of awareness of the potential impact of MIC on the community by the people operating the plant; lack of communication with community officials before and during the incident;

- inadequate community planning, allowing a large population to live near a hazardous manufacturing plant. This situation was not unusual in the chemical industry in the early 1980s, and one major impact of Bhopal was to warn all chemical plants about the importance of these considerations in the siting and operation of facilities.

1.1.3 The Plant Layout 1984

Shown in Fig. 3, Fig. 4, and Fig. **5**, are engineering details related to the storage of MIC at the Bhopal plant. The plant was "backwards" integrated from a formulation plant to a fully integrated chemical plant in the late 1970's. Fig. 3 is a photograph looking North where the 3 MIC storage tanks were surrounded with earth for insulation (on the right side, or towards the East side of the plant layout). Fig. 4 shows a cross section of the storage tanks with piping provided for internal cooling, MIC flow in and out, and attachment to the process vent system via a relief valve should the tanks be overpressurized. Finally, Fig. **5** shows the effluent treatment system downstream of the relief valves, for treatment of any material that might be discharged if the relief valve opened, either due to overpressurization of an MIC tank or because of a relief valve leak.

Fig. 3. Photograph taken shortly after the accident. A pipe rack is shown on the left and the partially buried storage tanks (three total) for MIC are located mid photo right.

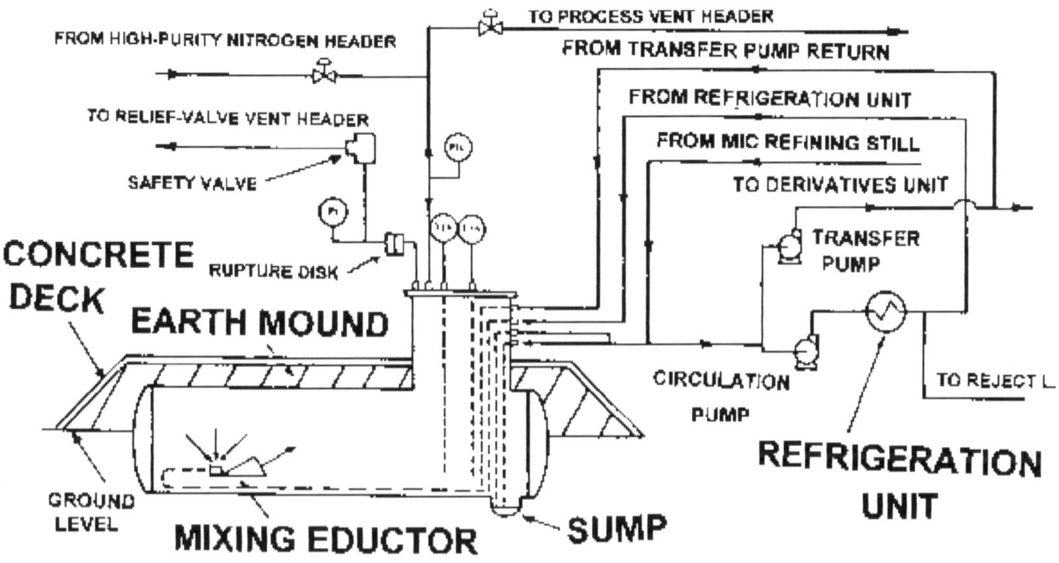

Fig. 4. Engineering piping details around the storage tanks. Storage tanks could hold up to 40 metric tons of liquid MIC.

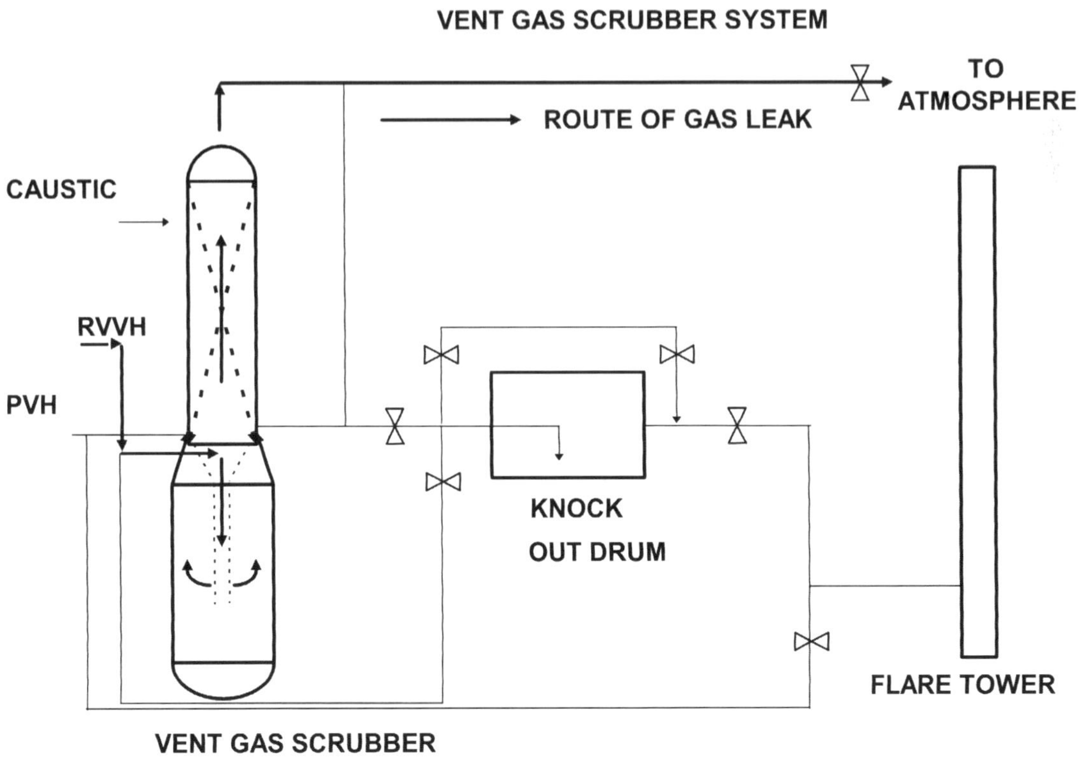

Fig. 5. Schematic of emergency relief effluent treatment system that included a scrubber and flare tower in series.

1.2 The 20th Anniversary of the Bhopal Gas Tragedy

On December 1 to 3, 2004, Prof. J. Gupta of the Indian Institute of Technology – Kanpur organized the "International Conference on the 20th Anniversary of the Bhopal Gas Tragedy," in Kanpur, India [16]. The focus of the conference was the Bhopal tragedy and its effects on process safety. Process Safety representatives from around the world attended. Papers from this conference are available in print [9] and on the Internet [17]. Perhaps the most memorable presentations were the first hand witness stories from Bhopal residents and UCIL workers. For example, Swaraj Puri, the chief of police of Bhopal described how, exposed to MIC vapors, he risked his life seeking answers in the dark night to manage the chaos of evacuation, obtain medical help for the victims, and eventually oversee the removal and disposal of bodies. . Poor communications (less than 10,000 telephones for 900,000 residents at the time) hampered the discovery of what was affecting the multitudes.

A plant operator spoke. He told how workers prided themselves working at the plant. However, change was underway, driven by Union Carbide's decision to close its Bhopal operations. Pressure to eliminate costs resulted in personnel layoffs, and decisions mentioned above compromised what could have been a safely run plant. Other speakers discussed the long-term health effects, and epidemiology studies and monitoring of the long term health impacts of the disaster have unexpectedly ceased. Finally, there were many papers related to process safety and how this accident influenced process safety practice across the world.

2. The site 20 years later

Our most astonishing discovery was the current condition of the Bhopal plant site. As one approaches, it looks like a jungle. Plants and trees have taken over (Fig. 6). Tank 610, the tank that over pressured and released the MIC, was removed from its underground vault, and is now above ground and totally surrounded by brush and overgrowth. Even more astonishing, sadly, bags of chemicals have been left behind, as shown in Figure 7. For 20 years! Fig. 8. is a photograph very near the filter area where a water washing cycle was underway the night of the accident. We were able to identify the gate valve that fed the process vent header located above in the pipe rack. Its stem indicated full closure, although this provides no evidence of its condition more than 20 years earlier during the incident. Fig. 9 is a photograph of the top entry to the concrete vault containing Tank 611, an identical storage tank to 610. Much of the piping is in place, undisturbed after 20 years. The rupture disk tag is still readable, and somewhat hidden, the spring relief valve that leads to the relief vent system lines. In studying the piping up close, we conclude that it is quite sophisticated. If sabotage was committed, the person who did it had to have intimate knowledge of piping into and out of the tank. Further, they had to have identified exactly how to remove a pressure indicator, and attached a quick disconnect hose connection, then after the event remove this quick disconnect hose connection (the condition it was reported to have been found by an eye witness). Fig. 10 shows the scrubber system. Manufactured of stainless steel, it is in reasonably good condition after more than 20 years. This plant had state-of-the-art equipment in place. Some motors appear as if they could operate today; simply flip the switch. Pumps, however, do show signs of corrosion after sitting in the elements for 20 years. Fig. 11 is a photograph of the flare system. This flare system went out service several weeks earlier due to a corroded pipe replacement project [14]. Little priority was placed on repairing it because the plant at the time was not manufacturing MIC. Fig. 12 shows the control room. By early 1980's standards, this was a typical control room with state-of-the-art equipment. Much of the paper that litters the floor are the remains of operating instructions and safe work permits. Its location from the plant was about 100 m to the West, well isolated from the process area. Fig. 13 made a sobering tour somber. We heard complaints from the local engineers who served as our hosts that mercury was contaminating the ground. Initially, we thought this was an exaggeration. Regrettably, as Fig. 13 shows, mercury droplets are quite prevalent near certain units. Fig. 14 shows a tank that has corroded completely, unknown contents lying below on the ground. Disturbing.

Such surface pollution is thought to have contaminated groundwater and rendered the drinking water in Bhopal hazardous. It is important for us to learn that when environmental activists speak of the deaths caused by the Bhopal Plant, they regard the MIC release as minor relative to the long-term environmental effects.

Fig. 6. Twenty years later, the flora covers Tank 610 which has been removed from the earthen mound.

Fig. 7. Former Plant operator T. R. Chauhan with bags of chemicals left behind in the background.

Fig. 8. Photograph near the filter area. Some of the critical components are identified in the photograph.

Fig. 9. Photograph of Tank 611 entry piping. Rupture disk is in center, spring relief valve is behind flora that has grown in over the years.

Fig. 10. Photograph of the scrubber system.

Fig. 11. Photograph of the disabled flare tower.

Fig. 12. Photograph of the control room 20 years later.

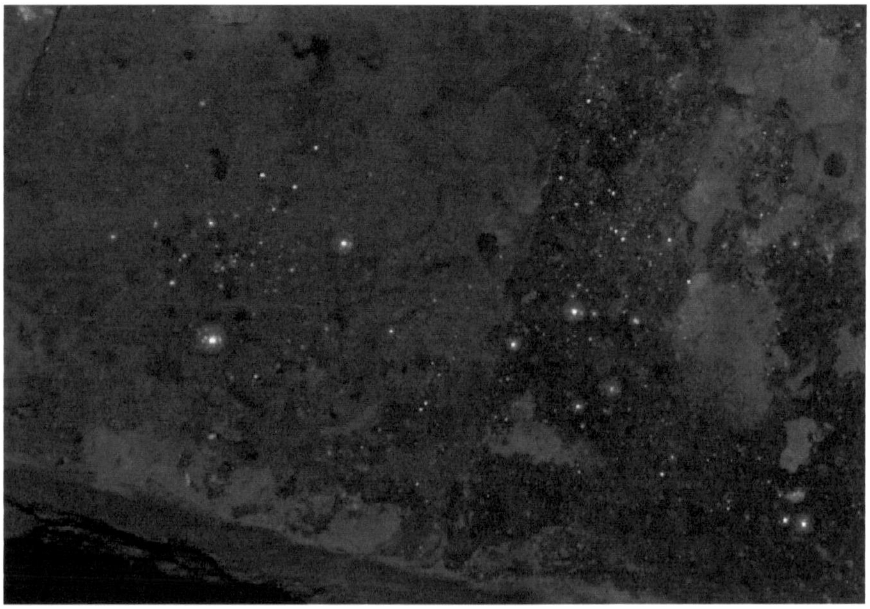

Fig. 13. Mercury droplets spotted on tour.

Fig. 14. Corroded tank with unknown contents lying below.

3. Discussion

The Bhopal disaster began long before the actual event, and its effects continue today, twenty-one years later. Although many of the early investigation efforts were directed to the initiating event – improper line blinding during filter washing, or intentional contamination by attachment of a water hose – it is clear today that the initiating event is basically irrelevant.

Clues to this accident were available in the days and months before it happened, and from the perspective of our experience twenty-one years later, perhaps even years before. If even one of the basic protections, i.e. the refrigeration system, the scrubber, the flare, the pressure gauge, the water curtain, and community emergency procedures, had been in place and functioning properly, many, many lives could have been saved. If communications between the plant and corporate management had been stronger – and perhaps given the state of communication technology in 1984, communications were as strong as they could have been – local management may have had better information and support and therefore not have made flawed decisions about fundamental safety principles. Likewise, corporate management would have had the opportunity to hasten the already-planned shutdown of the plant.

Roots of this accident extend even further back, well before the involvement of Union Carbide's manufacturing and safety personnel. Optimistic market-size expectations led to an oversized plant – and therefore MIC storage capacity oversized by a factor of three. Failure of state and local government to control the shantytown growth near the plant meant a considerably larger impacted population unable to shelter in place.

The consequences of Bhopal extend well beyond December 3, 1984 and the days immediately following. Thousands of people injured that day continue to suffer from symptoms caused by exposure to MIC, including respiratory distress. And, because Union Carbide was banished so abruptly from the site, the chemicals remaining on site were never properly removed, leading to significant environmental and health impacts, including a large number of stillbirths, which continue to this day.

Union Carbide paid an unprecedented large settlement to the Indian government and suffered other significant business and market losses in the years following 1984. Forced to sell of business after business in order to maintain its core operation, Union Carbide was eventually sold to The Dow Chemical Company in 1999, marking the sad end of a chemical industry pioneer.

Serious accidents are not common, so it is easy to fall into the mindset that they cannot happen. This is a common human failing, and it explains many risky behaviors from driving too fast on the highway to failure to maintain critical safety systems in a chemical plant. Bhopal clearly disproves this. In no circumstances does "unlikely" mean "impossible," and some possible consequences are clearly so significant that "unlikely" still leaves one with unacceptable risk that must be addressed more aggressively.

If there is a silver lining to this dark cloud, Bhopal has lead to improved process safety practices, through leading companies, through AIChE's Safety and Health Division and Center for Chemical Process Safety, and through many other organizations around the world. Fundamental principles such as Management of Change, Mechanical Integrity, Hazard Analysis, and Layers of Protection are now in the toolbox of most practicing chemical engineers around the world. It would be a stretch to say that an accident like Bhopal could never happen today, but as we work together to build a global culture of process safety, the use of the process safety tools – and the strength of process safety practices – will help prevent future Bhopals.

Finally, to keep our industry moving forward on this course and to prevent future tragedies like Bhopal, we as chemical engineers must all commit to never forgetting what happened on December 3, 1984. It is distressing how many chemical engineering students have never heard of Bhopal – we must be sure that this is one of their very first engineering lessons.

4.Conclusions

1. Every business decision has safety consequences.
2. A negative safety outcome is a negative business outcome.
3. In order to do the right thing, politics and the local community must be assessed, understood, and protected.

6. References

[1]P. Shrivastava, *Bhopal: Anatomy of a Crisis (2nd Ed.)*, Paul Chapman Publishing Ltd, London, UK 1992.

[2] T. R. Chouhan and others, *Bhopal: The Inside Story*, The Apex Press, New York, NY 2005.

[3] D. LaPierre and J. Moro, *Five Past Midnight in Bhopal*, Warner Books, Inc., New York, NY 2002.

[4] Chemical Engineering News, See February 11, 1985, December 2, 1985, and December 19, 1994 issues.

[5] R. J. Willey, *The Bhopal Disaster - a Case History*, AIChE-CCPS, New York, New York 1998.

[6] Shop A&E and The History Channel : Episode #11 VHS, 2004, The Most: Episode #11 VHS , Item Number: AAE-43088 http://store.aetv.com/

[7] S. Condie, 2004, "One Night in Bhopal ", BBC NEWS Programmes, Bhopal, How a dream turned into a nightmare, news.bbc.co.uk/1/hi/programmes/bhopal/4037703.stm

[8] A. S. Kalelkar, *Investigation of Large-magnitude Incidents: Bhopal as a Case Study*, Institution of Chemical Engineers Conference on Preventing Major Chemical Accidents, London, UK, 1988.

[9] J.P.Gupta, Ed., "*Selected Papers Presented at the International Conference on Bhopal Gas Tragedy and its Effects on Process Safety*," J. Loss Prevention, Vol 18, 2005, pp. 195-558.

[10] Bhopal Information Center, http://www.bhopal.com/, Accessed 30 Dec 2005.

[11] Bhopal Gas Tragedy Relief and Rehabilitation Department, Bhopal, Government of Madhya Pradesh, http://www.mp.nic.in/bgtrrdmp/profile.htm, Accessed 30 Dec 2005.

[12] International Campaign for Justice in Bhopal, http://www.bhopal.net/index1.html Accessed 30 Dec 2005.

[13] Bhopal Gas Tragedy Relief and Rehabilitation Department, Bhopal, Government of Madhya Pradesh, http://www.mp.nic.in/bgtrrdmp/back.htm Accessed 30 Dec 2005.

[14] F. P. Lees, Loss Prevention in *Process Industries: Hazard Identification, Assessment, and Control 2nd ed.*, Butterworth-Heinemann, Oxford, GB 1996.

[15] S. Dodson, *Week in Business,* The New York Times, February 19, 1989, **3,** 14.

[16] International Conference on the 20th Anniversary of the Bhopal Gas Tragedy "Bhopal Gas Tragedy and its Effects on Process Safety" Indian Institute of Technology Kanpur, India, http://www.iitk.ac.in/che/jpg/bhopal2.htm, 1-3 Dec 2005, Accessed: 12 Dec 2005.

[17] http://www.iitk.ac.in/che/jpg/papersb/Titlesm.htm, (accessed 13 Dec 2005).

American Institute of Chemical Engineers
3 Park Avenue
New York, NY 10016-5991